Nutritional Support for Chronic Disease

Nutritional Support for Chronic Disease

Editor

Sareen Gropper

MDPI • Basel • Beijing • Wuhan • Barcelona • Belgrade • Manchester • Tokyo • Cluj • Tianjin

Editor
Sareen Gropper
Florida Atlantic University
USA

Editorial Office
MDPI
St. Alban-Anlage 66
4052 Basel, Switzerland

This is a reprint of articles from the Special Issue published online in the open access journal *Nutrients* (ISSN 2072-6643) (available at: https://www.mdpi.com/journal/nutrients/special_issues/Nutritional_Chronic).

For citation purposes, cite each article independently as indicated on the article page online and as indicated below:

LastName, A.A.; LastName, B.B.; LastName, C.C. Article Title. *Journal Name* **Year**, *Volume Number*, Page Range.

ISBN 978-3-0365-7062-4 (Hbk)
ISBN 978-3-0365-7063-1 (PDF)

Contents

nutrients

Editorial

The Role of Nutrition in Chronic Disease

Sareen S. Gropper

Christine E. Lynn College of Nursing, Florida Atlantic University, Boca Raton, FL 33431, USA; sgropper@fau.edu

According to the Centers for Disease Control and Prevention, six out of every ten adults in the United States have at least one chronic disease, and about four in ten have two or more chronic diseases [1]. Chronic diseases, i.e., conditions that occur for at least one or more years and necessitate ongoing medical care, include diseases such as cardiovascular conditions, cancers, diabetes mellitus, and Alzheimer's disease. These conditions are also among the leading causes of death globally, accounting for 70% of all deaths around the world [2–4].

Diet, often considered as a lifestyle factor, contributes to the development of many chronic conditions including obesity, cardiovascular disease, hypertension, stroke, type 2 diabetes, metabolic syndrome, some cancers, and perhaps some neurological diseases. Moreover, one medical condition, when present, often contributes to the development of other medical conditions, such as the impact of obesity or excess body weight/fat as a risk factor for conditions including type 2 diabetes, hypertension, metabolic syndrome, and some cancers, among others. This Special Issue features research conducted by Ding and coworkers [5], which demonstrated significant associations between weight change during the different phases of adulthood and the risk of non-alcoholic fatty liver disease. The authors reported that their findings, if causal, could translate to the prevention of about 73% of incident non-alcoholic fatty liver disease if individuals maintained a healthy body mass index across adulthood [5].

Studies employing modifications of "usual" dietary practices are sometimes used as a means of evaluating diet's impact on disease risk and/or on specific disease risk factors. Two papers in this Special Issue focus on the impact of diet on chronic disease risk. First, the study by Kim and Giovannucci [6] examined the long-term impact of plant-based diets and disease risk in an Asian population. Their findings indicated that healthier plant-based diets are associated with a lower incidence of hypertension and type-2 diabetes, especially among those with a family history of the disease [6]. Moreover, in this Special Issue, under the area of diet and disease risk, is a systematic review written by Giosue and colleagues [7]. These researchers evaluated published studies that examine dairy product consumption and disease risk with a focus on cardiovascular disease, while also addressing its association with factors that contribute to disease risks such as body weight, fasting blood glucose, glycated hemoglobin, blood pressure, and inflammation to name a few. Their findings may indeed contribute to changes in dietary guidelines regarding dairy product consumption [7].

Dietary guidelines in many countries have more recently focused on overall dietary patterns versus individual nutrient intake and disease risk. Yet, it is well established that nutrient deficiencies or suboptimal nutritional status can contribute to the development of diseases and/or health problems (for example inadequate vitamin D and/or calcium status and their impact on bone development/maintenance, as well as inadequate vitamin K status and its impact on blood coagulation, among other things). The study by Li and colleagues [8] evaluated serum 25-hydroxyvitamin D concentrations in adults who had previously experienced a stroke to assess the vitamin's associations with risk for a recurrent stoke. A J-shaped relationship was found between serum 25-hydroxyvitamin D concentrations and risk of recurrent stroke in these adults with a stroke history, and as the

Citation: Gropper, S.S. The Role of Nutrition in Chronic Disease. *Nutrients* **2023**, *15*, 664. https://doi.org/10.3390/nu15030664

Received: 10 January 2023
Accepted: 19 January 2023
Published: 28 January 2023

author's point out, there is a clear need for further studies to examine if there is a possible cause–effect relationship [8]. Just as impairments in vitamin D status can negatively impact critical body functions and be negatively associated with disease risk, other nutrients play a multitude of other roles in the intricate workings of the cells of the body. Das [9] explored the roles and mechanisms of the fatty acid arachidonic acid as a mechanotransducer of renin cell baroreceptor.

While dietary modifications can help to prevent the development of many chronic diseases, once a condition has developed, changes to a person's usual diet are often needed to assist with disease (or symptom) management. The article by Dynka, Kowalcze, and Paziewska [10] reviewed the effectiveness of ketogenic diets in the management of epilepsy and other neurological conditions, as well as the diet's possible action mechanisms.

In addition to the role of dietary modifications in the treatment of disease or its symptoms, disease management may also include the use of dietary supplements. The effectiveness of oral nutritional supplements for individuals with diabetes/prediabetes was examined in two studies in this Special Issue. In the paper by López-Gómez and colleagues [11], the researchers demonstrated the effectiveness of a diabetes-specific oral supplement on reducing the prevalence of malnutrition and sarcopenia in patients with diabetes/prediabetes. Similarly, the effectiveness of nutrition supplementation and education (versus standard care for wound treatment) on inflammatory biomarkers was assessed in patients with diabetic foot ulcers by Basiri and coworkers [12]. Supplementation and education were found to positively control inflammation in the patients.

While an array of oral nutritional products on the market provides nutritional support to individuals with chronic diseases, investigations into the effectiveness of vast numbers of nutraceuticals targeting disease prevention and treatment are also exploding in the scientific literature. One such condition is age-related macular degeneration, as explored by Luján and colleagues [13]. The paper provides a detailed examination of the role of mitochondrial dysfunction in the progression of the disease, as well as an examination of the nutraceuticals/drugs that may be able to up-regulate mitophagy and mitochondrial biogenesis to enable the possible prevention or control of the disease. Some of the nutraceuticals/drugs that are examined include ferulic acid, melatonin, urolithin A and glucosamine, metformin and berberine, lipoic acid and broccoli sprout extract, and fibrate drugs and astaxanthin. Moreover, the effectiveness of extra-virgin olive oil as a means of reducing gut-permeability-derived low-grade endotoxemia in adults with impaired fasting glucose is presented in a study by Bartimoccia and colleagues [14].

Finally, the role of nutritional support for chronic disease is not complete without the inclusion of some studies investigating the complex relationships of diet/food and of disease with the gut microbiota. Zhang and colleagues [15] studied the associations between habitual diet patterns and gut microbiota in Chinese adults. The authors found that the intake of specific foods or food groups, such as whole grains, vegetables, and red meats, among others, was associated with changes in the abundance of specific genera and species of gut microbiota. Moreover, the review by Araujo, Borges-Canha, and Pimentel-Nunes [16] explored differences in the gut microbiome among individuals with metabolic syndrome (versus healthy adults) as well as the potential for probiotics and/or synbiotics to modulate the microbiome in an effort to mitigate some of the metabolic disturbances in some individuals with metabolic syndromes.

The studies featured in this Special Issue, as well as in the wider scientific literature at large, are critical in furthering the knowledge of diet and nutrition support of chronic disease. As biological mechanisms underlying chronic diseases continue to be elucidated and the causes and consequences of diet-related conditions are better characterized, new intervention strategies can be implemented, studied, and evaluated. These findings will not only assist in the formation of additional evidence-based dietary guidelines, but may also help healthcare professionals educate their patients and facilitate their adoption of healthful eating behaviors [17].

Conflicts of Interest: The authors declare no conflict of interest.

References

1. Centers for Disease Control and Prevention. National Center for Chronic Disease Prevention and Health Promotion. About Chronic Diseases. Available online: https://www.cdc.gov/chronicdisease/about/index.htm#risks (accessed on 3 January 2023).
2. Kimokoti, R.W.; Millen, B.E. Nutrition for the Prevention of Chronic Diseases. *Med. Clin. N. Am.* **2016**, *100*, 1185–1198. [CrossRef]
3. Global Burden of Disease Study 2013 Collaborators. Global, regional, and national incidence, prevalence, and years lived with disability for 301 acute and chronic diseases and injuries in 188 countries, 1990–2013: A systematic analysis for the Global Burden of Disease Study 2013. *Lancet* **2015**, *386*, 743–800. [CrossRef]
4. Global Burden of Disease 2013 Mortality and Causes of Death Collaborators. Global, regional, and national age-sex specific all-cause and cause-specific mortality for 240 causes of death, 1990–2013: A systematic analysis for the Global Burden of Disease Study 2013. *Lancet* **2015**, *385*, 117–171. [CrossRef]
5. Ding, Y.; Xu, X.; Tian, T.; Yu, C.; Ge, X.; Gao, J.; Lu, J.; Ge, Z.; Jiang, T.; Jiang, Y.; et al. Weight Change across Adulthood in Relation to Non-Alcoholic Fatty Liver Disease among Non-Obese Individuals. *Nutrients* **2022**, *14*, 2140. [CrossRef] [PubMed]
6. Kim, J.; Giovannucci, E. Healthful Plant-Based Diet and Incidence of Type 2 Diabetes in Asian Population. *Nutrients* **2022**, *14*, 3078. [CrossRef] [PubMed]
7. Giosuè, A.; Calabrese, I.; Vitale, M.; Riccardi, G.; Vaccaro, O. Consumption of Dairy Foods and Cardiovascular Disease: A Systematic Review. *Nutrients* **2022**, *14*, 831. [CrossRef] [PubMed]
8. Li, G.; Li, L.; Adachi, J.D.; Wang, R.; Ye, Z.; Liu, X.; Thabane, L.; Lip, G.Y.H. Relationship between Serum 25-Hydroxyvitamin D Level and Risk of Recurrent Stroke. *Nutrients* **2022**, *14*, 1908. [CrossRef] [PubMed]
9. Das, U.N. Arachidonic Acid as Mechanotransducer of Renin Cell Baroreceptor. *Nutrients* **2022**, *14*, 749. [CrossRef] [PubMed]
10. Dyńka, D.; Kowalcze, K.; Paziewska, A. The Role of Ketogenic Diet in the Treatment of Neurological Diseases. *Nutrients* **2022**, *14*, 5003. [CrossRef] [PubMed]
11. López-Gómez, J.J.; Gutiérrez-Lora, C.; Izaola-Jauregui, O.; Primo-Martín, D.; Gómez-Hoyos, E.; Jiménez-Sahagún, R.; De Luis-Román, D.A. Real World Practice Study of the Effect of a Specific Oral Nutritional Supplement for Diabetes Mellitus on the Morphofunctional Assessment and Protein Energy Requirements. *Nutrients* **2022**, *14*, 4802. [CrossRef] [PubMed]
12. Basiri, R.; Spicer, M.; Levenson, C.; Ledermann, T.; Akhavan, N.; Arjmandi, B. Improving Dietary Intake of Essential Nutrients Can Ameliorate Inflammation in Patients with Diabetic Foot Ulcers. *Nutrients* **2022**, *14*, 2393. [CrossRef] [PubMed]
13. Luján, L.M.L.; McCarty, M.F.; Di Nicolantonio, J.J.; Ruiz, J.C.G.; Rosas-Burgos, E.C.; Plascencia-Jatomea, M.; Assanga, S.B.I. Nutraceuticals/Drugs Promoting Mitophagy and Mitochondrial Biogenesis May Combat the Mitochondrial Dysfunction Driving Progression of Dry Age-Related Macular Degeneration. *Nutrients* **2022**, *14*, 1985. [CrossRef] [PubMed]
14. Bartimoccia, S.; Cammisotto, V.; Nocella, C.; Del Ben, M.; D'Amico, A.; Castellani, V.; Baratta, F.; Pignatelli, P.; Loffredo, L.; Violi, F.; et al. Extra Virgin Olive Oil Reduces Gut Permeability and Metabolic Endotoxemia in Diabetic Patients. *Nutrients* **2022**, *14*, 2153. [CrossRef] [PubMed]
15. Zhang, Y.; Chen, H.; Lu, M.; Cai, J.; Lu, B.; Luo, C.; Dai, M. Habitual Diet Pattern Associations with Gut Microbiome Diversity and Composition: Results from a Chinese Adult Cohort. *Nutrients* **2022**, *14*, 2639. [CrossRef] [PubMed]
16. Araujo, R.; Borges-Canha, M.; Pimentel-Nunes, P. Microbiota Modulation in Patienxts with Metabolic Syndrome. *Nutrients* **2022**, *14*, 4490. [CrossRef] [PubMed]
17. *Dietary Guidelines for Americans, 2020–2025*, 9th ed.; U.S. Department of Agriculture: Washington, DC, USA; U.S. Department of Health and Human Services: Washington, DC, USA, 2020. Available online: https://www.dietaryguidelines.gov/sites/default/files/2021-03/Dietary_Guidelines_for_Americans-2020-2025.pdf (accessed on 3 January 2023).

nutrients

MDPI

Article

Weight Change across Adulthood in Relation to Non-Alcoholic Fatty Liver Disease among Non-Obese Individuals

Yuqing Ding [1,†], Xin Xu [1,†], Ting Tian [1], Chengxiao Yu [1], Xinyuan Ge [1], Jiaxin Gao [1], Jing Lu [1,2], Zijun Ge [3], Tao Jiang [1], Yue Jiang [1], Hongxia Ma [1,4], Ci Song [2,4,*] and Zhibin Hu [1,*]

1 Department of Epidemiology, School of Public Health, Nanjing Medical University, Nanjing 211166, China; dyqdyq226@163.com (Y.D.); xuxin0061@163.com (X.X.); tianting0628@njmu.edu.cn (T.T.); yuchengxiao@njmu.edu.cn (C.Y.); gxy_iris499@163.com (X.G.); jiaxingao131@163.com (J.G.); lj_123456@126.com (J.L.); tao.chiang0923@njmu.edu.cn (T.J.); jiangyue@njmu.edu.cn (Y.J.); hongxiama@njmu.edu.cn (H.M.)

2 Department of Health Promotion Center, Jiangsu Province Hospital and the First Affiliated Hospital of Nanjing Medical University, Nanjing 210029, China

3 Office of Infection Management, Jiangsu Province Hospital and the First Affiliated Hospital of Nanjing Medical University, Nanjing 210029, China; gzj0408@126.com

4 Research Units of Cohort Study on Cardiovascular Diseases and Cancers, Chinese Academy of Medical Sciences, Beijing 100730, China

* Correspondence: songci@njmu.edu.cn (C.S.); zhibin_hu@njmu.edu.cn (Z.H.); Tel.: +86-025-86868291 (C.S.); +86-025-86868471 (Z.H.)

† These authors contributed equally to this work.

Abstract: Background: To investigate the associations of weight change patterns across adulthood with the risk of non-alcoholic fatty liver disease (NAFLD). Methods: Using data from the National Health and Nutrition Examination Survey (NHANES) 2017–2018 cycle, we performed a retrospective cohort study with 2212 non-obese participants aged 36 years old over. Weight change patterns were categorized as "stable non-obese", "early adulthood weight gain", "middle and late adulthood weight gain" and "revert to non-obese" according to the body mass index (BMI) at age 25, 10 years prior and at baseline. Vibration-controlled transient elastography (VCTE) was performed to diagnose NAFLD. Modified Poisson regression was used to quantify the associations of weight change patterns with NAFLD. Results: Compared with participants in the "stable non-obese" group, those who gained weight at early or middle and late adulthood had an increased risk of NAFLD, with an adjusted rate ratio (RR) of 2.19 (95% CI 1.64–2.91) and 1.92 (95% CI 1.40–2.62), respectively. The risk of NAFLD in "revert to the non-obese" group showed no significant difference with the stable non-obese group. If the association of weight change and NAFLD was causal, we estimated that 73.09% (95% CI 55.62–82.93%) of incident NAFLD would be prevented if the total population had a normal BMI across adulthood. Conclusions: Weight gain to obese at early or middle and late adulthood was associated with an evaluated risk of NAFLD. A large proportion would have been prevented with effective weight intervention.

Keywords: weight gain; NAFLD; population attributable fraction

Citation: Ding, Y.; Xu, X.; Tian, T.; Yu, C.; Ge, X.; Gao, J.; Lu, J.; Ge, Z.; Jiang, T.; Jiang, Y.; et al. Weight Change across Adulthood in Relation to Non-Alcoholic Fatty Liver Disease among Non-Obese Individuals. *Nutrients* **2022**, *14*, 2140. https://doi.org/10.3390/nu14102140

Academic Editor: Naoki Tanaka

Received: 26 April 2022
Accepted: 17 May 2022
Published: 20 May 2022

1. Introduction

Non-alcoholic fatty liver disease (NAFLD) is the most prevalent form of liver disease, affecting nearly one billion adults worldwide [1–3]. This multisystem condition is associated with an increased risk of liver-related and cardiovascular extrahepatic diseases [4,5]. Therefore, NAFLD prevention of a high-risk population is the cornerstone of lowering the occurrence of related morbidity and mortality.

Obesity is an established risk factor for NAFLD [6]; nearly 80% of obese adults will progress to NAFLD [7]. Despite this, around 10–40% of NAFLD occurs in those with a normal body mass index (BMI), often called non-obese NAFLD [8,9]. A series of risk factors has been described related to non-obese NAFLD, such as visceral obesity, high fructose intake, cholesterol intake, and genetic risk factors (e.g., palatin-like phospholipase domain-containing 3) [10]. Longitudinal studies found that short-term weight gain remained an independent risk factor for NAFLD development in non-obese individuals [11,12]. By contrast, tracking long-term weight may accurately and integrally reflect the dynamic change of excess body fat through an entire life, thereby improving the proposal of health guidance or recommendation on weight control for different age phases. Weight gain through early and middle adulthood may be relevant to NAFLD because of growth of age and excess adiposity deposition during this period [13,14].

To date, few studies have solely investigated the impact of early and midlife adulthood weight gain on NAFLD risk. A recent study based on the Nurses' Health Study II cohort found that both early and lifetime adulthood weight gain trajectories were independently associated with an excess risk of developing NAFLD [15]. Although based on a well-designed cohort, the generalizability of the findings was constrained by the fact that the samples were female and not nationally representative (selection bias), and NAFLD ascertainment was self-reported (information bias). The National Health and Nutrition Examination Survey (NHANES) is a nationally representative designed survey. The questionnaire contains weight history questions, including recalled weight and height at age 25 and recalled weight at 10 years prior to baseline, alongside measured weight and height at the baseline survey. NAFLD It also diagnosed by using an accurate and quantified application of vibration-controlled transient elastography (VCTE). Taking advantage of the NHANES 2017–2018 database, we have an opportunity to investigate the association between histories of weight gain and the incidence of NAFLD.

Thus, using the nationally representative NHANES 2017–2018 data, we aimed to (1) determine whether non-obese individuals who gained weight during early and middle adulthood were at an increased risk of NAFLD (the risk raise hypothesis); (2) determine whether non-obese individuals who converted from obese to non-obese through adulthood were at a reduced risk of NAFLD (the risk elimination hypothesis); (3) qualify the amount of NAFLD that would be prevented if the total population maintained a normal BMI during a life course.

2. Materials and Methods

2.1. Study Population

We used data from the 2017–2018 cycle of NHANES, which exclusively assessed hepatic steatosis and liver fibrosis by VCTE. The NHANES is a cross-sectional survey program well represented for the civilian United States population with a complex stratified, multistage and clustered probability sampling design [16–18]. We used recalled questions on weight history and NAFLD diagnosed by VCTE at baseline (after excluding self-reported fatty liver) to establish a retrospective cohort based on the cross-sectional data. The survey was approved by the National Center for Health Statistics Research Ethics Review Board (protocol number: 2018-01), and written informed consent was obtained from all participants.

In this study, participants aged $\geq$ 36 years old (n = 4218) at baseline were enrolled. Excluded from the study were pregnant women (n = 11), those who did not receive physical examination (n = 227), with BMI $\geq$ 30 kg/m^2 at age 25 or missing BMI at age 25 or baseline (n = 287), and those who partially completed, were ineligible, or did not perform VCTE (n = 458). Then, we excluded participants with excessive alcohol consumption ($\geq$2 drinks/day in men or $\geq$1 drink/day in women) (n = 150), exposure to hepatitis virus infection (n = 132), self-reported fatty liver and other liver conditions (n = 75), and self-reported malignancy (n = 366). In total, 2212 participants who were non-obese at age 25 were included in the analysis (Figure S1).

2.2. Measurement of Exposure: Weight Change Patterns

Baseline weight and height were measured during physical examination. Weight at age 25 and 10 years prior to baseline were recalled at in-person interview. BMI was calculated as weight (kg) divided by the square of height (m^2). BMI at age 25 ($BMI_{age\ 25}$) was calculated using recalled height and weight, considering the possibility of height declined with age. BMI at 10 years prior to baseline ($BMI_{10\ prior}$, mean age 45.74 years old) was calculated using recalled weight and measured height. BMI at baseline ($BMI_{baseline}$, mean age 55.74 years old) was calculated using measured height and weight.

BMI change patterns were generated according to BMI at the three timepoints: "stable non-obese" ($BMI_{age\ 25} < 30$ kg/m^2, $BMI_{10\ prior} < 30$ kg/m^2 and $BMI_{baseline} < 30$ kg/m^2), "early adulthood weight gain" ($BMI_{age\ 25} < 30$ kg/m^2, $BMI_{10\ prior} \geq 30$ kg/m^2 and $BMI_{baseline} \geq 30$ kg/m^2), "middle and late adulthood weight gain" ($BMI_{age\ 25} < 30$ kg/m^2, $BMI_{10\ prior} < 30$ kg/m^2 and $BMI_{baseline} \geq 30$ kg/m^2), "revert to non-obese" ($BMI_{age\ 25} < 30$ kg/m^2, $BMI_{10\ prior} \geq 30$ kg/m^2 and $BMI_{baseline} < 30$ kg/m^2) (Figure 1).

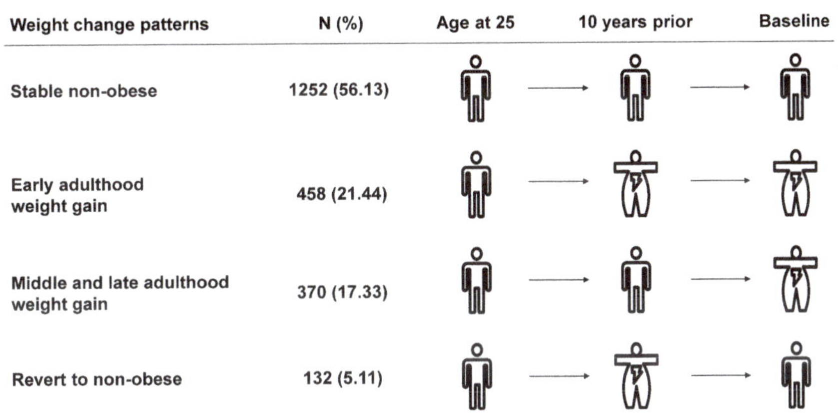

Figure 1. The definition of weight change patterns.

2.3. Measurement of Outcome: Fatty Liver

In the 2017–2018 cycle, FibroScan® model 502 V2 Touch (Echosens™, Nevada, North America) was performed, equipped with a medium (M) (74% of participants) or extra-large (XL) probe. Controlled attenuation parameter (CAP) scores were derived as the indicator of hepatic steatosis, and liver stiffness measurement (LSM) scores were measured to detect liver fibrosis.

One was defined as NAFLD if he/she carried a median CAP score ≥ 280 dB/m (S3, severe fatty liver, the cutoff of sensitivity fixed at 82% [19]). Participants with a median LSM value of 9.7 kPa or more were considered with advanced fibrosis [20].

2.4. Covariates

Information on covariates was obtained through baseline questionnaires, including age (years), sex (male and female), race and ethnicity (non-Hispanic White, non-Hispanic Black, non-Hispanic Asian, Hispanic and other), education (less than high school, high school or equivalent, college or above), family income–poverty ratio level (0–1.0, 1.1–3.0, >3.0) and marital status (married, separated, never married). Alcohol consumption was defined as a non-drinker and low to moderate drinker (<2 drinks/day in men and <1 drink/day in women). Smoking status was grouped into never smoker, former smoker and current smoker. We defined leisure time physical activity level as 0 times/week, 1–2 times/week and $\geq$3 times/week. Healthy eating index scores (HEI-2015) were calculated to represent dietary quality. Diabetes was defined as fasting plasma glucose ≥ 126 mg/dL, HbA1c $\geq 6.5\%$, and/or currently taking insulin or diabetic pills [21]. Hypertension was defined as average blood pressure ≥ 140/90 mmHg, and/or receipt of an anti-hypertensive medication [22]. Dyslipidemia was defined as total cholesterol ≥ 240 mg/dL, and/or taking prescription for cholesterol [23].

2.5. *Statistical Analysis*

2.5.1. Baseline Description

Appropriate sampling weights, clusters, and stratums were applied to this study. Continuous variables were presented as weighted mean ± standard errors (SE), and categorical variables were presented as numbers with percentages. We compared baseline characteristics by weight change patterns using linear regression adjusted for sampling weights for continuous variables and Rao-Scott χ^2 test for categorical variables.

2.5.2. Association Analysis

We qualified the association of weight change patterns with NAFLD using the multivariable modified Poisson regression model with a robust variance estimator to estimate rate ratios (RRs) and 95% confidence intervals (CI) directly, recognizing that odds ratios (ORs) may overestimate effect differences and do not give a good approximation of the RRs when the outcome is common (affecting > 10% of population). We adjusted baseline age, sex and race/ethnicity in model 1. Model 2 was additionally adjusted for education level, family income–poverty ratio level, marital status, alcohol consumption, smoking status, and chronic diseases. Model 3 was further adjusted for leisure time physical activity level and HEI. We also assessed the association between weight change patterns and advanced fibrosis in the secondary analysis.

2.5.3. Calculation of Population Attributable Fraction (PAF)

We calculated PAF to estimate the percentage of NAFLD cases that could be prevented under the four hypothetical scenarios: (1) "weight loss", defined as the scenario in which individuals who gained weight from early adulthood could have lost weight in later adulthood ("early adulthood weight gain" group vs. "revert to non-obese" group); (2) "weight maintenance", defined as the scenario in which individuals who gained weight to obese could have maintained non-obese during adulthood ("early or middle and late adulthood weight gain" group vs. "stable non-obese" group); (3) "partial prevention", defined as the scenario in which the total population maintained non-obese across adulthood (BMI < 30 kg/m^2, "stable non-obese" vs. other counterparts); and (4) "comprehensive prevention", defined as the scenario in which the total population had a normal BMI during the life course (BMI < 25 kg/m^2, "stable normal" vs. other counterparts).

Consistent with prior studies [24], PAFs were calculated according to the following equation: $PAF = \sum_i pd_i\left(\frac{RR_i-1}{RR_i}\right)$, where pd_i is the proportion of total NAFLD cases observed in the ith weight change pattern category and RR_i is the adjusted rate ratio for the ith exposure category. A category-specific attributable fraction is the percentage of NAFLD cases that would be eliminated if individuals in that weight change category were to be shifted to another lower risk category, assuming a causal relation.

2.5.4. Sensitivity Analysis

We performed a series of sensitivity analyses to access the stability of the results. First, we defined NAFLD as a median CAP value of 248 dB/m or more (≥S1, mild fatty liver) and a median CAP value of 268 dB/m or more (≥S2, moderate fatty liver). Second, we removed participants with baseline diabetes, hypertension and dyslipidemia to avoid inverse association. Third, we classified absolute weight change into five groups (weight loss of at least 2.5 kg, weight change within 2.5 kg, weight gain of at least 2.5 kg but less than 10.0 kg, weight gain of at least 10 kg but less than 20.0 kg and weight gain of at least 20.0 kg) and examined the association between absolute weight change group and NAFLD. Finally, the association between absolute weight change (continuous variable) and

NAFLD was executed to explore the possible non-linear relation by using multivariable linear regression models based on restricted cubic splines with 4 knots.

All analyses were performed by Stata 16.0 (StataCorp, College Station, TX, USA), using Taylor series linearization. A two-tailed p value less than 0.05 was considered statistically significant.

3. Results

3.1. Baseline Characteristics

The correlation coefficients between $BMI_{age\ 25}$, $BMI_{10\ prior}$, and $BMI_{baseline}$ ranged from 0.397 to 0.668 (Table S1). The mean age was 55.74 years old at baseline and 46.22% were male. Throughout the whole adulthood, 1252 (56.13%) participants remained non-obese across adulthood (classified as "stable non-obese" group), 458 (21.44%) became obese from early adulthood (classified as "early adulthood weight gain" group), and 370 (17.33%) moved to obesity from middle and late adulthood (classified as "middle and late adulthood weight gain" group). There were also 132 (5.11%) participants who moved to obese at 10 years prior but reverted to the non-obese at baseline (classified as "revert to non-obese" group) (Figure 1).

Table 1 shows baseline characteristics of study participants. The "early adulthood weight gain" group had a higher proportion of non-Hispanic White, and the "middle and late adulthood weight gain" group had a higher proportion of non-Hispanic Black. Compared with the "stable non-obese" group, the weight gaining groups were more likely to be low to moderate drinkers, current smokers, physically inactive and have a lower healthy eating index score. Participants in the "revert to the non-obese" group were more likely to be former smokers, physically active and have healthier eating habit.

Table 1. Baseline characteristics of study participants in NHANES 2017–2018.

Characteristics	Total (n = 2212, 100%)	Weight Change Patterns				p Value
		Stable Non-Obese (n = 1252, 56.13%)	Early Adulthood Weight Gain (n = 458, 21.44%)	Middle and Late Adulthood Weight Gain (n = 370, 17.33%)	Revert to Non-Obese (n = 132, 5.11%)	
Age, years, mean ± SE	55.74 ± 0.47	55.24 ± 0.50	58.25 ± 0.90	52.58 ± 0.89	61.43 ± 1.30	<0.001
Sex, n (%)						0.009
Male	1056 (46.22)	611 (44.03)	223 (56.10)	148 (38.12)	74 (56.23)	
Female	1156 (53.78)	641 (55.97)	235 (43.90)	222 (61.88)	58 (43.77)	
BMI, kg/m^2, mean ± SE						
At age 25 years	22.85 ± 0.12	21.77 ± 0.12	25.05 ± 0.31	23.56 ± 0.21	23.17 ± 0.36	<0.001
10 years prior	27.38 ± 0.21	24.20 ± 0.07	34.65 ± 0.30	27.10 ± 0.16	33.17 ± 0.66	<0.001
At baseline	29.11 ± 0.28	25.16 ± 0.14	35.95 ± 0.42	33.92 ± 0.16	27.47 ± 0.24	<0.001
Absolute weight change from age 25 years old to baseline, kg, mean ± SE	15.72 ± 0.56	7.95 ± 0.42	28.66 ± 0.95	26.68 ± 0.71	9.52 ± 1.42	<0.001
Waist circumference, cm, mean ± SE	100.32 ± 0.76	90.91 ± 0.40	117.87 ± 1.04	109.78 ± 0.65	98.13 ± 0.84	<0.001
Race and ethnicity, n (%)						<0.001
Non-Hispanic White	727 (64.00)	381 (61.52)	181 (71.55)	115 (61.73)	50 (67.34)	
Non-Hispanic Black	533 (11.03)	248 (9.24)	144 (12.39)	117 (15.93)	24 (8.42)	
Non-Hispanic Asian	385 (6.90)	330 (10.49)	14 (1.16)	32 (3.63)	9 (2.58)	
Hispanic	460 (13.54)	245 (14.29)	86 (9.14)	89 (15.41)	40 (17.44)	
Other	107 (4.53)	48 (4.46)	33 (5.76)	17 (3.30)	9 (4.23)	
Education, n (%)						0.158
Less than high school	416 (10.85)	230 (11.36)	81 (8.46)	68 (10.83)	37 (15.41)	
High school or equivalent	498 (25.63)	265 (23.55)	100 (24.34)	96 (32.51)	37 (30.64)	
College or above	1293 (63.52)	754 (65.10)	277 (67.19)	205 (56.66)	57 (53.95)	
Family income–poverty ratio level, n (%)						0.169
0–1.0	303 (10.06)	159 (9.21)	63 (7.71)	65 (15.04)	16 (11.88)	
1.1–3.0	838 (33.00)	472 (32.40)	183 (32.10)	132 (34.18)	51 (39.15)	
>3.0	806 (56.94)	480 (58.39)	157 (60.19)	130 (50.78)	39 (48.97)	

Table 1. *Cont.*

Characteristics	Weight Change Patterns					
	Total (*n* = 2212, 100%)	Stable Non-Obese (*n* = 1252, 56.13%)	Early Adulthood Weight Gain (*n* = 458, 21.44%)	Middle and Late Adulthood Weight Gain (*n* = 370, 17.33%)	Revert to Non-Obese (*n* = 132, 5.11%)	*p* Value
Marital status, *n* (%)						0.352
Married	1293 (64.95)	786 (67.45)	245 (62.35)	191 (61.68)	71 (59.47)	
Separated	581 (23.47)	286 (21.35)	154 (27.86)	103 (24.53)	38 (24.76)	
Never married	335 (11.58)	178 (11.20)	59 (9.80)	75 (13.78)	23 (15.77)	
Alcohol consumption, *n* (%)						0.005
Non-drinker	747 (27.84)	428 (29.48)	167 (27.98)	97 (18.39)	55 (41.72)	
Low to moderate drinker	1340 (72.16)	737 (70.52)	274 (72.02)	259 (81.61)	70 (58.28)	
Smoking status, *n* (%)						0.113
Never smoker	1284 (59.21)	752 (60.44)	242 (54.74)	216 (60.30)	74 (60.85)	
Former smoker	335 (12.71)	195 (13.05)	51 (9.45)	62 (13.19)	27 (20.90)	
Current smoker	593 (28.08)	305 (26.51)	165 (35.81)	92 (26.50)	31 (18.25)	
Leisure time physical activity level, *n* (%)						0.003
0 times/week	1199 (45.91)	633 (41.16)	275 (49.85)	207 (54.60)	84 (51.96)	
1–2 times/week	707 (38.06)	421 (38.55)	135 (40.34)	122 (36.03)	29 (29.90)	
≥3 times/week	304 (16.04)	196 (20.28)	48 (9.82)	41 (9.37)	19 (18.14)	
Healthy eating index score, *n* (%)						0.021
Quarter 1	381 (20.38)	174 (15.79)	106 (25.25)	76 (27.94)	25 (23.57)	
Quarter 2	484 (25.60)	257 (24.54)	115 (28.51)	79 (26.24)	33 (22.40)	
Quarter 3	514 (24.30)	291 (25.85)	107 (23.95)	92 (22.67)	24 (14.58)	
Quarter 4	667 (29.72)	422 (33.83)	108 (22.29)	99 (23.16)	38 (39.45)	
Chronic diseases, *n* (%)						
Diabetes	488 (15.31)	188 (8.254)	174 (32.15)	77 (13.46)	49 (28.35)	<0.001
Hypertension	1013 (39.03)	478 (30.92)	281 (52.56)	174 (42.63)	80 (57.13)	<0.001
Dyslipidemia	812 (34.13)	433 (31.06)	202 (44.02)	120 (30.09)	57 (39.15)	0.035

In total, 5, 265, 3, 125, 2, 166, 54 and 58 participants had missing information for baseline education, family income–poverty ratio, marital status, alcohol consumption, leisure time physical activity level, healthy eating index score, hypertension and dyslipidemia, respectively.

3.2. Relations of Weight Change Patterns with NAFLD

When evaluating the weight status at each time point (Table S2), we found that overweight and obesity were significantly associated with increased risks of NAFLD (p for trend < 0.001). Substantially rising curves were observed when we assessed the association between absolute weight changes and CAP value during the three-time intervals (Figure S2A). When describing the distribution of CAP value across the weight change patterns (Figure S2B), the "early adulthood weight gain" group had the highest median CAP value, the "middle and late adulthood weight gain" group came a close second, whereas the "revert to non-obese" group sharply declined, and the "stable non-obese" group came lowest ($p < 0.001$).

The association of weight change patterns across adulthood with NAFLD was presents in Table 2. The age-adjusted incidence rate of NAFLD among the four groups from the highest to the lowest is below: 67.20% (95% CI 58.75–75.64%) in the "early adulthood weight gain" group, 54.57% (95% CI 46.82–62.31%) in the "middle and late adulthood weight gain" group, 37.40% (95% CI 27.85–46.95%) in the "revert to the non-obese" group and 23.82% (95% CI 19.66–27.98%) in the "stable non-obese" group. Compared with the stable non-obese participants, those who gained weight in early adulthood were associated with a 119% higher risk of NAFLD (RR 2.19, 95% CI 1.64–2.91) and those who gained weight at middle and late adulthood had 92% higher risk of NAFLD (RR 1.92, 95% CI 1.40–2.62). Notably, the "revert to non-obese" group showed a null association with NAFLD, with RR of 1.01 (95% CI 0.62–1.64).

Table 2. Association of weight change patterns across adulthood with NAFLD in NHANES 2017–2018.

	Weight Change Patterns			
	Stable Non-Obese	**Early Adulthood Weight Gain**	**Middle and Late Adulthood Weight Gain**	**Revert to Non-Obese**
Number of NAFLD	355	298	226	45
Age adjusted incidence rate (%)	23.82 (19.66, 27.98)	67.20 (58.75, 75.64)	54.57 (46.82, 62.31)	37.40 (27.85, 46.95)
Model 1, RR (95% CI)	1.00 (ref)	2.77 (2.21, 3.49)	2.33 (1.78, 3.05)	1.37 (0.95, 1.98)
Model 2, RR (95% CI)	1.00 (ref)	2.23 (1.73, 2.88)	1.98 (1.45, 2.71)	1.09 (0.70, 1.68)
Model 3, RR (95% CI)	1.00 (ref)	2.19 (1.64, 2.91)	1.92 (1.40, 2.62)	1.01 (0.62, 1.64)

Model 1 was adjusted for age, sex, and race/ethnicity. Model 2 was additionally adjusted for education level, family income–poverty ratio level, marital status, alcohol consumption and smoking status, and chronic diseases. Model 3 was further adjusted for leisure time physical activity level and HEI. Incidence rates were directly standardized to age distribution of entire study population.

Regarding the risk raise hypothesis, participants who gained weight through adulthood had a 107% higher risk of NAFLD (RR 2.07, 95% CI 1.60–2.67) compared with those stable non-obese participants (Table 3). In the risk elimination hypothesis, we observed that the participants who reverted to non-obese had a 52% lower risk of NAFLD (RR 0.48, 95% CI 0.37–0.61) when compared with those who gained weight at early adulthood.

Table 3. Risk raise and risk elimination: RRs for weight change pattern and incident NAFLD.

Hypothesis	Weight Change Pattern	RR (95% CI)	*p* Value
Risk raise	Stable non-obese	1.00 (ref)	
	Early or middle and late adulthood weight gain	2.07 (1.60, 2.67)	<0.001
Risk elimination	Early adulthood weight gain	1.00 (ref)	
	Revert to non-obese	0.48 (0.37, 0.61)	<0.001

Risk estimates were adjusted for baseline age, sex, race/ethnicity, education level, family income–poverty ratio level, marital status, alcohol consumption and smoking status, chronic diseases, baseline leisure time physical activity level and HEI.

In the stratified analysis, we found that the associations were stronger among participants who were younger (p for interaction = 0.015, Figure S3A). The overall patterns were broadly comparable when stratified by sex (Figure S3B). We observed that the effects of weight gain were stronger in non-Hispanic Black when stratified by race and ethnicity; however, there were no significant interactions with race (p for interaction = 0.844, Figure S4).

In the secondary analysis, we reported the association of weight change patterns with advanced fibrosis in Table S3. Similarly, compared with the "stable non-obese" participants, those who gained at early adulthood or middle and late adulthood had a higher incidence risk of advanced fibrosis, with an RR of 3.20 (95% CI 1.19–8.62) and 2.46 (95% CI 0.97–6.20). We observed a null association between "revert to non-obese" with advanced fibrosis, with an RR of 0.70 (95% CI 0.14–3.54).

3.3. Assessment of the Public Health Impact of Weight Change in Populations

PAFs for population counterfactuals are reported in Table 4. In the weight loss scenario, if those who gained weight from early adulthood could have lost weight in later adulthood, 16.27% (95% CI 12.16–19.50%) of observed NAFLD cases in the total population might have been averted. In the weight maintenance scenario, if those who gained weight could have maintained non-obese during adulthood, 26.74% (95% CI 19.45–32.38%) of the incident NAFLD might have been prevented. In the partial prevention scenario, if the total population maintained non-obese (BMI < 30 kg/m^2) during the life course, 26.67% (95% CI 18.17–33.23%) of NAFLD cases would have been prevented. In the comprehensive prevention scenario, keeping a normal weight (BMI < 25 kg/m^2) across adulthood would have prevented 73.09% (95% CI 55.62–82.93%) of NAFLD cases. As for advanced fibrosis, if the total population had a normal BMI, 80.26% (95% CI 53.14–91.23%) of the case could have been averted (Table S4).

Table 4. PAFs for population counterfactuals of NAFLD.

Scenario	Definition	PAF (%), 95% CI of Non-Obese Population	PAF (%), 95% CI of Total Population
Weight loss	If those who gained weight from early adulthood could have lost weight in later adulthood ("early adulthood weight gain" group vs. "revert to non-obese" group).	18.98 (14.18, 22.74)	16.27 (12.16, 19.50)
Weight maintenance	If those who gained weight could have maintained non-obese during adulthood ("early or middle and late adulthood weight gain" group vs. "stable non-obese" group).	31.18 (22.69, 37.76)	26.74 (19.45, 32.38)
Partial prevention	If the total population maintained non-obese (BMI < 30 kg/m^2) across adulthood ("stable non-obese" vs. other counterparts).	31.10 (21.19, 38.76)	26.67 (18.17, 33.23)
Comprehensive prevention	If the total population had a normal BMI (BMI < 25 kg/m^2) across adulthood ("stable normal" vs. other counterparts).	73.32 (55.03, 82.05)	73.09 (55.62, 82.93)

Risk estimates were adjusted for baseline age, sex, race/ethnicity, education level, family income–poverty ratio level, marital status, alcohol consumption and smoking status, chronic diseases, baseline leisure time physical activity level and HEI.

3.4. Sensitivity Analysis

Similar patterns of results were observed for mild fatty liver (Table S5) and moderate fatty liver (Table S6). The results were substantially unchanged when we removed participants with chronic diseases at baseline (Table S7). When evaluating the absolute weight changes (categorical variables), there was a dose–response association with risk of NAFLD (Table S8). Furthermore, we identified a J-shaped relationship between NAFLD and absolute weight change (continuous variables) during the three intervals (p for non-linear < 0.001, Figure S5A).

4. Discussion

Based on the nationally representative US adults' cohort, we revealed the association between long-term weight change patterns across adulthood and NAFLD identified by VCTE among non-obese individuals. The lowest NAFLD risk was observed in stable non-obese individuals, whereas weight gain from early adulthood and middle and late adulthood were both at significantly elevated risk for NAFLD. Additionally, the risk of individuals who reverted to non-obese showed no significant difference compared with stable non-obese individuals. We further observed that a large percentage of NAFLD and advanced fibrosis cases would have been prevented with effective weight intervention in early adulthood.

We found strong evidence in support of the "risk raise" hypothesis. Those who gained weight to obese had an elevated risk of NAFLD. The association between absolute weight gain in midlife and NAFLD has been explicitly studied in previous longitudinal cohorts [11,25]. However, most studies were designed to assess the short-term effect of weight gain on incident NAFLD, and linear association was approximately reported. In this study, we reached a similar conclusion by using the long-term weight change through the adulthood as the exposure. In addition, several studies with the long-term track of weight have reported the linear relationship between the magnitude of weight growth and NAFLD risk [26,27], which were also seen in our current findings. In the study here, weight gain from a non-obese to an obese pattern from young to middle and middle to late adulthood had a 119% and 92% higher NAFLD risk, respectively. Moderate weight gain (ranging from 10.0 to 20.0 kg) and extreme weight gain (≥20.0 kg) from age 25 years to baseline were associated with a 166% and 279% higher NAFLD risk. Hence, our findings were generally consistent with previous studies.

We found that the effect of weight gain was generally stronger from young to middle adulthood than from middle to late adulthood. Previous studies have reported that weight gain from early adulthood is a risk factor for metabolic disorder [12,28], diabetes [17], hypertension [29], cardiovascular diseases [30]. Additionally, weight gain from earlier adulthood is more strongly associated with unfavorable metabolic indicators (e.g., adiponectin, C-peptide, HbA1c and gamma-glutamyl transferase) than weight gain from later adulthood [28,31]. Perennial accumulation of adipose tissue accumulation may result in abnormalities of free fatty acid [14]. Excessive free fatty acid transport to the liver and skeletal muscle leads to an increase in intrahepatic triglyceride and causes accumulation of liver fat [32,33]. Although the potential mechanism needs to be further studied, our results indicate that early prevention of weight gain would be an effective strategy to reduce future NAFLD risk.

Our findings were in support of the "risk elimination" hypothesis. In our study, participants who reverted to non-obese in midlife had a 52% lower risk of NAFLD when compared with the "early adulthood weight gain" group and even had a similar incidence rate with the "stable non-obese" group. The results were in line with previous prospective cohort and trial studies that weight loss through lifestyle interventions improve biomarkers of NAFLD [34,35]. A systematic review and meta-analysis also found a dose–response relationship between absolute weight loss and resolution of NAFLD [36]. In the current study, we estimated the PAF to explore the potential effect of weight loss intervention. As evaluated, a proportion of 16.27% NAFLD cases in the total population would be prevented if those who gained weight from early adulthood took measures to lose weight. We also found that 26.67% and 73.09% of NAFLD cases would be prevented if the total population maintained a non-obese or normal BMI throughout their life course, respectively. In total, a normal weight is associated with a greater improvement of NAFLD.

This study had several strengths. Taking advantage of NHANES, a large nationwide representative survey, our study had the unique feature of evaluating weight change across the life course and the development of NAFLD, and the results were more broadly generalizable to the US population. Our study shows the importance of weight change surveillance through an entire life because the underlying mechanisms of weight gain on NAFLD in certain life periods might differ. In early adulthood, weight gain was due in large part to accumulation of fat mass, whereas in later adulthoods, it was generally attributed to a decrease in lean mass [18]. Another strength of our research is the application of VCTE in fatty liver diagnosis, which further reduced information bias due to its accurate and quantitative features compared with abdomen ultrasound. Furthermore, we adjusted for adequate potential confounders to improve the stability of results.

Our study also had several limitations. First, we used recalled data on weight at age 25 and 10 years prior to baseline survey such that the misclassification bias could not be ignored. However, several validation studies suggested that self-reported weight was strongly correlated with anthropometric measures and could be used in life course epidemiology studies [37,38]. Second, we could not distinguish whether the weight change was intentional or unintentional, especially for those in the revert to non-obese group. Third, we did not capture the exact time of NAFLD onset, which may confound the effect of weight change on NAFLD.

5. Conclusions

In the non-obese individuals, weight gain across entire adulthood was independently associated with an increased risk of developing NAFLD. A larger reduction in NAFLD risk can be expected with losing weight in later life. Taken together, this study provides critical quantitative estimates that strategies for maintaining a healthy weight at the individual level may help reduce NAFLD progression and its associated consequences. At the national level, policies and programs developing to control the prevalence of obesity would be beneficial for the primary prevention of NAFLD.

Supplementary Materials: The following supporting information can be downloaded at: https://www.mdpi.com/article/10.3390/nu14102140/s1. Figure S1. Flowchart for the selection of study participants; Figure S2. The relationship between weight change and CAP value; Figure S3. Associations between weight change patterns across adulthood and risk of NAFLD S3 stratified by age and sex; Figure S4. Associations between weight change patterns across adulthood and risk of NAFLD S3 stratified by race/ethnicity; Figure S5. Dose–response association between absolute weight change across adulthood and risk of NAFLD S3 and advanced fibrosis; Table S1. Pearson correlation coefficients for BMI at three time points and absolute weight change during three intervals; Table S2. Incidence rate ratios (95% confidence intervals) of NAFLD S3 with BMI at three time points in the NHANES 2017–2018; Table S3. Association between weight change patterns across adulthood and advanced fibrosis in NHANES 2017–2018; Table S4. Population attributable fractions (PAF) for population counterfactuals of advanced fibrosis; Table S5. Sensitivity analyses of the association between weight change patterns across adulthood and NAFLD S1 in NHANES 2017–2018; Table S6. Sensitivity analyses of the association between weight change patterns across adulthood and NAFLD S2 in NHANES 2017–2018; Table S7. Sensitivity analyses of the association between weight change patterns across adulthood and NAFLD S3 with exclusion of chronic diseases at baseline; Table S8. Association between absolute weight change across adulthood and NAFLD S3 in NHANES 2017–2018.

Author Contributions: Conceptualization, X.X., Y.D., T.T., C.Y., C.S. and Z.H.; methodology, Y.D., X.X. and T.T.; software, Y.D. and C.Y.; validation, X.X., H.M., T.J. and Y.J.; formal analysis, Y.D. and X.X.; investigation, X.X., Z.G., J.G. and X.G.; resources, C.S. and Z.H.; data curation, C.S. and Z.H.; writing—original draft preparation, Y.D. and X.X.; writing—review and editing, all authors; visualization, Y.D. and J.L.; supervision, C.S. and Z.H.; project administration, C.S. and Z.H.; funding acquisition, C.S. All authors have read and agreed to the published version of the manuscript.

Funding: This research was funded by the National Natural Science Foundation of China (grant numbers: 81903382); Natural Science Foundation of Jiangsu Province (grant numbers: BK20190652); Science and Technology Young Scientific and Technological Talents Project of Jiangsu Province (grant numbers: 2021-50); China Postdoctoral Science Foundation (grant numbers: General Program, 2019M651900), and National Science Foundation for Post-doctoral Scientists of China (grant numbers: 2018M640466).

Institutional Review Board Statement: This study was approved by the National Center for Health Statistics Research Ethics Review Board (Protocol number: 2018-01).

Informed Consent Statement: Written informed consent was obtained from all participants.

Data Availability Statement: Detailed survey operation manuals, consent documents, and brochures are available on the NHANES website (https://www.cdc.gov/nchs/nhanes/about_nhanes, accessed on 9 May 2022).

Acknowledgments: The most important acknowledgement is to the participants in NHANES and officers for data collection and review, as well as to the project development and management teams.

Conflicts of Interest: The authors declare no conflict of interest.

References

1. Younossi, Z.M.; Koenig, A.B.; Abdelatif, D.; Fazel, Y.; Henry, L.; Wymer, M. Global epidemiology of nonalcoholic fatty liver disease-Meta-analytic assessment of prevalence, incidence, and outcomes. *Hepatology* **2016**, *64*, 73–84. [CrossRef] [PubMed]
2. Younossi, Z.; Tacke, F.; Arrese, M.; Chander Sharma, B.; Mostafa, I.; Bugianesi, E.; Wai-Sun Wong, V.; Yilmaz, Y.; George, J.; Fan, J.; et al. Global Perspectives on Nonalcoholic Fatty Liver Disease and Nonalcoholic Steatohepatitis. *Hepatology* **2019**, *69*, 2672–2682. [CrossRef] [PubMed]
3. Younossi, Z.M. Non-alcoholic fatty liver disease—A global public health perspective. *J. Hepatol.* **2019**, *70*, 531–544. [CrossRef] [PubMed]
4. Byrne, C.D.; Targher, G. NAFLD: A multisystem disease. *J. Hepatol.* **2015**, *62* (Suppl. S1), S47–S64. [CrossRef]
5. Adams, L.A.; Anstee, Q.M.; Tilg, H.; Targher, G. Non-alcoholic fatty liver disease and its relationship with cardiovascular disease and other extrahepatic diseases. *Gut* **2017**, *66*, 1138–1153. [CrossRef]
6. Li, L.; Liu, D.W.; Yan, H.Y.; Wang, Z.Y.; Zhao, S.H.; Wang, B. Obesity is an independent risk factor for non-alcoholic fatty liver disease: Evidence from a meta-analysis of 21 cohort studies. *Obes. Rev.* **2016**, *17*, 510–519. [CrossRef]
7. Williams, C.D.; Stengel, J.; Asike, M.I.; Torres, D.M.; Shaw, J.; Contreras, M.; Landt, C.L.; Harrison, S.A. Prevalence of nonalcoholic fatty liver disease and nonalcoholic steatohepatitis among a largely middle-aged population utilizing ultrasound and liver biopsy: A prospective study. *Gastroenterology* **2011**, *140*, 124–131. [CrossRef]
8. Younossi, Z.M.; Corey, K.E.; Lim, J.K. AGA Clinical Practice Update on Lifestyle Modification Using Diet and Exercise to Achieve Weight Loss in the Management of Nonalcoholic Fatty Liver Disease: Expert Review. *Gastroenterology* **2021**, *160*, 912–918. [CrossRef]
9. Kwon, Y.M.; Oh, S.W.; Hwang, S.S.; Lee, C.; Kwon, H.; Chung, G.E. Association of nonalcoholic fatty liver disease with components of metabolic syndrome according to body mass index in Korean adults. *Am. J. Gastroenterol.* **2012**, *107*, 1852–1858. [CrossRef]
10. Kim, D.; Kim, W.R. Nonobese Fatty Liver Disease. *Clin. Gastroenterol. Hepatol.* **2017**, *15*, 474–485. [CrossRef]
11. Yamada, G.; Hagiwara, Y.; Kimura, T.; Takeuchi, Y.; Oba, K.; Masuda, K.; Matsuyama, Y. Impact of Body Weight Gain on the Incidence of Nonalcoholic Fatty Liver Disease in Nonobese Japanese Individuals. *Am. J. Gastroenterol.* **2021**, *116*, 733–740. [CrossRef] [PubMed]
12. Zheng, Y.; Manson, J.E.; Yuan, C.; Liang, M.H.; Grodstein, F.; Stampfer, M.J.; Willett, W.C.; Hu, F.B. Associations of Weight Gain From Early to Middle Adulthood with Major Health Outcomes Later in Life. *JAMA* **2017**, *318*, 255–269. [CrossRef] [PubMed]
13. Adams, K.F.; Leitzmann, M.F.; Ballard-Barbash, R.; Albanes, D.; Harris, T.B.; Hollenbeck, A.; Kipnis, V. Body mass and weight change in adults in relation to mortality risk. *Am. J. Epidemiol.* **2014**, *179*, 135–144. [CrossRef] [PubMed]
14. Fabbrini, E.; Sullivan, S.; Klein, S. Obesity and nonalcoholic fatty liver disease: Biochemical, metabolic, and clinical implications. *Hepatology* **2010**, *51*, 679–689. [CrossRef]
15. Kim, M.N.; Lo, C.H.; Corey, K.E.; Liu, P.H.; Ma, W.; Zhang, X.; Jovani, M.; Song, M.; Chan, A.T.; Simon, T.G. Weight gain during early adulthood, trajectory of body shape and the risk of nonalcoholic fatty liver disease: A prospective cohort study among women. *Metabolism* **2020**, *113*, 154398. [CrossRef]
16. Centers for Disease Control and Prevention (CDC). About the National Health and Nutrition Examination Survey. 2017. Available online: https://www.cdc.gov/nchs/data/nhanes/nhanes_13_14/NHANES_Overview_Brochure.pdf (accessed on 5 January 2022).
17. Stokes, A.; Collins, J.M.; Grant, B.F.; Scamuffa, R.F.; Hsiao, C.W.; Johnston, S.S.; Ammann, E.M.; Manson, J.E.; Preston, S.H. Obesity Progression Between Young Adulthood and Midlife and Incident Diabetes: A Retrospective Cohort Study of U.S. Adults. *Diabetes Care* **2018**, *41*, 1025–1031. [CrossRef]
18. Chen, C.; Ye, Y.; Zhang, Y.; Pan, X.F.; Pan, A. Weight change across adulthood in relation to all cause and cause specific mortality. Prospective cohort study. *BMJ* **2019**, *367*, l5584. [CrossRef]
19. Karlas, T.; Petroff, D.; Sasso, M.; Fan, J.G.; Mi, Y.Q.; de Ledinghen, V.; Kumar, M.; Lupsor-Platon, M.; Han, K.H.; Cardoso, A.C.; et al. Individual patient data meta-analysis of controlled attenuation parameter (CAP) technology for assessing steatosis. *J. Hepatol.* **2017**, *66*, 1022–1030. [CrossRef]
20. Eddowes, P.J.; Sasso, M.; Allison, M.; Tsochatzis, E.; Anstee, Q.M.; Sheridan, D.; Guha, I.N.; Cobbold, J.F.; Deeks, J.J.; Paradis, V.; et al. Accuracy of FibroScan Controlled Attenuation Parameter and Liver Stiffness Measurement in Assessing Steatosis and Fibrosis in Patients With Nonalcoholic Fatty Liver Disease. *Gastroenterology* **2019**, *156*, 1717–1730. [CrossRef]
21. Wang, L.; Li, X.; Wang, Z.; Bancks, M.P.; Carnethon, M.R.; Greenland, P.; Feng, Y.Q.; Wang, H.; Zhong, V.W. Trends in Prevalence of Diabetes and Control of Risk Factors in Diabetes Among US Adults, 1999–2018. *JAMA* **2021**, *326*, 1–13. [CrossRef]
22. Muntner, P.; Hardy, S.T.; Fine, L.J.; Jaeger, B.C.; Wozniak, G.; Levitan, E.B.; Colantonio, L.D. Trends in Blood Pressure Control Among US Adults with Hypertension, 1999–2000 to 2017–2018. *JAMA* **2020**, *324*, 1190–1200. [CrossRef] [PubMed]
23. Decker, S.L.; Kostova, D.; Kenney, G.M.; Long, S.K. Health status, risk factors, and medical conditions among persons enrolled in Medicaid vs uninsured low-income adults potentially eligible for Medicaid under the Affordable Care Act. *JAMA* **2013**, *309*, 2579–2586. [CrossRef] [PubMed]
24. Rockhill, B.; Newman, B.; Weinberg, C. Use and misuse of population attributable fractions. *Am. J. Public Health* **1998**, *88*, 15–19. [CrossRef] [PubMed]

25. Zelber-Sagi, S.; Lotan, R.; Shlomai, A.; Webb, M.; Harrari, G.; Buch, A.; Nitzan Kaluski, D.; Halpern, Z.; Oren, R. Predictors for incidence and remission of NAFLD in the general population during a seven-year prospective follow-up. *J. Hepatol.* **2012**, *56*, 1145–1151. [CrossRef]
26. VanWagner, L.B.; Khan, S.S.; Ning, H.; Siddique, J.; Lewis, C.E.; Carr, J.J.; Vos, M.B.; Speliotes, E.; Terrault, N.A.; Rinella, M.E.; et al. Body mass index trajectories in young adulthood predict non-alcoholic fatty liver disease in middle age: The CARDIA cohort study. *Liver Int.* **2018**, *38*, 706–714. [CrossRef]
27. Du, S.; Wang, C.; Jiang, W.; Li, C.; Li, Y.; Feng, R.; Sun, C. The impact of body weight gain on nonalcoholic fatty liver disease and metabolic syndrome during earlier and later adulthood. *Diabetes Res. Clin. Pract.* **2016**, *116*, 183–191. [CrossRef]
28. Montonen, J.; Boeing, H.; Schleicher, E.; Fritsche, A.; Pischon, T. Association of changes in body mass index during earlier adulthood and later adulthood with circulating obesity biomarker concentrations in middle-aged men and women. *Diabetologia* **2011**, *54*, 1676–1683. [CrossRef]
29. Zhou, Y.; Wang, T.; Yin, X.; Sun, Y.; Seow, W.J. Association of Weight Loss from Early to Middle Adulthood and Incident Hypertension Risk Later in Life. *Nutrients* **2020**, *12*, 2622. [CrossRef]
30. Liu, M.; Zhang, Z.; Zhou, C.; He, P.; Zhang, Y.; Li, H.; Li, Q.; Liu, C.; Wang, B.; Li, J.; et al. Relationship of Weight Change Patterns From Young to Middle Adulthood With Incident Cardiovascular Diseases. *J. Clin. Endocrinol. Metab.* **2021**, *106*, e812–e823. [CrossRef]
31. Kvaerner, A.S.; Hang, D.; Giovannucci, E.L.; Willett, W.C.; Chan, A.T.; Song, M. Trajectories of body fatness from age 5 to 60 y and plasma biomarker concentrations of the insulin-insulin-like growth factor system. *Am. J. Clin. Nutr.* **2018**, *108*, 388–397. [CrossRef]
32. Shoelson, S.E.; Herrero, L.; Naaz, A. Obesity, inflammation, and insulin resistance. *Gastroenterology* **2007**, *132*, 2169–2180. [CrossRef] [PubMed]
33. Boden, G.; Shulman, G.I. Free fatty acids in obesity and type 2 diabetes: Defining their role in the development of insulin resistance and beta-cell dysfunction. *Eur. J. Clin. Investig.* **2002**, *32* (Suppl. S3), 14–23. [CrossRef] [PubMed]
34. Cho, I.Y.; Chang, Y.; Sung, E.; Kang, J.H.; Shin, H.; Wild, S.H.; Byrne, C.D.; Ryu, S. Weight Change and the Development of Nonalcoholic Fatty Liver Disease in Metabolically Healthy Overweight Individuals. *Clin. Gastroenterol. Hepatol.* **2021**, *20*, E505–E507. [CrossRef]
35. Koutoukidis, D.A.; Jebb, S.A.; Tomlinson, J.W.; Cobbold, J.F.; Aveyard, P. Association of Weight Changes With Changes in Histological Features and Blood Markers in Nonalcoholic Steatohepatitis. *Clin. Gastroenterol. Hepatol.* **2021**, *20*, E538–E547. [CrossRef] [PubMed]
36. Koutoukidis, D.A.; Koshiaris, C.; Henry, J.A.; Noreik, M.; Morris, E.; Manoharan, I.; Tudor, K.; Bodenham, E.; Dunnigan, A.; Jebb, S.A.; et al. The effect of the magnitude of weight loss on non-alcoholic fatty liver disease: A systematic review and meta-analysis. *Metabolism* **2021**, *115*, 154455. [CrossRef] [PubMed]
37. De Rubeis, V.; Bayat, S.; Griffith, L.E.; Smith, B.T.; Anderson, L.N. Validity of self-reported recall of anthropometric measures in early life: A systematic review and meta-analysis. *Obes. Rev.* **2019**, *20*, 1426–1440. [CrossRef]
38. Stokes, A.; Ni, Y. Validating a summary measure of weight history for modeling the health consequences of obesity. *Ann. Epidemiol.* **2016**, *26*, 821–826.e2. [CrossRef]

nutrients

Article

Healthful Plant-Based Diet and Incidence of Type 2 Diabetes in Asian Population

Jihye Kim [1,*] and Edward Giovannucci [2]

1 Department of Genetics and Biotechnology, College of Life Sciences, Kyung Hee University, Yongin 17104, Korea
2 Departments of Epidemiology and Nutrition, Harvard T.H. Chan School of Public Health and Harvard Medical School, Boston, MA 02115, USA; egiovann@hsph.harvard.edu
* Correspondence: kjhye@khu.ac.kr; Tel.: +82-31-201-3497

Abstract: Plant-based diets have been suggested to be beneficial for type 2 diabetes (T2D). However, studies investigating the association between the healthiness of a plant-based diet and T2D risk are limited. This study explored the prospective association between scores from three different plant-based diet indices and risk of T2D and investigated whether associations differ by demographic and lifestyle factors in the Korean population. Data were derived from the Korean Genome and Epidemiology Study (KoGES), a prospective cohort study initiated between 2001 and 2002. Dietary intakes were assessed using a validated food frequency questionnaire. Scores for three plant-based diet indices (overall plant-based diet index (PDI), healthful plant-based diet index (hPDI), and unhealthful plant-based diet index (uPDI)) were measured. A total of 7363 Korean adults aged 40–69 years without T2D and related chronic diseases at baseline were included. Incident T2D was defined as elevated plasma glucose (≥126 mg/dL), self-report of a doctor's diagnosis of T2D, or use of oral hypoglycemic drug. Multivariable Cox proportional hazards models were used to estimate hazard ratios (HRs) and 95% CIs for T2D risk. During a follow-up period of 14 years, 977 participants developed T2D. A 10-point higher score in hPDI was associated with a 14% lower risk of T2D (HR: 0.86, 95% CI, 0.77–0.95), adjusting for potential confounders. In subgroup analysis, inverse associations between hPDI and T2D risk were stronger in participants with a family history of T2D (HR: 0.58, 95% CI, 0.44 0.76) or history of hypertension (HR: 0.73, 95% CI, 0.60 0.89) than those without a family history of T2D (p interaction = 0.01) or history of hypertension (p interaction = 0.04). Considering the quality of the plant foods may be important for the prevention of T2D in the Korean population, which habitually consumes diets rich in plant foods.

Keywords: plant-based diets; plant food quality; type 2 diabetes; prospective study; Asian

Citation: Kim, J.; Giovannucci, E. Healthful Plant-Based Diet and Incidence of Type 2 Diabetes in Asian Population. *Nutrients* **2022**, *14*, 3078. https://doi.org/10.3390/nu14153078

Academic Editor: Antonio Brunetti

Received: 30 June 2022
Accepted: 25 July 2022
Published: 27 July 2022

1. Introduction

Type 2 diabetes (T2D) is a major metabolic disorder, which contributes substantially to morbidity and mortality in the world [1]. Diet is a modifiable risk factor in the development of T2D [2]. Plant-based diets have been known to be beneficial for the prevention and management of T2D [3]. Several plant foods, such as fruits, vegetables, whole grains, and legumes, are favorable for the prevention of T2D [4–6], but not all plant foods are healthy. For instance, plant foods such as refined grains, sweets, and sugar-sweetened beverages have unfavorable effects on the development of T2D [7–9]. Moreover, some animal foods, such as dairy and fish may be beneficial for health outcomes [10–12].

The 2015 Dietary Guidelines for Americans recommends gradually moving to diets rich in plant foods and progressively decrease animal food consumption [13]. The plant-based diet indices including overall plant-based diet index (PDI), healthful plant-based diet index (hPDI), and unhealthful plant-based diet index (uPDI), assess intakes of both plant foods and animal foods, taking the quality of plant foods into account [14,15]. In these indices, animal foods are negatively weighed but differ with respect to how plant foods are weighed. Existing studies reported that the scores from PDI or hPDI were associated with a lower risk of cardiovascular disease, T2D, and related chronic diseases [14–17]. On the contrary, the score from uPDI was positively associated with the risk of these metabolic diseases [14,15,18,19]. However, the study on the association of newly established plant-based diet indices with T2D risk is limited in the Asian population although their eating patterns and metabolic responses may be different from the Western population [20,21]. Asians consume higher amounts of grains and vegetables and lower amounts of meat than Western populations [21]. Different types of grains may affect glucose metabolism differently [22,23]. Previous studies reported ethnic differences in T2D risk associated with scores of a priori-defined dietary patterns [24,25]. One prospective study found that the scores of PDI and hPDI was associated with a reduced risk of type 2 diabetes in Singapore Chinese [16].

Thus, this study evaluated the associations between scores from three different plant-based diet indices (PDI, hPDI, and uPDI) and risk of T2D and investigated whether associations differed by demographic and lifestyle factors in a large community-based cohort of Korean middle-aged and older adults.

2. Materials and Methods

2.1. Study Population

We used the data from the Korean Genome and Epidemiology Study (KoGES), a population-based cohort study, which aimed to explore the genetic and lifestyle factors of T2D and hypertension among Koreans [26]. A total of 10,030 community residents (40–69 years of age) living in Ansan and Ansung city, near Seoul, were enrolled. The KoGES was initiated between 2001 and 2002 (baseline) and participants were followed up biennially until 2016. The follow-up rate in the KoGES was over 90%. The Institutional Review Boards of the Korea Disease Control and Prevention Agency and Kyung Hee University (KHGIRB-19-398) approved the study protocol, and participants provided written informed consent.

In the analysis, exclusion criteria included individuals with implausible energy intake at baseline (<500 kcal/d or >5000 kcal/d) ($n = 410$), had cardiovascular disease or cancer at baseline ($n = 304$), did not visit in follow-up examinations ($n = 841$), who had T2D at baseline ($n = 595$), and who had missing data on the outcome of T2D or covariates including education level, physical activity, cigarette smoking, alcohol consumption, baseline body mass index (BMI), total energy intake, family history of T2D, and history of hypertension ($n = 487$). The final analysis included 7393 (3466 men and 3927 women) (Figure 1).

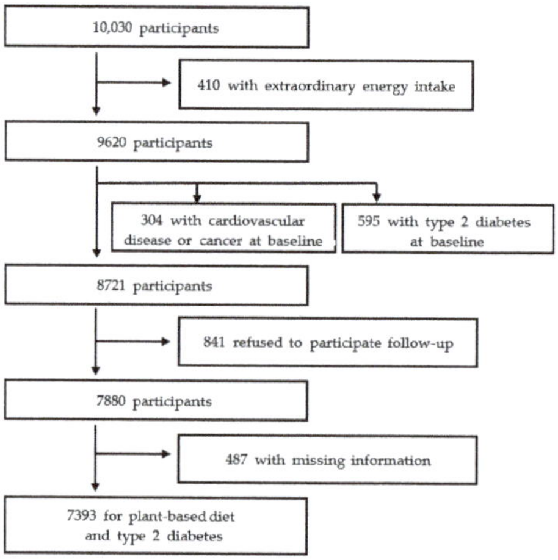

Figure 1. Flow diagram of participant selection.

2.2. Assessment of Plant-Based Diet Index Score

Participants were asked for their usual food intake with a 106-item semi-quantitative food frequency questionnaire (FFQ). Validity and reproducibility for FFQ have been previously evaluated [27]. The correlation coefficient of nutrient density between the two FFQs examined at a 1-year interval was between 0.22 (vitamin A) and 0.51 (calcium) (average: 0.39). The median value of correlation coefficients for nutrients between the FFQ and the 12-day diet records was 0.39. The FFQ was assessed at baseline and visit 3, which is the second follow-up (visit 3: 2005–2006). We applied the average of dietary intake from two FFQs for the calculation of plant-based diet indices. When participants developed T2D before visit 3 or did not complete the questionnaire at visit 3, we used only baseline dietary intake. Participants were asked to report the frequency and the portion size of food consumption over the past year. The FFQ had nine answers for frequency of consumption, ranging from "almost never" to "3 times or more per day," and three answers for portion size (small, medium, or large) [28].

We applied previously reported processes for calculating three plant-based diet index scores [15,18] (Table S1). Briefly, the food items were classified into 17 food groups based on nutrient and culinary similarities and then the food groups were categorized into three larger groups, which are healthy, less healthy plant and animal food group. We distinguished between healthy plant foods and less healthy plant foods depending on the associations of food items with disease risk [14,15,18,19]. Healthy plant foods include whole grains, fruits, vegetables, nuts, legumes, tea/coffee, and less healthy plant foods include refined grains, potatoes, sugar-sweetened beverages, sweets and desserts, salty foods. Animal foods include animal fat, dairy, eggs, fish, meat, and miscellaneous animal foods. We classified salty foods (i.e., kimchi) as less healthy plant foods due to high sodium content. We did not separate vegetable oil and fruit juices included in the original indices as food groups [14,15], because the oil intake was not queried in the FFQ, and fruits and fruit juices were queried together. Alcoholic beverages were excluded from the calculation of indices due to unclear directions of association for various health outcomes. Some mixed dishes, such as pizza and hamburgers/sandwiches, were queried individually in the FFQ and were categorized into miscellaneous animal foods.

After we formulated group foods, participants were ranked into energy-adjusted quintiles. For the PDI score, participants in the highest quintile of each of all plant foods were scored 5 while those in the lowest quintile of each of all plant foods were scored 1. For the hPDI, participants in the highest quintile of only healthy plant foods were scored 5 while those in the highest quintile of less healthy plant foods were scored 1. For the uPDI, participants in the highest quintile of only less healthy plant foods were scored 5 while those in the highest quintile of healthy plant foods were scored 1. For all plant-based diet indices, animal foods were adversely scored. For instance, participants in the highest quintile of animal fat consumption were scored 1, and those in the lowest quintile of animal fat consumption were scored 5. Thus, higher scores in all plant-based diet indices represented lower consumption of animal foods. After summing up the scores across these categories for plant and animal foods, the overall diet scores were divided into quintiles for analysis. In the present study, the Spearman correlation coefficients between PDI and hPDI were the highest (0.44), and the correlations were −0.15 between hPDI and uPDI, and 0.14 between PDI and uPDI.

2.3. Ascertainment of Type 2 Diabetes

T2D incidence was defined as having one or more of the following criteria [29]: elevated fasting plasma glucose, use of the oral hypoglycemic drug, or current treatment with insulin. Biochemical assessment, medical history, and medication use were identified at biennial follow-up visits. Elevated plasma glucose was considered as ≥126 mg/dL. Blood samples were collected after ≥8 h of fasting and the samples were stored at −80 °C until analyses. An auto-analyzer (ADVIA 1650, Bayer HealthCare) was used to measure the glucose concentration enzymatically using a standardized protocol. In a reliability study, the laboratory value of this biomarker is highly reproducible (Pearson's correlation > 0.99) [30].

2.4. Assessment of Covariates

Information on demographic and lifestyle factors at baseline were investigated using structured questionnaires, administered by trained interviewers. Educational level was divided into ≤6, 7 to 12, and >12 years. Cigarette smoking was queried as pack-years of cigarettes. Alcohol intake was assessed among former and current drinkers who had consumed alcohol within one year. Physical activity was calculated using the metabolic equivalent of task based on the types and intensity of physical activity [31]. Baseline height and weight of participants were measured by trained staff. Height was measured to the nearest 0.1 cm without shoes using a stadiometer (Samhwa Instrument, Seoul, Korea) and body weight was measured to 0.1 kg in light clothes without shoes. Body mass index (BMI) was calculated from measured weight (kg) divided by height squared (m^2). We calculated total energy intake using a food composition table from the Korean Nutrition Society [32]. History of hypertension at baseline was defined as a self-report of a doctor's diagnosis of hypertension.

2.5. Statistical Analysis

Baseline characteristics of participants are described as mean and standard deviation (SD) or number and percentage (categorical variables). Food group consumption was compared between the lowest and highest quintile of each plant-based diet indices and was expressed as serving size per 1000 kcal.

The risk of T2D per a 10-point higher score of plant-based diet indices were tested using multivariable Cox proportional hazards models. Age (year, continuous) and sex (men/women) were adjusted in Model 1. Additionally, residential location (rural/urban), education (≤6, 7–12, >12 years), physical activity (MET/day, continuous), cigarette smoking (continuous), alcohol consumption (g/day, continuous), BMI (kg/m^2, continuous), total energy intake (kcal/day, continuous), family history of diabetes (yes/no), and history of hypertension at baseline (yes/no) were adjusted in Model 2. We selected the potential confounding factors from the previous literature [14,18].

Next, the restricted cubic splines with 4 knots were applied to examine the shape of the associations for plant-based diet indices. Effect modification by sex, baseline BMI, family history of diabetes, and history of hypertension were tested with cross-product terms.

Person year was calculated as the time from baseline examination until the date of T2D event or censoring. Censoring was defined as those who did not develop T2D until the end date of the study or those who developed cancer or cardiovascular disease before developing diabetes during follow–up visits. We did not consider the death of participants, due to the unavailability of the data.

We tested the proportional hazard assumption using Schoenfeld's residuals, which was met [33]. All data were analyzed using SAS software, version 9.4 (SAS Institute, Cary, NC, USA) [34]. $p < 0.05$ was considered significant for two-sided tests.

3. Results

During a follow-up of 82,351 person-years, 977 (13%) T2D cases were identified. Table 1 summarizes participants' characteristics at baseline. Individuals in the Q5 (highest quintile) of PDI and hPDI were more likely to be older, women, to live in rural areas, had lower education levels and cigarette smoking (modest for PDI), consuming less alcohol, had higher BMI, were more physically active, and were more likely to have hypertension compared to those in the Q1 (lowest quintile) of PDI and hPDI. Individuals in the Q5 of uPDI were more likely to be older, men, to live in rural areas, had lower education levels, higher cigarette smoking, had lower BMI, were more physically active, and were more likely to have hypertension compared to those in the Q1 of uPDI.

Table 1. Baseline characteristics of study participants in the lowest versus highest quintile of plant-based diet indices.

	PDI		hPDI		uPDI	
	Quintile 1	**Quintile 5**	**Quintile 1**	**Quintile 5**	**Quintile 1**	**Quintile 5**
Sample size, *n*	1451	1438	1506	1452	1586	1186
Median score (range)	44 (31–46)	58 (56–69)	43 (28–45)	59 (57–74)	43 (29–46)	62 (60–76)
Female, *n* (%)	673 (46.4)	875 (60.9)	542 (36.0)	1033 (71.1)	1014 (63.9)	581 (49.0)
Age, years	49.7 (8.4)	53.8 (8.9)	49.2 (8.2)	54.2 (8.7)	49.1 (7.8)	55.4 (8.9)
Residential location, *n* (%)						
Rural, Ansung	618 (42.6)	830 (57.7)	496 (32.9)	834 (57.4)	394 (24.8)	939 (79.2)
Urban, Ansan	833 (57.4)	608 (42.3)	1010 (67.1)	618 (42.6)	1192 (75.2)	247 (20.8)
Education level, *n* (%)						
≤6 years	334 (23.0)	607 (42.2)	315 (20.9)	616 (42.4)	233 (14.7)	627 (52.9)
7–12 years	842 (58.0)	709 (49.3)	886 (58.8)	716 (49.3)	1009 (63.6)	499 (42.1)
>12 years	275 (19.0)	122 (8.5)	305 (20.3)	120 (8.3)	344 (21.7)	60 (5.0)
Cigarette smoking (pack-year)	9.5 (14.9)	8.4 (15.6) *	12.6 (16.4)	5.7 (13.1)	6.3 (13.1)	11.0 (17.1)
Alcohol intake (g/day)	11.8 (23.1)	7.1 (20.1)	13.7 (26.3)	5.7 (17.8)	8.7 (20.7)	8.4 (19.0)
Body Mass Index (kg/m^2)	24.4 (3.0)	24.8 (3.2)	24.5 (3.1)	24.8 (3.3)	24.7 (3.0)	24.3 (3.2)
Physical activity (MET/day)	21.7 (14.3)	25.9 (15.6)	21.7 (14.0)	24.8 (15.5)	20.2 (12.0)	27.2 (16.8)
Family history of diabetes, *n* (%)	179 (12.3)	140 (9.7) *	160 (10.6)	143 (9.9) *	223 (14.1)	91 (7.7)
History of hypertension, *n* (%)	151 (10.4)	227 (15.8)	160 (10.6)	282 (19.4)	200 (12.6)	203 (17.1)

Data are expressed as mean ± SD or n (%). PDI, overall plant-based diet index; hPDI, healthful plant-based diet index, MET, Metabolic equivalent of task; uPDI, unhealthful plant-based diet index. * Values are not significantly different between Q1 and Q5 group. Except for that, all values are significantly different between Q1 and Q5 group.

Table 2 shows the food group consumption of participants in the lowest (Q1) versus highest quintile (Q5) of plant-based diet indices. Individuals in the Q5 of PDI consumed greater amounts of whole grains, fruits, vegetables, legumes, tea and coffee, potatoes, sweets and desserts, and salty foods and less amounts of refined grains, dairy, eggs, fish, and meat compared to those in the Q1 of PDI. Individuals in the Q5 of hPDI consumed greater amounts of whole grains, fruits, vegetables, and legumes and consumed less amounts of refined grains, potatoes, sugar-sweetened beverages, sweets and desserts, and all animal foods compared to those in the Q1 of hPDI. On the contrary, individuals in the Q5 of uPDI consumed less amounts of all healthy plant foods and greater amounts of refined grains and salty foods compared to those in the Q1 of uPDI.

Table 2. Dietary consumption of participants in the lowest versus highest quintile of plant-based diet indices.

	PDI		hPDI		uPDI	
	Quintile 1	**Quintile 5**	**Quintile 1**	**Quintile 5**	**Quintile 1**	**Quintile 5**
Total energy intake, kcal/day	2048 (676)	1836 (584)	2016 (594)	1850 (617)	2064 (584)	1717 (584)
Food group intake (servings/1000 kcal/week)						
Whole grain	4.6 (4.4)	7.0 (4.5)	2.5 (3.2)	9.9 (3.7)	7.5 (3.5)	3.6 (4.5)
Fruits	8.2 (7.0)	13.2 (8.3)	7.4 (5.3)	13.7 (9.2)	14.0 (7.7)	7.0 (6.7)
Vegetables	11.9 (6.7)	18.2 (8.4)	13.5 (6.4)	16.1 (9.3)	18.5 (8.0)	10.6 (6.4)
Nuts	0.1 (0.3)	0.3 (0.5)	0.1 (0.4)	0.3 (0.6)	0.4 (0.7)	0.0 (0.2)
Legumes	1.7 (1.5)	3.8 (3.0)	1.9 (1.6)	3.7 (3.0)	3.3 (2.0)	1.7 (2.2)
Tea and coffee	4.3 (3.8)	6.5 (4.9)	6.2 (4.7)	4.6 (4.4)	6.4 (4.4)	3.9 (4.8)
Refined grains	3.3 (4.6)	1.8 (2.9)	4.7 (5.0)	0.9 (1.6)	1.0 (1.7)	4.6 (5.4)
Potatoes	0.6 (0.6)	1.3 (1.2)	1.0 (0.9)	0.8 (1.0)	0.9 (0.9)	0.9 (1.5) *
Sugar-sweetened beverages	0.6 (1.0)	0.7 (1.0)	1.1 (1.1)	0.3 (0.6)	0.6 (0.8)	0.6 (1.0) *
Sweets and desserts	3.4 (3.0)	4.8 (3.8)	5.6 (3.5)	2.6 (2.9)	3.9 (2.9)	3.8 (4.2) *
Salty foods	11.6 (6.2)	19.6 (8.2)	16.0 (7.5)	15.2 (8.3)	13.3 (6.2)	18.8 (9.3)
Animal fat	3.5 (3.8)	3.5 (4.4) *	5.3 (4.3)	1.9 (3.2)	4.1 (3.9)	2.8 (4.3)
Dairy	3.4 (2.7)	1.8 (2.1)	3.0 (2.3)	2.1 (2.5)	3.8 (2.6)	1.3 (1.9)

Table 2. *Cont.*

	PDI		hPDI		uPDI	
	Quintile 1	**Quintile 5**	**Quintile 1**	**Quintile 5**	**Quintile 1**	**Quintile 5**
Eggs	1.1 (1.0)	0.6 (0.8)	1.1 (0.9)	0.6 (0.8)	1.2 (0.9)	0.4 (0.7)
Fish	4.7 (2.7)	3.7 (2.9)	5.0 (2.7)	3.5 (3.1)	5.9 (2.9)	2.0 (1.7)
Meat	0.9 (1.1)	0.4 (0.5)	1.2 (1.3)	0.3 (0.5)	0.6 (0.7)	0.5 (0.8)
Miscellaneous animal foods	0.2 (0.3)	0.1 (0.2)	0.2 (0.3)	0.1 (0.1)	0.2 (0.2)	0.1 (0.2)
Healthy plant foods	30.8 (12.0)	48.9 (13.6)	31.7 (10.8)	48.3 (14.7)	50.2 (12.5)	26.8 (10.7)
Less healthy plant foods	19.5 (8.4)	28.2 (9.7)	28.3 (9.9)	19.7 (8.8)	19.7 (6.9)	28.8 (11.9)
Animal foods	13.8 (5.7)	10.1 (5.7)	15.8 (5.5)	8.5 (5.3)	15.8 (5.0)	7.0 (4.9)

Values are means (SD). * Values are not significantly different between Q1 and Q5 group. Except for that, all values are significantly different between Q1 and Q5 group.

Table 3 reports the hazard ratios and 95% confidence intervals for developing T2D according to a 10-point higher score of plant-based diet indices. In the age and sex-adjusted model, the hPDI was not significantly associated with the risk of T2D (HR = 0.92, 95% CI, 0.83–1.02). However, in the multivariable-adjusted model, a 10-point higher score of hPDI was associated with a 14 % lower risk of T2D after adjustment for potential confounders (HR: 0.86, 95% CI, 0.77–0.95). When we additionally adjusted for dietary fiber or calcium, the results remained the same. This association was reflected when the relation between hPDI with incident T2D was visually depicted (Figure 2). However, neither PDI nor uPDI was significantly associated with the risk of T2D.

Table 3. Hazard ratios and 95% confidence intervals for incident type 2 diabetes according to a 10-point higher score of plant-based diet indices.

Model	PDI	hPDI	uPDI
Median (SD)	51 (5.2)	51 (6.4)	52 (7.0)
Model 1	1.04 (0.92–1.18)	0.92 (0.83–1.02)	1.06 (0.97–1.16)
Model 2	0.99 (0.88–1.12)	0.86 (0.77–0.95)	1.06 (0.96–1.18)

Model 1 was adjusted for age (year, continuous) and sex (men/women). Model 2 was additionally adjusted for residential area (rural/urban), education (≤6, 7–12, >12 years), physical activity (MET-hour/day, continuous), smoking cigarettes (continuous), alcohol intake (g/day, continuous), baseline body mass index (kg/m^2, continuous), total energy intake (kcal/day, continuous), family history of diabetes (yes/no), and history of hypertension at baseline (yes/no).

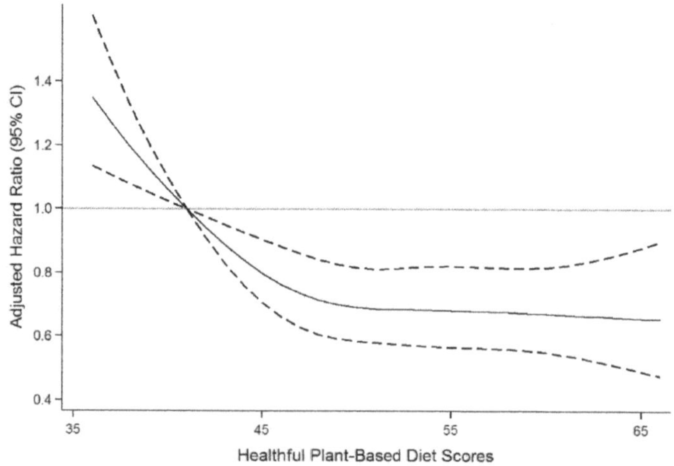

Figure 2. Association of healthful plant-based diet score with incident type 2 diabetes according to the continuous plant-based diet score among the middle-aged and older Korean population.

The solid lines represent the multivariable-adjusted hazard ratios for incident type 2 diabetes, modelled using restricted cubic splines with 4 knots (5th, 35th, 65th, and 95th percentiles). The reference was set at the 5th percentile of the score. The dashed lines represent 95% confidence intervals. The model adjusted for age (year, continuous) and sex (men/women), residence area (rural/urban), education (≤6, 7–12, >12 years), physical activity (MET/day, continuous), cigarette smoking (continuous), alcohol intake (g/day, continuous), baseline body mass index (kg/m^2, continuous), total energy intake (kcal/day, continuous), family history of diabetes (yes/no), and history of hypertension at baseline (yes/no).

Additionally, when we conducted the subgroup analysis stratified by sex, baseline BMI, family history of T2D, and history of hypertension, an inverse relationships between hPDI and T2D risk were stronger in participants with a family history of T2D (HR: 0.58, 95% CI, 0.44 0.76) or history of hypertension (HR: 0.73, 95% CI, 0.60 0.89) than those without a family history of T2D (p interaction = 0.01) or history of hypertension (p interaction = 0.04). (Table 4)

Table 4. Hazard ratios (HR) and 95% confidence intervals (CI) for incident type 2 diabetes according to the continuous hPDI score, stratified by selected characteristics.

	Healthful Plant-Based Diet Index	p Interaction
Sex		
Men	0.89 (0.77–1.04) *	0.85
Women	0.84 (0.72–0.97)	
Baseline body mass index		
≥25 kg/m^2	0.84 (0.73–0.96)	0.19
<25 kg/m^2	0.90 (0.76–1.07)	
Family history of T2D		
Yes	0.58 (0.44–0.76)	0.01
No	0.92 (0.82–1.03)	
History of hypertension		
Yes	0.73 (0.60–0.89)	0.04
No	0.92 (0.81–1.04)	

* per a 10-point higher score of healthful plant-based diet index. Adjusted for age (years, continuous), residence area (rural/urban), education (≤6, 7–12, >12 years), physical activity (MET/day, continuous), cigarette smoking (continuous), alcohol intake (g/day, continuous), body mass index (kg/m^2, continuous), total energy intake (kcal/day, continuous), family history of diabetes (yes/no), and history of hypertension at baseline (yes/no). MET, metabolic equivalent task; T2D, type 2 diabetes.

4. Discussion

Greater adherence to a healthful plant-based diet (captured by hPDI) was associated with a 14% lower risk of T2D in South Korean adults after adjustment for demographic factors, lifestyle factors, and BMI. In subgroup analyses, the inverse relationship was stronger among participants with a family history of T2D or history of hypertension than those without a family history of T2D or history of hypertension. However, overall plant-based diets (captured by PDI) and unhealthy plant-based diets (captured by uPDI) were not significantly associated with the risk of T2D. The inverse association of hPDI with the risk of T2D highlights that greater adherence to diets higher in healthy plant foods and lower in less healthy plant foods and animal foods may be important for the prevention of T2D.

Several studies have explored the association between plant-based diets and T2D risk. Three prospective cohort studies have reported that higher score of PDI and hPDI were associated with 49% and 45% lower risk of T2D, respectively, whereas higher score of uPDI was associated with a 16% higher risk of T2D in the US adults [14]. The current study also showed a stronger association of the hPDI with T2D. A Singapore Chinese study showed that higher scores of PDI and hPDI were associated with 17% and 19% lower risk of T2D, respectively, in the middle-aged and older adults [16]. In the Rotterdam study, a higher adherence to an overall plant-based diet was associated with a 13% lower T2D risk after adjustment for potential confounders in Dutch adults [3]. Kim et al. showed that great adherence to PDI was associated with a 20% lower risk of elevated fasting glucose (≥100 mg/dL) as a component of metabolic syndrome, but not with hPDI or uPDI among Korean adults [18]. The differences in findings between the two Korean studies may be due to the differences in exclusion criteria of participants, diagnosis criteria (e.g., blood glucose cut point), and covariates for adjustment, although the same cohort data were used.

Unlike US adults [14], an overall plant-based diet captured by PDI was not associated with the incidence of T2D in the Korean population. No association of PDI with T2D shows that increasing the amount of plant foods without consideration of the quality of plant foods may not be beneficial for the prevention of T2D in a population that habitually consumes diets high in plant foods and low in animal foods. Based on our data, Korean adults consumed half the amount of animal foods and more plant foods than US adults [14]. In addition, the differences in non-dietary lifestyle risk factors for T2D may be attributable to differential effect size in T2D risk between US adults and Korean adults. Korean adults have healthier indicators on average including lower BMI, more physical activity, and less alcohol consumption than US adults [14].

In the current study, participants in the highest quintile of hPDI consumed greater amounts of whole grains, fruits, vegetables, and legumes, and lesser amounts of refined grains, sugar sweetened beverages, animal fat, and meat compared to those in the lowest quintile of hPDI. Combinations of these food groups would contribute to reducing the risk of T2D. In a meta-analysis, whole grains, fruits, vegetables, and soy products were associated with a reduced risk of T2D, while refined grains, red meat, processed meat, and sugar sweetened beverages were associated with an increased risk of T2D [35,36]. Several mechanisms have been suggested for how a healthful plant-based diet lowers the risk of T2D. Diets rich in healthy plant foods include various favorable food components and nutrients such as dietary fiber, antioxidants, micronutrients, and unsaturated fatty acids [37,38], and the diet is low in saturated fatty acids and cholesterol. A meta-analysis of clinical trials has shown that soluble fiber intake improved glycemic control in patients with T2D [39]. Epidemiologic studies have shown flavonoids/polyphenols may have anti-diabetic effects by increasing glucose metabolism, improving vascular functions as well as reducing insulin resistance [40]. Polyunsaturated fatty acids are associated with improved glycemic control and with anti-inflammatory effects [41]. In addition, another mechanism may be related to the gut microbiome. A healthy plant-based diet may improve gut microbiota profiles that facilitate the metabolism of dietary components, such as fiber and polyphenols, and diminish the metabolism of microbial metabolites, such as trimethylamine N-oxide, which are primarily found in animal originated foods [42,43]. This microbial change may induce a reduction of T2D risk.

Interestingly, an inverse association between the healthier version of the plant-based diet and T2D was stronger in participants who have a family history of T2D or history of hypertension. This population may be at a higher risk of T2D, possibly due to abnormal obesity-related body composition (higher BMI and waist-hip ratio) and genetic factors [44]. This suggests that a healthier dietary pattern may have more benefits for a vulnerable population who are at a higher risk of T2D.

This study explored the prospective associations between different plant-based diet indices assessed by the healthiness of plant-based diet and the risk of T2D in the Korean population, which has traditionally consumed diets rich in plant foods. Strengths of this study include the use of data from a community-based cohort, validated FFQ, repeated dietary assessments, and a relatively long follow-up period. The current study also focused on the Korean population which may have different eating habits from the population in Western countries.

However, several limitations should be noted. Reporting of dietary intakes can be subject to measurement error although the FFQ was validated in South Korean adults [45]. The FFQ has some mixed dishes that combine healthy plant food and less healthy plant food and combines plant food and animal food such as Bibimbap. We queried fruit (healthy) and fruit juice (less healthy) together in the FFQ and data on vegetable oil intake was not collected from the FFQ. Furthermore, the processing/cooking method was not considered in the differentiation of healthy plant and less healthy plant foods. These food classifications might not completely distinguish the consumption of healthy plant or less healthy plant foods and, thus, they may have led to an attenuation of the association between uPDI and incident T2D. Although we used the average of dietary intakes from two FFQs, there is possibility that dietary habits of participants might have changed over time. T2D cases might be missing because HbA1c was not included as the criteria of diagnosis. Some person's time may be misclassified because data on death or moving were not available. Missing data on covariates were not imputed in the current analysis. Lastly, there may still be residual confounding although important confounders were adjusted.

5. Conclusions

In a community-based cohort of Korean men and women, greater adherence to diets high in healthy plant foods and low in unhealthy plant foods and animal foods in the context of a plant-based diet was associated with a lower risk of incident T2D. These results emphasize the importance of considering the quality of plant foods for the prevention of T2D in Koreans. Further research confirming the associations between plant-based diet indices and T2D in diverse ethnic populations with different dietary habits are warranted.

Supplementary Materials: The following are available online at https://www.mdpi.com/article/10.3390/nu14153078/s1, Table S1: Classification of food items in the Korean Genome and Epidemiology Study (KoGES).

Author Contributions: Conceptualization, methodology, formal analysis, funding acquisition, and writing—original draft preparation, J.K.; Conceptualization and writing—review and editing, E.G. All authors have read and agreed to the published version of the manuscript.

Funding: This research was funded by the National Research Foundation of Korea (NRF) of the Korea government (Ministry of Science and ICT), grant number 2021R1A2C1003211, to J.K.

Institutional Review Board Statement: The study was conducted according to the guidelines of the Declaration of Helsinki, and approved by the Institutional Review Board of the Korea Disease Control and Prevention Agency and Kyung Hee University (KHGIRB-19-398 and 19 December 2019).

Informed Consent Statement: Informed consent was obtained from all subjects involved in the study.

Data Availability Statement: Data underlying the results of our study are not publicly available due to KoGES data policy. Data are available from the Division of Genetic Epidemiology and Health Index, NIH, Korea Disease Control and Prevention Agency (contact via Mi-Jin Cho at whalwls0227@korea.kr) for researchers who meet the criteria for access to confidential data.

Acknowledgments: Data used in this study were obtained from the Korean Genome and Epidemiology Study (KoGES; 4851-302), National Research Institute of Health, Korea Disease Control and Prevention Agency, Ministry for Health and Welfare, Republic of Korea. We would like to thank the study participants and staff.

Conflicts of Interest: The authors declare no conflict of interest. The funders had no role in the design of the study; in the collection, analyses, or interpretation of data; in the writing of the manuscript; or in the decision to publish the results.

References

1. Kos, K. Cardiometabolic morbidity and mortality with smoking cessation, review of recommendations for people with diabetes and obesity. *Curr. Diab. Rep.* **2020**, *20*, 82. [CrossRef]
2. Ley, S.H.; Hamdy, O.; Mohan, V.; Hu, F.B. Prevention and management of type 2 diabetes: Dietary components and nutritional strategies. *Lancet* **2014**, *383*, 1999–2007. [CrossRef]
3. Chen, Z.; Zuurmond, M.G.; van der Schaft, N.; Nano, J.; Wijnhoven, H.A.H.; Ikram, M.A.; Franco, O.H.; Voortman, T. Plant versus animal based diets and insulin resistance, prediabetes and type 2 diabetes: The Rotterdam Study. *Eur. J. Epidemiol.* **2018**, *33*, 883–893. [CrossRef]

4. Aune, D.; Norat, T.; Romundstad, P.; Vatten, L.J. Whole grain and refined grain consumption and the risk of type 2 diabetes: A systematic review and dose–response meta-analysis of cohort studies. *Eur. J. Epidemiol.* **2013**, *28*, 845–858. [CrossRef] [PubMed]
5. Cooper, A.J.; Forouhi, N.G.; Ye, Z.; Buijsse, B.; Arriola, L.; Balkau, B.; Barricarte, A.; Beulens, J.W.; Boeing, H.; Büchner, F.L. Fruit and vegetable intake and type 2 diabetes: EPIC-InterAct prospective study and meta-analysis. *Eur. J. Clin. Nutr.* **2012**, *66*, 1082–1092. [CrossRef] [PubMed]
6. Muraki, I.; Imamura, F.; Manson, J.E.; Hu, F.B.; Willett, W.C.; van Dam, R.M.; Sun, Q. Fruit consumption and risk of type 2 diabetes: Results from three prospective longitudinal cohort studies. *BMJ* **2013**, *347*, f5001. [CrossRef] [PubMed]
7. Halton, T.L.; Willett, W.C.; Liu, S.; Manson, J.E.; Stampfer, M.J.; Hu, F.B. Potato and french fry consumption and risk of type 2 diabetes in women. *Am. J. Clin. Nutr.* **2006**, *83*, 284–290. [CrossRef]
8. Malik, V.S.; Popkin, B.M.; Bray, G.A.; Després, J.-P.; Willett, W.C.; Hu, F.B. Sugar-sweetened beverages and risk of metabolic syndrome and type 2 diabetes: A meta-analysis. *Diabetes Care* **2010**, *33*, 2477–2483. [CrossRef]
9. Soriguer, F.; Colomo, N.; Olveira, G.; García-Fuentes, E.; Esteva, I.; de Adana, M.S.R.; Morcillo, S.; Porras, N.; Valdés, S.; Rojo-Martínez, G. White rice consumption and risk of type 2 diabetes. *Clin. Nutr.* **2013**, *32*, 481–484. [CrossRef]
10. Gijsbers, L.; Ding, E.L.; Malik, V.S.; De Goede, J.; Geleijnse, J.M.; Soedamah-Muthu, S.S. Consumption of dairy foods and diabetes incidence: A dose-response meta-analysis of observational studies. *Am. J. Clin. Nutr.* **2016**, *103*, 1111–1124. [CrossRef]
11. Mozaffarian, D. Dietary and policy priorities for cardiovascular disease, diabetes, and obesity: A comprehensive review. *Circulation* **2016**, *133*, 187–225. [CrossRef] [PubMed]
12. Zhang, M.; Picard-Deland, E.; Marette, A. Fish and marine omega-3 polyunsatured fatty acid consumption and incidence of type 2 diabetes: A systematic review and meta-analysis. *Int. J. Endocrinol.* **2013**, *2013*, 501015. [CrossRef] [PubMed]
13. US Department of Agriculture, US Department of Health and Human Services. Scientific Report of the 2015 Dietary Guidelines Advisory Committee: Advisory Report to the Secretary of Health and Human Services and the Secretary of Agriculture. Washington (District of Columbia). 2015. Available online: https://health.gov/our-work/nutrition-physical-activity/dietary-guidelines/previous-dietary-guidelines/2015/advisory-report (accessed on 20 July 2022).
14. Satija, A.; Bhupathiraju, S.N.; Rimm, E.B.; Spiegelman, D.; Chiuve, S.E.; Borgi, L.; Willett, W.C.; Manson, J.E.; Sun, Q.; Hu, F.B. Plant-based dietary patterns and incidence of type 2 diabetes in US men and women: Results from three prospective cohort studies. *PLoS Med.* **2016**, *13*, e1002039. [CrossRef] [PubMed]
15. Satija, A.; Bhupathiraju, S.N.; Spiegelman, D.; Chiuve, S.E.; Manson, J.E.; Willett, W.; Rexrode, K.M.; Rimm, E.B.; Hu, F.B. Healthful and unhealthful plant-based diets and the risk of coronary heart disease in US adults. *J. Am. Coll. Cardiol.* **2017**, *70*, 411–422. [CrossRef] [PubMed]
16. Chen, G.-C.; Koh, W.-P.; Neelakantan, N.; Yuan, J.-M.; Qin, L.-Q.; van Dam, R.M. Diet quality indices and risk of type 2 diabetes mellitus: The Singapore Chinese Health Study. *Am. J. Epidemiol.* **2018**, *187*, 2651–2661. [CrossRef] [PubMed]
17. Kim, J.; Kim, H.; Giovannucci, E.L. Quality of plant-based diets and risk of hypertension: A Korean genome and examination study. *Eur. J. Nutr.* **2021**, *60*, 3841–3851. [CrossRef]
18. Kim, H.; Lee, K.; Rebholz, C.M.; Kim, J. Plant-based diets and incident metabolic syndrome: Results from a South Korean prospective cohort study. *PLoS Med.* **2020**, *17*, e1003371. [CrossRef]
19. Lee, K.; Kim, H.; Rebholz, C.M.; Kim, J. Association between different types of plant-based diets and risk of dyslipidemia: A prospective cohort study. *Nutrients* **2021**, *13*, 220. [CrossRef]
20. Afshin, A.; Micha, R.; Khatibzadeh, S.; Fahimi, S.; Shi, P.; Powles, J.; Singh, G.; Yakoob, M.Y.; Abdollahi, M.; Al-Hooti, S.; et al. Global Burden of Diseases I, Risk Factors Study N, Chronic Diseases Expert G, Metabolic Risk Factors of Chronic Diseases Collaborating Group. The impact of dietary habits and metabolic risk factors on cardiovascular and diabetes mortality in countries of the Middle East and North Africa in 2010: A comparative risk assessment analysis. *BMJ Open* **2015**, *5*, e006385. [PubMed]
21. Micha, R.; Khatibzadeh, S.; Shi, P.; Andrews, K.G.; Engell, R.E.; Mozaffarian, D. Global Burden of Diseases N, Chronic Diseases Expert G. Global, regional and national consumption of major food groups in 1990 and 2010: A systematic analysis including 266 country-specific nutrition surveys worldwide. *BMJ Open* **2015**, *5*, e008705. [CrossRef]
22. Hopping, B.N.; Erber, E.; Grandinetti, A.; Verheus, M.; Kolonel, L.N.; Maskarinec, G. Dietary fiber, magnesium, and glycemic load alter risk of type 2 diabetes in a multiethnic cohort in Hawaii. *J. Nutr.* **2010**, *140*, 68–74. [CrossRef] [PubMed]
23. Kolonel, L.N.; Henderson, B.E.; Hankin, J.H.; Nomura, A.M.; Wilkens, L.R.; Pike, M.C.; Stram, D.O.; Monroe, K.R.; Earle, M.E.; Nagamine, F.S. A multiethnic cohort in Hawaii and Los Angeles: Baseline characteristics. *Am. J. Epidemiol.* **2000**, *151*, 346–357. [CrossRef] [PubMed]
24. Jacobs, S.; Harmon, B.E.; Boushey, C.J.; Morimoto, Y.; Wilkens, L.R.; Le Marchand, L.; Kröger, J.; Schulze, M.B.; Kolonel, L.N.; Maskarinec, G. A priori-defined diet quality indexes and risk of type 2 diabetes: The Multiethnic Cohort. *Diabetologia* **2015**, *58*, 98–112. [CrossRef] [PubMed]
25. Liese, A.D.; Nichols, M.; Sun, X.; D'Agostino, R.B.; Haffner, S.M. Adherence to the DASH diet is inversely associated with incidence of type 2 diabetes: The insulin resistance atherosclerosis study. *Diabetes Care* **2009**, *32*, 1434–1436. [CrossRef]
26. Kim, Y.; Han, B.-G. KoGES Group. Cohort profile: The Korean genome and epidemiology study (KoGES) consortium. *Int. J. Epidemiol.* **2017**, *46*, 1350. [CrossRef]
27. Ahn, Y.; Kwon, E.; Shim, J.; Park, M.; Joo, Y.; Kimm, K.; Park, C.; Kim, D. Validation and reproducibility of food frequency questionnaire for Korean genome epidemiologic study. *Eur. J. Clin. Nutr.* **2007**, *61*, 1435–1441. [CrossRef]

28. Korea Disease Control and Prevention Agency (KDCA). Manual of Korean Genome and Epidemiology Study. Osong: KDCA, National Institute of Health, Center for Genomics. 2011. Available online: https://nih.go.kr/contents.es?mid=a50401010300#1 (accessed on 20 July 2022).
29. American Diabetes Association. Standards of medical care in diabetes 2010. *Diabetes Care* **2010**, *33* (Suppl. S1), S11–S61. [CrossRef]
30. Yang, J.J.; Yang, J.H.; Kim, J.; Cho, L.Y.; Park, B.; Ma, S.H.; Song, S.H.; Min, W.-K.; Kim, S.S.; Park, M.S. Reliability of quadruplicated serological parameters in the Korean genome and epidemiology study. *Epidemiol. Health* **2011**, *33*, e2011004. [CrossRef]
31. Ainsworth, B.E.; Haskell, W.L.; Whitt, M.C.; Irwin, M.L.; Swartz, A.M.; Strath, S.J.; OBrien, W.L.; Bassett, D.R.; Schmitz, K.H.; Emplaincourt, P.O. Compendium of physical activities: An update of activity codes and MET intensities. *Med. Sci. Sports Exerc.* **2000**, *32* (Suppl. S1), S498–S504. [CrossRef]
32. The Korean Nutrition Society. Food Composition Table. In *Recommended Dietary Allowances for Koreans, 7th ed*; The Korean Nutrition Society: Seoul, Korea, 2000.
33. Schoenfeld, D. Partial residuals for the proportional hazards regression model. *Biometrika* **1982**, *69*, 239–241. [CrossRef]
34. *SAS [Computer Program]*; Version 9.4; SAS Institue Inc.: Cary, NC, USA, 2020.
35. Li, W.; Ruan, W.; Peng, Y.; Wang, D. Soy and the risk of type 2 diabetes mellitus: A systematic review and meta-analysis of observational studies. *Diabetes Res. Clin. Pract.* **2018**, *137*, 190–199. [CrossRef] [PubMed]
36. Schwingshackl, L.; Hoffmann, G.; Lampousi, A.M.; Knuppel, S.; Iqbal, K.; Schwedhelm, C.; Bechthold, A.; Schlesinger, S.; Boeing, H. Food groups and risk of type 2 diabetes mellitus: A systematic review and meta-analysis of prospective studies. *Eur. J. Epidemiol.* **2017**, *32*, 363–375. [CrossRef] [PubMed]
37. Schwingshackl, L.; Schwedhelm, C.; Hoffmann, G.; Lampousi, A.-M.; Knüppel, S.; Iqbal, K.; Bechthold, A.; Schlesinger, S.; Boeing, H. Food groups and risk of all-cause mortality: A systematic review and meta-analysis of prospective studies. *Am. J. Clin. Nutr.* **2017**, *105*, 1462–1473. [CrossRef]
38. Ma, Y.; Hébert, J.R.; Li, W.; Bertone-Johnson, E.R.; Olendzki, B.; Pagoto, S.L.; Tinker, L.; Rosal, M.C.; Ockene, I.S.; Ockene, J.K. Association between dietary fiber and markers of systemic inflammation in the Women's Health Initiative Observational Study. *Nutrition* **2008**, *24*, 941–949. [CrossRef]
39. Xie, Y.; Gou, L.; Peng, M.; Zheng, J.; Chen, L. Effects of soluble fiber supplementation on glycemic control in adults with type 2 diabetes mellitus: A systematic review and meta-analysis of randomized controlled trials. *Clin. Nutr.* **2020**, *40*, 1800–1810. [CrossRef]
40. Cao, H.; Ou, J.; Chen, L.; Zhang, Y.; Szkudelski, T.; Delmas, D.; Daglia, M.; Xiao, J. Dietary polyphenols and type 2 diabetes: Human study and clinical trial. *Crit. Rev. Food Sci. Nutr.* **2019**, *59*, 3371–3379. [CrossRef] [PubMed]
41. Poreba, M.; Rostoff, P.; Siniarski, A.; Mostowik, M.; Golebiowska-Wiatrak, R.; Nessler, J.; Undas, A.; Gajos, G. Relationship between polyunsaturated fatty acid composition in serum phospholipids, systemic low-grade inflammation, and glycemic control in patients with type 2 diabetes and atherosclerotic cardiovascular disease. *Cardiovasc. Diabetol.* **2018**, *17*, 29. [CrossRef] [PubMed]
42. Rinninella, E.; Cintoni, M.; Raoul, P.; Lopetuso, L.R.; Scaldaferri, F.; Pulcini, G.; Miggiano, G.A.D.; Gasbarrini, A.; Mele, M.C. Food components and dietary habits: Keys for a healthy gut microbiota composition. *Nutrients* **2019**, *11*, 2393. [CrossRef] [PubMed]
43. Tomova, A.; Bukovsky, I.; Rembert, E.; Yonas, W.; Alwarith, J.; Barnard, N.D.; Kahleova, H. The effects of vegetarian and vegan diets on gut microbiota. *Front. Nutr.* **2019**, *6*, 47. [CrossRef] [PubMed]
44. Choi, J.; Choi, J.-Y.; Lee, S.-A.; Lee, K.-M.; Shin, A.; Oh, J.; Park, J.; Song, M.; Yang, J.J.; Lee, J.-k. Association between family history of diabetes and clusters of adherence to healthy behaviors: Cross-sectional results from the Health Examinees-Gem (HEXA-G) study. *BMJ Open* **2019**, *9*, e025477. [CrossRef]
45. Watanabe, D.; Yoshida, T.; Yoshimura, E.; Nanri, H.; Goto, C.; Ishikawa-Takata, K.; Ebine, N.; Fujita, H.; Kimura, M.; Yamada, Y. Doubly labelled water–calibration approach attenuates the underestimation of energy intake calculated from self reported dietary assessment data in Japanese older adults. *Public Health Nutr.* **2021**, *25*, 1893–1903. [CrossRef] [PubMed]

nutrients

Review

Consumption of Dairy Foods and Cardiovascular Disease: A Systematic Review

Annalisa Giosuè [1,†], Ilaria Calabrese [1,†], Marilena Vitale [1], Gabriele Riccardi [1] and Olga Vaccaro [2,*]

1 Department of Clinical Medicine and Surgery, "Federico II" University of Naples, 80131 Naples, Italy; annalisa.giosue@gmail.com (A.G.); ilariacalabrese@live.it (I.C.); marilena.vitale@unina.it (M.V.); riccardi@unina.it (G.R.)
2 Department of Pharmacy, "Federico II" University of Naples, 80131 Naples, Italy
* Correspondence: ovaccaro@unina.it; Tel.: +39-081-7463665
† These authors contributed equally to this work.

Abstract: Limited consumption of dairy foods and use of low-fat products is recommended for cardiovascular (CV) prevention; however, other features besides fat content modulate their metabolic effects. We analyze updated evidence on the relationship of different dairy products (low/full-fat dairy, milk, cheese, yogurt) with CVD by reviewing meta-analyses of cohort studies and individual prospective cohort studies with CV hard endpoints (CVD/CHD incidence/mortality), together with meta-analyses of randomized controlled trials exploring the effect of dairy on major CV risk factors. The analyses provide evidence that moderate dairy consumption (up to 200 g/day, globally) has no detrimental effects on CV health and that their effect depends more on the food type (cheese, yogurt, milk) than on the fat content. These data expand current knowledge and may inform revision of current guidelines for CVD prevention.

Keywords: dairy foods; cheese; yogurt; cardiovascular disease; cardiovascular risk factors

Citation: Giosuè, A.; Calabrese, I.; Vitale, M.; Riccardi, G.; Vaccaro, O. Consumption of Dairy Foods and Cardiovascular Disease: A Systematic Review. *Nutrients* **2022**, *14*, 831. https://doi.org/10.3390/nu14040831

Academic Editor: Alaa El-Din A. Bekhit

Received: 28 December 2021
Accepted: 10 February 2022
Published: 16 February 2022

1. Introduction

Cardiovascular diseases (CVDs) constitute the leading cause of global mortality and are a major contributor to reduced quality of life, and thus are a key challenge to health care systems [1,2]. Dietary patterns and particularly fat intake have long been implicated in the modulation of the cardiovascular (CV) risk [3]. A high intake of saturated fatty acids (SFAs) and trans-fatty acids (TFAs) has been linked to an enhanced risk of CVD, and this effect is thought to be mediated predominantly by increased plasma LDL cholesterol levels and their proatherogenic effect [4–6]. The most recent guidelines of the European Society of Cardiology (ESC) and American Heart Association (AHA) recommend <10% of total energy intake from SFAs, yet most developed countries currently exceed this recommendation [7,8].

Recent research has emphasized the importance of focusing on whole foods rather than single nutrients when exploring the relationship of nutrition with disease risk [9].

This is justified by the observation that the physical form of the food, the specific combination of macro/micronutrients and non-nutrient bioactive compounds within each food can play an important role in modulating the metabolic effect of foods and their impact on health beyond their nutrient's composition. Furthermore, detailed information on the associations between each food item and health outcomes can also facilitate the translation of the evidence derived from nutritional research into clinical recommendations, thus improving adherence [10–13].

Among the foods that represent a major source of dietary SFAs, dairy foods deserve consideration since they are largely consumed worldwide [14] and contribute a relevant proportion to the global SFAs intake (i.e., one fifth in the USA, 17–41% in Europe) [15]. A substantial reduction in the intake of dairy and a preferential consumption of low-fat products has been advocated as a strategy for CVD prevention, although evidence in this regard is scant and inconsistent. Dairy foods are a highly heterogeneous food group comprising foods with different biochemical composition, nutritional characteristics (i.e., fat, micronutrients, and salt content), preparation techniques (i.e., fermentation, pasteurization, processed by enzymatic procedures) which may all impact on their nutritional properties and metabolic effects. Partly due to this heterogeneity, the relationship of dairy consumption with cardiovascular diseases remains controversial. The large geographical variation of patterns of dairy food consumption, deeply rooted in sociocultural behaviors and, therefore, influenced by the background diet, further adds to the complexity of the relationship of dairy consumption with health outcomes.

It is therefore appropriate to evaluate the updated evidence on the relationship between dairy food consumption and CVD taking into account the available information on the specific food items included in the dairy food group; in particular, it is relevant to evaluate the impact of dairy foods on established and emerging cardiovascular risk factors in order to substantiate possible mechanisms trough which dairy foods may impact on CV health and differentiate population groups that might be more prone to their possible untoward effects.

Against this background, to shed light on the complex relationship between dairy intake and CV health, we have reviewed the literature on the relationship between the consumption of total dairy foods, or single dairy food items, and CV hard endpoints (i.e., CVD and coronary heart disease (CHD) incidence/mortality) as well as all-cause mortality, focusing on meta-analyses of prospective studies, since they represent a comprehensive and weighted synthesis of the available evidence. Furthermore, to substantiate possible cause–effect relationships and investigate biologically plausible mechanisms through which dairy may have an impact on CVD risk, we have reviewed the available meta-analyses of randomized controlled trials (RCTs) exploring the effect of the various dairy products (i.e., low/full-fat dairy, milk, butter, cheese, yogurt) on major cardiovascular risk factors. By fulfilling the study aims we expect to contribute to a better understanding of the complexity of the relationship between nutrition and CV health.

2. Materials and Methods

2.1. Meta-Analyses of Prospective Cohort Studies on Cardiovascular Hard Endpoints and All-Cause Mortality: Literature Search Strategy and Data Extraction

We performed a systematic literature search in PubMed, Embase, Scopus, and Cochrane Library databases for meta-analyses of prospective cohort studies examining the association between dairy food consumption and all-cause mortality as well as CVD/CHD incidence and mortality according to PRISMA (Preferred Reporting Items for Systematic reviews and Meta-Analyses) guidelines [16], up to 30 April 2021. We used various combinations of the following keywords: "dairy", "dairy products", "total dairy", "full fat dairy", "low fat dairy", "milk", "fermented dairy", "cheese", "yogurt", "all-cause mortality", "coronary heart disease incidence", "coronary heart disease mortality", "cardiovascular disease incidence", "cardiovascular disease mortality", "coronary artery disease", "acute myocardial infarction". The full details on the search strategy are presented in the Online Supplementary Materials (Supplemental Methods S1). We also performed an additional manual search through the reference lists of original publications to identify further pertinent studies. The search was limited to meta-analyses of human studies and was restricted to papers written in English.

Studies were considered for inclusion in the present systematic review if they met the following criteria: (1) the authors reported data from an original, peer-reviewed study (not reviews, conferences, and letters); (2) the study had a prospective design; (3) the authors reported RRs, HRs, or ORs with 95% CIs for dairy food consumption; (4) the investigators reported $\geq$1 of the outcomes of CVD risk, including incidence of total CVD and CHD, or CVD and CHD mortality, or all-cause mortality. We included only prospective cohorts to minimize recall and selection bias. Exclusion criteria were meta-analyses of cross-sectional and/or case-control studies and meta-analyses of prospective cohort studies conducted on populations with special dietary habits (i.e., vegetarians, vegans) or with physiological or pathological conditions requiring specific dietary treatment (i.e., pregnancy or breastfeeding, childhood, diabetes, dyslipidemia, hypertension, etc.).

Two investigators (IC and AG) independently conducted a 2-stage selection process to identify eligible studies: an initial screening of titles and abstracts, followed by an evaluation of all potentially relevant full-length articles. Any discrepancy was resolved by discussions with another investigator (MV). Studies were excluded if they failed to meet the criteria detailed above.

The reason to focus on prospective studies was that this study design allows a more successful control for confounding factors and represents the best source of evidence when RCTs are not available. Furthermore, the utilization of meta-analyses allows a comprehensive and weighted summary of the available evidence. Results of prospective studies not included in the reviewed meta-analyses were also evaluated and reported when relevant for the aims of this review.

The association of dairy consumption with the outcomes was generally estimated by comparing the highest with the lowest level of consumption since a minority of studies report dose–response analyses; where available, these data were collected in order to identify the serving size associated with the best outcomes or representing the threshold of intake above or below which the relationship curve departs from linearity.

2.2. Meta-Analyses of RCTs on the Effect of Dairy Products on Major CV Risk Factor: Literature Search Strategy and Data Extraction

One of the biologically plausible mechanisms through which the consumption of dairy products can influence cardiovascular outcomes is by modulating major cardiometabolic risk factors. Therefore, we reviewed the literature on this topic.

We performed a systematic search in PubMed, Embase, Scopus, and Cochrane Library databases for meta-analyses of randomized controlled trials examining the impact of dairy food consumption on the following parameters known to be relevant in relation to cardiovascular risk: body weight and waist circumference, plasma glucose, glycated hemoglobin, insulin resistance (HOMA-IR), blood pressure (systolic and diastolic), plasma lipids (triglycerides, total cholesterol, LDL and HDL cholesterol), and markers of subclinical inflammation (C-reactive protein, TNF-α, IL-6, adiponectin), according to PRISMA (Preferred Reporting Items for Systematic reviews and Meta-Analyses) guidelines [16], up to 30 April 2021. The full details on the search strategy are presented in the Online Supplementary Materials (Supplemental Methods S1). We also performed an additional manual search through the reference lists of original publications to identify further pertinent studies. The search was limited to meta-analyses of human studies and was restricted to papers written in English.

The following inclusion criteria were applied: (1) study design: meta-analyses of randomized controlled trials; (2) subjects: general adult population without prior cardiovascular events; (3) interventions: common dairy foods or dairy supplemented with probiotics. Meta-analyses of randomized controlled trials conducted on populations with special dietary habits (i.e., vegetarians, vegans, child) or with pre-existing cardiovascular diseases and/or cancer were excluded.

Two investigators (IC and AG) independently reviewed each eligible study, and the following data were extracted: first author's name, publication year, number of studies, the number and type of participants, intervention, comparison, weighted mean difference (Supplemental Tables S1–S5).

3. Results

The results from the literature search and study selection process are shown in Supplemental Figure S1. We identified 36,428 articles (18,666 meta-analyses of prospective cohort studies and 17,762 meta-analyses of RCT) from PubMed, Embase, Scopus, and

Cochrane Library databases by 30 April 2021. After two rounds of review and searching citations of retained articles, 332 potentially relevant studies were initially selected (181 meta-analyses of prospective cohort studies and 151 meta-analyses of RCT). After evaluating the full texts, we further excluded 295 studies: 10 for inappropriate study design, 117 for wrong exposure (i.e., milk proteins, vitamin fortified products, etc.), 47 for wrong outcomes (i.e., cerebrovascular diseases, diabetes, etc.), 57 for special population (i.e., pregnant, vegetarians, infant, etc.), 64 inappropriate study setting (breast feeding, animal feeding, pharmacological setting).

3.1. Association of Dairy Products with All-Cause Death, CVD or CHD

3.1.1. Total Dairy Foods

The association between globally considered dairy products (i.e., milk, cheese, yogurt) and all-cause death has been investigated by four meta-analyses, two comparing high vs. low dairy intake [17,18] and two exploring the dose–response relationship (Table 1) [19,20]. None of them showed a significant association with all-cause death. As for CVD incidence and mortality, three meta-analyses [19,21,22] are available for the former and one [17] for the latter endpoint. Overall, they indicated that the total consumption of dairy foods was not associated with increased CVD risk, whereas Qin et al. [21] reported a statistically significant 12% reduction in the incidence of CVD for high vs. low dairy intake.

Six meta-analyses have focused on CHD incidence [19,21–25], five of which explored the dose–response relationship [19,22–25] and consistently showed that a total dairy food consumption up to 200 g per day was not associated with a higher CHD incidence. In one meta-analysis in which dairy intake was quantified in servings/day, rather than g/day, a significant inverse association with CHD incidence emerged for a consumption of three or more servings per day [22]. The only meta-analysis on CHD mortality reported no significant association with dairy consumption [26].

In summation, the available evidence indicates that the consumption of total dairy foods not exceeding 200 g per day is not associated with all-cause deaths, nor with CVD/CHD incidence and mortality. Above this amount, there seems to a trend towards an increased risk of CHD [25].

Table 1. Summary of the available meta-analyses of prospective cohort studies on the relation between dairy products and CV hard endpoints or all-cause mortality.

	Meta-Analysis	Neutral Relation	Inverse Relation (% Risk Reduction)	Positive Relation (% Risk Increase)
TOTAL DAIRY				
All-cause mortality	O'Sullivan 2013 [17]	✓		
	Guo 2017 [19]	✓		
	Eleftheriou 2018 [18]	✓		
	Schwingshackl 2018 [20]	✓		
CVD incidence	Qin 2015 [21]		✓ (−12% high vs. low intake)	
	Alexander 2016 [22]	✓		
	Guo 2017 [19]	✓		
CVD mortality	O' Sullivan 2013 [17]	✓		

Table 1. Cont.

	Meta-Analysis	Neutral Relation	Inverse Relation (% Risk Reduction)	Positive Relation (% Risk Increase)
CHD incidence	Soedamah-Muthu 2011 [23]	✓		
	Qin 2015 [21]	✓		
	Alexander 2016 [22]	✓ (high vs. low intake)	✓ (−14% per > 3 s/d)	
	Guo 2017 [19]	✓		
	Soedamah-Muthu 2018 [24]	✓		
	Bechthold 2019 [25]	✓		
CHD mortality	Mazidi 2019 [26]	✓		
FULL-FAT DAIRY				
All-cause mortality	Guo 2017 [19]	✓		
CVD incidence	Guo 2017 [19]	✓		
CHD incidence	Soedamah-Muthu 2011 [23]	✓		
	Qin 2015 [21]	✓		
	Alexander 2016 [22]	✓		
	Guo 2017 [19]	✓		
CHD mortality	Mazidi 2019 [26]	✓		
LOW-FAT DAIRY				
All-cause mortality	Guo 2017 [19]	✓		
CVD incidence	Guo 2017 [19]	✓		
CHD incidence	Soedamah-Muthu 2011 [23]	✓		
	Qin 2015 [21]	✓		
	Alexander 2016 [22]		✓ (−10% high vs. low intake)	
	Guo 2017 [19]	✓		
MILK				
All-cause mortality	Soedamah-Muthu 2011 [23]	✓		
	O'Sullivan 2013 [17]	✓		
	Mullie 2016 [27]	✓		
	Guo 2017 [19]	✓		
CVD incidence	Soedamah-Muthu 2011 [23]		✓ (−6% per 200 mL/d)	
	Alexander 2016 [22]	✓		
	Guo 2017 [19]	✓		
CVD mortality	O'Sullivan 2013 [17]	✓		

Table 1. Cont.

	Meta-Analysis	Neutral Relation	Inverse Relation (% Risk Reduction)	Positive Relation (% Risk Increase)
CHD incidence	Soedamah-Muthu 2011 [23]	✓		
	Alexander 2016 [22]	✓		
	Mullie 2016 [27]	✓		
	Guo 2017 [19]	✓		
	Soedamah-Muthu 2018 [24]	✓		
	Jakobsen 2021 [28]	✓		
CHD mortality	Mazidi 2019 [26]			✓ (+4% high vs. low intake)
FERMENTED DAIRY PRODUCTS				
All-cause mortality	Guo 2017 [19]		✓ (−2% per 20 g/d)	
CVD incidence	Guo 2017 [19]		✓ (−2% per 20 g/d)	
	Zhang 2020 [29]		✓ (−20% high vs. low intake)	
CVD mortality	Zhang 2020 [29]	✓		
CHD incidence	Guo 2017 [19]	✓		
	Zhang 2020 [29]	✓		
CHEESE				
All-cause mortality	O'Sullivan 2013 [17]	✓		
	Guo 2017 [19]	✓		
	Tong 2017 [30]	✓		
CVD incidence	Alexander 2016 [22]	✓		
	Chen 2017 [31]	✓ per 50 g/d	✓ (−10% high vs. low intake)	
	Guo 2017 [19]		✓ (−2% per 10 g/d)	
	Zhang 2020 [29]		✓ (−13% high vs. low intake)	
CVD mortality	O'Sullivan 2013 [17]	✓		
CHD incidence	Qin 2015 [21]		✓ (−16% high vs. low intake)	
	Alexander 2016 [22]		✓ (−14% per 50 g/d)	

Table 1. Cont.

	Meta-Analysis	Neutral Relation	Inverse Relation (% Risk Reduction)	Positive Relation (% Risk Increase)
	Chen 2017 [31]		✓ (−10% per 50 g/d)	
	Guo 2017 [19]	✓		
	Jakobsen 2021 [28]		✓ (−4% per 20 g/d)	
YOGURT				
All-cause mortality	Guo 2017 [19]	✓		
	Gao 2020 [32]	✓ high vs. low intake	✓ (−5% per 200 g/d)	
CVD incidence	Alexander 2016 [22]	✓		
	Guo 2017 [19]	✓		
	Wu 2017 [33]	✓ high vs. low intake	✓ (−8% per ≥ 200 g/d)	
	Zhang 2020 [29]		✓ (−22% high vs. low intake)	
CVD mortality	Gao 2020 [32]	✓ high vs. low intake	✓ (−8% per 200 g/d)	
CHD incidence	Qin 2015 [21]	✓		
	Alexander 2016 [22]	✓		
	Wu 2017 [33]	✓		
	Guo 2017 [19]	✓		
	Jakobsen 2021 [28]	✓		

3.1.2. Full-Fat and Low-Fat Dairy Foods

The only meta-analysis that has explored the consumption of full-fat or low-fat dairy foods in relation to all-cause mortality and CVD incidence found no significant association for consumption of up to 200 g per day of either full-fat or low-fat products (Table 1) [19]. Coherent with this finding, two dose–response meta-analyses reported no association with CHD incidence for the same amount of consumption [19,23]. Two meta-analyses which compared high vs. low intake of full fat or low fat products consistently reported no association between full-fat dairy and CHD incidence [21,22], while for low-fat products a significant 10% risk reduction was reported by one of them [22]. A recent meta-analysis on CHD mortality confirmed a neutral association with full-fat or low-fat dairy products [26].

According to the reviewed evidence, consumption of up to 200 g/day of either full-fat or low-fat dairy products is not associated with all-cause death or cardiovascular outcomes in healthy people. For a consumption above this amount the data do not allow a clear trend to be identified.

3.1.3. Milk

Four dose–response meta-analyses that have explored the relationship between milk intake and all-cause death consistently showed a neutral association for a daily consumption of approximately 200 mL (Table 1) [17,19,23,27]. This result is coherent with all but one [23] of the available meta-analyses on milk consumption and CVD outcomes, and with all the meta-analyses on CHD incidence [19,22–24,27,28]. The only available meta-analysis on milk consumption and CHD mortality reported a marginal increased risk in the highest versus the lowest category of milk intake, without assessing the dose–response relationship [26].

In a meta-analysis published in 2021, which investigated the association of low-fat and full-fat milk consumption with fatal and non-fatal CHD events, Jakobsen et al. [28] showed a neutral relationship for a daily consumption of up to 200 g of low-fat milk, but a significant 8% higher risk associated with the same amount of full-fat milk (RR: 1.08, 95% CI: 1.00–1.16). More recent studies—not included in this meta-analysis—in which the different fat content in milk was evaluated in relation to death from all-cause, CVD or CHD, showed no significant association with any of the considered endpoints irrespective of the fat content of milk [34–36].

Overall, the evidence indicates that for milk consumption the association with all-cause mortality and cardiovascular endpoints is neutral for a daily intake up to 200 mL. Above this quantity, data are too scanty to allow a meaningful evaluation of a trend.

3.1.4. Fermented Dairy Foods, Cheese and Yogurt

In a recent dose–response meta-analysis [19] the consumption of fermented dairy products (i.e., sour milk products, yogurt, cheese) was associated with a marginal, but statistically significant, risk reduction (−2%) of all-cause mortality and CVD incidence for each 20 g increase in fermented dairy consumption (Table 1).

This finding was confirmed by a later meta-analysis, which showed a significant 20% lower risk of total CV events [29]. As for fatal CVD, the only available meta-analysis found no significant association with fermented dairy intake [29]. Similar results were reported in the two available meta-analyses on CHD incidence [19,29]. It is, however, notable that in the paper by Zhang et al. [29], high vs. low consumption of fermented dairy was associated with a significant 18% lower risk of myocardial infarction (MI). An inverse association between fermented dairy intake and MI was also reported by some recently published prospective studies not included in the published meta-analyses [37,38].

Cheese and yogurt, the two main components of the fermented dairy food group, were also analyzed. Cheese consumption was not associated with all-cause mortality [17,19,30]; for CVD incidence, either a neutral association [22] or a statistically significant 10 and 13% risk reduction [29,31] were reported for high vs. low cheese intake, whereas one out of two available dose–response analyses [19,31] showed a marginal but significant 2% risk reduction associated with the consumption of 10 g of cheese per day [19]. No relationship with CVD mortality was reported [17]. A daily amount of 50 g of cheese (a standard serving of hard and semi-hard types) was associated with a statistically significant 10 and 14% lower risk of CHD in two meta-analyses including a large number of subjects [22,31]. This finding is shared by the analyses comparing high vs. low cheese intakes, but for lower amounts of cheese consumption the magnitude of the risk reduction is smaller (i.e., −4% for 20 g per day) or null (i.e., for 10 g per day) [19,28]. Similarly, there is no association with CV endpoints for higher intakes [31].

With regard to yogurt, the consumption of 200 g per day was associated with a significant 5% lower total mortality and with an 8% risk reduction of total CVD events [32,33]. A further confirmation of the inverse relationship between a high yogurt consumption and CVD comes from a recent meta-analysis of 10 cohort studies conducted by Zhang et al. [29] that found a statistically significant 22% lower risk of CV events in the high vs. low consumption group.

These findings expand prior knowledge provided by a dose–response analysis, which reported no association between yogurt–but at a lower amount of consumption (i.e., 50 g)—and all-cause death or CVD events [19]. As for CHD incidence, data from five meta-analyses consistently showed a neutral relationship with yogurt consumption [19,21,22,28,33]; this has been confirmed by Jakobsen et al. [28] for both low-fat and full-fat yogurt.

In summation, the available evidence indicates that the consumption of fermented dairy foods is inversely associated with all-cause death and CVD risk. In particular, a generous serving of yogurt (≥200 g per day) is associated with a lower CVD risk, while a moderate cheese consumption (50 g/day) is associated with a reduced risk of CHD.

3.2. Effects of Dairy Foods on Cardiovascular Risk Factors

Among the mechanisms by which the consumption of dairy products can influence cardiovascular outcomes, the modulation of the CV risk factors profile may certainly play a relevant role. Since dairy products have different nutrients composition and a different food matrix, they may also have different effects on major CV risk factors. Here, we review the evidence from meta-analyses of clinical trials relative to body weight and body fat distribution, fasting glucose and glycated hemoglobin, insulin resistance, blood pressure, plasma lipids, and markers of subclinical inflammation (Table 2).

Table 2. Summary of the available meta-analyses of randomized controlled trials on the effect of dairy products on cardiovascular risk factors.

	No. of Meta-Analyses Reporting No Effect	No. of Meta-Analyses Reporting a Significant Reduction	No. of Meta-Analyses Reporting a Significant Increase
	TOTAL DAIRY		
Body weight	3	5 *	2
Waist circumference	3	3	0
Fasting glucose	0	0	2
Glycated hemoglobin	0	1	0
Insulin resistance (HOMA-IR)	2	1	0
Systolic blood pressure	1	0	0
Diastolic blood pressure	1	0	0
Total cholesterol	1	0	0
LDL cholesterol	2	0	0
HDL cholesterol	1	0	0
Triglycerides	1	0	0
C-reactive protein	1	1	0
TNF-α	0	1	0
IL-6	0	1	0
Adiponectin	0	0	1
	FULL-FAT DAIRY PRODUCTS		
Body weight	0	0	1
Waist circumference	1	0	0
Fasting glucose	0	0	1
Systolic blood pressure	1	0	0
Diastolic blood pressure	1	0	0
LDL cholesterol	1	0	0
HDL cholesterol	1	0	0
C-reactive protein	1	0	0
	LOW-FAT DAIRY PRODUCTS		
Body weight	0	0	1
Waist circumference	1	0	0

Table 2. *Cont.*

	No. of Meta-Analyses Reporting No Effect	No. of Meta-Analyses Reporting a Significant Reduction	No. of Meta-Analyses Reporting a Significant Increase
	LOW-FAT DAIRY PRODUCTS		
Fasting glucose	1	0	1
Glycated hemoglobin	1	0	0
Insulin resistance (HOMA-IR)	1	0	0
Systolic blood pressure	1	0	0
Diastolic blood pressure	1	0	0
LDL cholesterol	1	0	0
HDL cholesterol	1	0	0
C-reactive protein	1	0	0
	MILK AND/OR YOGURT		
Fasting glucose	0	0	1
Glycated hemoglobin	0	1	0
Insulin resistance (HOMA-IR)	1	0	0
C-reactive protein	1	0	0
TNF-α	0	1	0
IL-6	1	0	0
Adiponectin	0	0	1
	CHEESE		
Fasting glucose	1	0	0
Insulin resistance (HOMA-IR)	1	0	0
Total cholesterol	0	1	0
LDL cholesterol	0	1	0
HDL cholesterol	0	1	0
Triglycerides	1	0	0
	FERMENTED DAIRIES OR DAIRIES PLUS PROBIOTICS		
Waist circumference	0	1	0
Fasting glucose	2	2	0
Glycated hemoglobin	1	1	0
Insulin resistance (HOMA-IR)	1	0	0
Systolic blood pressure	2	1	0
Diastolic blood pressure	3	1	0
Total cholesterol	0	7	0
LDL cholesterol	0	7	0
HDL cholesterol	5	0	1
Triglycerides	3	1	0

* 4 meta-analyses including studies where dairy supplementation was associated with energy restriction; 1 meta-analysis including studies where dairy supplementation was not associated with energy restriction. CV: cardiovascular.

3.2.1. Body Weight/Waist Circumference

Several meta-analyses have focused on the effect of total dairy on body weight [39–45]. A marginal weight reduction ranging from 0.6 to 1.2 kg was reported in some meta-analyses in which dairy supplementation was implemented within the context of an energy-restricted diet [40,44,45], whereas a neutral effect or a marginal weight increase—ranging from 0.36 to 0.60 kg—was shown by studies where dairy supplementation was not associated with energy restriction (Supplemental Table S1) [40,43,44]. The effect of low-fat or full-fat dairy foods on body weight was explored by Benatar et al. [43] in a meta-analysis showing a statistically significant increase for either low-fat or full-fat dairy products in the context of a dietary regimen without energy restriction. No information is available on the effects of individual foods such as cheese, milk, or yogurt on body weight and fat distribution.

The relation of total dairy consumption with waist circumference was investigated by four meta-analyses [39,40,43,44], three of which show a small, statistically significant reduction in waist circumference, ranging from −2.43 to −1.09 cm [39,40,44]; in these meta-analyses, dairy supplementation was implemented within the context of energy-restricted diets. As for dose–effect, Geng et al. [40] reported a non-significant effect for a supplement of 0.5 to 5.24 g/day of total dairy when evaluated within the context of a diet with or without energy restriction.

The effect of low-fat or full-fat dairy consumption on waist circumference has been explored by Benatar et al. [43] who report no significant association; no data are available to discern the effects of individual foods (i.e., cheese, milk, yogurt). Interventions including the consumption of dairy products supplemented with probiotics were associated with a marginal, but statistically significant, reduction in waist circumference [46].

Overall, the available evidence indicates that the effect of dairy foods on weight/waist circumference is marginal and largely driven by the energy content of the diet.

3.2.2. Fasting Glucose/Glycated Hemoglobin

The search retrieved four meta-analyses on the effects of dairy food consumption on fasting glucose (Supplemental Table S2) [43,46–48]. O'Connor et al. [48] reported a marginal but statistically significant increase in fasting glucose (1.3 mg/dL) with a high total dairy intake. When full-fat or low-fat products were analyzed separately, a marginal increase in fasting glucose was reported for full-fat dairy [43]. The two meta-analyses dealing with low-fat dairy provided conflicting findings: one reports a neutral effect [43], whereas the other shows a modest, but significant increase of 0.07 mmol/L (i.e., 1.26 mg/dL) [48]. Somewhat at variance with this latter observation, however, in a meta-analysis of only four studies the same author reported marginally lower glycated hemoglobin for high versus low consumption of total dairy (−0.09%; 95% CI: −0.16, −0.03) [48]; the finding was not confirmed when only low-fat products were analyzed.

As for specific dairy foods, no significant effect on glucose was reported for cheese consumption, whereas a marginal increase (1.44 mg/dL; 95% CI: 0.36, 2.52) was reported for a high consumption of milk/and or yogurt [48]; again, at variance with this finding, in an analysis of three studies the same author reports a lower value of glycated hemoglobin associated with higher milk and/or yogurt consumption.

Two recent meta-analyses have focused on the relationship of yogurt and other dairy foods enriched with probiotics with blood glucose, showing a moderate reduction ranging from 6 to 13 mg/dL [46,47]. Dixon et al. [47] have explored the effect of yogurt or milk enriched with probiotics: a significant reduction of 12.88 mg/dL in fasting glucose in the intervention group vs. placebo was reported for yogurt, but not for milk; they also showed a modest decrease (−0.55%; 95% CI: −1.03, −0.06) or no change in glycated hemoglobin after yogurt or milk intake, respectively, thus suggesting that probiotics might exert some beneficial effects on glucose homeostasis.

In summation, the evidence for a relationship between the consumption of dairy foods—total dairy, full-fat or low-fat products, or specific items such as cheese or milk—and glucose/glycated hemoglobin is weak; instead, fermented products added with probiotics might be beneficial.

3.2.3. Insulin/Insulin Resistance

Three meta-analyses have explored the relation of total dairy intake with insulin resistance as estimated by the HOMA index [39,43,48]; only one reports a significant improvement in insulin sensitivity associated with intervention of dairy supplementation with or without an energy deficit or caloric restriction (Supplemental Table S2) [39]. In this same meta-analysis, marginal reductions in waist circumference and weight were also reported, therefore it is not straightforward to extrapolate to what extent the improved HOMA-IR is mediated by changes in body weight or by the dietary intervention per se.

The effect of specific dairy food on insulin sensitivity has been extensively studied by O'Connor et al. [48] who have meta-analyzed the few available RCTs dealing with low-fat products (four studies), cheese (two studies), milk and yogurt (six studies), fermented dairy products (four studies), without finding any significant associations.

In summation, the available evidence is scant and overall does not support an association of dairy foods consumption with amelioration of insulin sensitivity.

3.2.4. Blood Pressure

Benatar et al. [43] have meta-analyzed the few studies on the effects of dairy intake on blood pressure; no significant effect was found for either total dairy or low- or full-fat products (Supplemental Table S3). The other available studies concern yogurt and dairy foods enriched with probiotics [47,49,50]. In particular, in a meta-analysis of 15 studies Usinger et al. [49] found a significant reduction in systolic blood pressure (−2.45 mmHg) in the group assigned to fermented milk consumption, whereas in a meta-analysis of three studies, Dixon et al. [47] found a reduction of 3.54 mmHg in diastolic blood pressure, associated with increased consumption of yogurt enriched with probiotics.

On the overall there is no evidence for a detrimental effect of dairy consumption on blood pressure; if anything, a slight improvement with increasing consumption of yogurt enriched with probiotics and fermented milk has been described.

3.2.5. Plasma Lipids

Dairy foods have long been considered unnecessary sources of saturated fats with adverse effects on plasm lipids; however, this concept is based on old studies and needs to be reconsidered in the light of the currently available evidence. The effects of dairy foods on plasma lipids have been extensively studied; data are available for total cholesterol, LDL cholesterol, HDL cholesterol and triglycerides (Supplemental Table S4).

With regard to total cholesterol, no detrimental effects were reported in a network meta-analysis in which total dairy foods were compared with fish, red meat, or sugar-sweetened beverages (SSBs) [51]. As for specific foods, the effect of cheese on total cholesterol was explored by one meta-analysis which shows an average reduction of 0.28 mmol/L (i.e., 10 mg/dL) for hard cheese intake vs. butter intake [52]. Seven meta-analyses of RCTs on the effects of yogurt and other dairy products enriched with probiotics consistently show a reduction in total cholesterol ranging from 6.8 to 16 mg/dL [46,47,53–57].

For LDL cholesterol, no significant effects were reported in a meta-analysis in which total dairy foods were compared with fish, red meat, or SSBs [51]. The only meta-analysis exploring the separate effect of full-fat or low-fat products showed a neutral effect for both [43]. The effects of cheese on plasma cholesterol were studied in one meta-analysis of five studies in which the intake of hard cheese was compared with that of butter with a similar polyunsaturated/saturated fatty acids ratio; an average reduction of 0.22 mmol/L (i.e., 7.9 mg/dL) was reported for LDL cholesterol in the cheese group [52].

Seven meta-analyses have focused on dairy products enriched with probiotics [46,47,53–57]; the results consistently show a reduction in LDL cholesterol ranging from 7.6 to 18 mg/dL.

Information on the dose is not available; the few existing data indicate a significant effect for a daily consumption of 80–600 mL of probiotic fermented milk [53].

The findings are very similar for HDL cholesterol. A neutral effect is reported for total, low-, or full-fat dairy, whereas de Goede et al. [52] report a reduction in HDL cholesterol of 0.05 mmol/L (i.e., 0.9 mg/dL) when hard cheese was compared with butter; this is in parallel with the reduction in total cholesterol and LDL cholesterol described in the same study. As for products enriched with probiotics, five out of six available meta-analyses show a neutral effect on HDL cholesterol [47,53–56]. The only exception is a meta-analysis of four RCTs showing that dairy foods enriched with probiotic increased HDL cholesterol by 0.26 mmol/L (i.e., 4.7 mg/dL) [46]; it is of note that in this same study a reduction in LDL cholesterol of 0.5 mmol/L (i.e., 18 mg/dL) was also reported; no data are available on the dosage.

Several meta-analyses have explored the effect of dairy consumption on plasma triglycerides [46,47,51,52,55,56]. No significant effect was reported by a meta-analysis in which dairy consumption was evaluated in comparison with fish, red meat, or SSBs [51]. The only available meta-analysis focusing on cheese shows a neutral effect [52]. Four meta-analyses have focused on dairy enriched with probiotics; three show a neutral effect [47,55,56], whereas a significant reduction of 0.46 mmol/L (8.3 mg/dL) was reported by Companys et al. [46] in a meta-analysis of three studies. In this same work, as already mentioned, a significant reduction in LDL cholesterol and a significant increase in HDL cholesterol was also reported.

Overall, the available evidence does not support a detrimental effect of dairy consumption on plasma lipids; conversely, there are consistent indications that consumption of dairy enriched with probiotics can ameliorate the plasma lipids profile and, in particular, reduce total and LDL cholesterol in people with hypercholesterolemia.

3.2.6. Subclinical Inflammation

Two meta-analyses have explored the effects on subclinical inflammation of total dairy with C-reactive protein (Supplemental Table S5) [43,58]. While Benatar et al. [43] have shown a neutral effect for total dairy and full-fat or low-fat products, a very recent meta-analysis based on a larger number of studies showed that a high total dairy intake could significantly reduce C-reactive protein (−0.24 mg/L) as well as other markers of subclinical inflammation such as TNF-α (−0.66 pg/mL) and IL-6 (−0.74 pg/mL) and increase serum adiponectin levels (+2.42 μg/mL), a cytokine with anti-inflammatory properties [58].

As for specific dairy foods, data are available only for milk and yogurt; for these products, a high vs. low consumption has a neutral effect on C-reactive protein and IL-6 and a reducing effect on TNF-α (−0.37 pg/mL), while increasing serum adiponectin levels (+14.28 μg/mL) [58].

Overall the evidence is scant, but suggestive of a beneficial effect of specific food items (yogurt and milk) on markers of subclinical inflammation.

4. Discussion

The meta-analyses of cohort studies reviewed here do not support a detrimental relation of dairy foods consumption with total mortality or cardiovascular diseases. Indeed, there is consistent evidence that a total consumption of dairy foods (i.e., including milk, cheese, and yogurt) up to 200 g per day has a neutral association with CVD risk, independent of whether full-fat or low-fat products are considered. As for specific dairy foods, a neutral association was found for milk, while fermented products—cheese and yogurt—were associated with a lower risk of total mortality and CV events. These results confirm and expand our previous observations on the relevance of food choices with regard to cardiovascular health [13], and highlight the wide heterogeneity existing among dairy foods with regard to their association with CVD [59]. This complexity is partly due to potential interrelated influences of different nutrients and non-nutrient bioactive compounds, as well as other food characteristics (i.e., fermentation physical features, processing and cooking procedures) which can modulate the bioavailability and the metabolic effects of nutrients and, consequently, their impact on health [60]. By focusing on foods rather than on nutrients (namely SFAs) the present work is aligned with the most recent research on nutrition and health, which emphasizes food choices and dietary patterns above the nutrients composition of the diet as major determinants of health [13,61]. This approach is also more suitable for the translation of the information into clinical recommendations, as it is more understandable to lay people.

Current nutritional recommendations for CVD prevention in adults are mainly informed by the evidence that SFAs, largely present in dairy, contribute to increasing plasma cholesterol levels which, in turn, increases CV risk; however, recent data indicate that not all SFAs have the same metabolic effects [62–64].

The knowledge that different sources of SFAs have a different impact on CVD is not new. Early prospective studies indicated that the consumption of dairy fat (mainly milk and butter) was associated with an increased mortality from CHD [65–67]; however, when cheese was included among the sources of dairy fat, the correlation coefficients were reduced and become less statistically significant [67].

A subsequent study based on quantities of 40 food items available for consumption from the 1977 Food Balance Sheets (FBSs) of the Food and Agriculture Organization (FAO), relative to 40 countries, has investigated the relationship between a dietary lipid score that combined the intakes of cholesterol and saturated fat (Cholesterol–Saturated Fat Index, CSI) and CHD mortality [68]. The results showed that in Finland CHD mortality was three to five times higher than in France, even though the CSIs of these two countries were almost the same; however, the quality of products consumed was different, since milk intake was 3.5 times greater in Finland than in France [68].

The difference between milk and cheese in relation to CVD outcomes has been ascribed, among others, to the higher calcium content of cheese. On one hand, calcium might partly limit the absorption of SFAs [67,69]; on the other hand, extracellular and intracellular calcium concentrations influence cell membrane potentials of excitable tissues, including the myocardium. In vitro and in silico studies have shown that cardiomyocyte calcium handling is a major determinant of excitation–contraction coupling [70]. However, the clinical implications of these findings remain unclear. Results of human studies on the role of calcium in CV health are incoherent: while dietary calcium does not seem associated with the incidence of CVD, calcium supplementation is reported to increase the risk of MI [71].

There is evidence that differences in the chain length of SFAs lead to different physico-chemical properties and biological effects [72]. The physical and nutritional structure of foods can further modulate their biological effects by influencing the digestion, absorption, and bioavailability of the various nutrients. This is particularly true for a complex food group such as dairy foods. For example, despite their high content in long-chain saturated fatty acids (60% of dairy fat), dairy products are an important source of potentially beneficial compounds, such as medium- and odd-chain saturated fats, natural trans fats, unsaturated and branched-chain fats, branched amino acids, vitamin K1 and K2, and calcium [72]. Probiotics and bioactive compounds naturally contained in fermented dairy further increase the complexity of this food group, since their presence can influence the composition and function of the gut microbiota of the host, which in turn modulates the cardiometabolic risk [64]. In more detail, the probiotics' activity favors the intestinal epithelial integrity and reduces the low-grade inflammation due to metabolic endotoxemia; moreover, it modulates host microbiota composition, thus improving the energy homeostasis, the intermediate metabolism, and the insulin sensitivity [72,73].

The relationship between the various categories of dairy products and cardiovascular disease is largely, though not completely, mediated by their impact on major CV risk factors. To further substantiate the findings on hard CV outcomes (i.e., fatal and non-fatal events), and investigate biologically plausible mechanisms through which dairy may impact on CVD risk, we have reviewed the available meta-analyses of RCTs exploring the effects of the various dairy foods (i.e., low/full-fat dairy, milk, cheese, yogurt) on major CV risk factors, namely: body weight/waist circumference, plasma glucose/glycated hemoglobin, insulin sensitivity, blood pressure, plasma lipids, subclinical inflammation.

For body weight/waist circumference, either total dairy foods or specific items showed a substantially neutral effect; very modest changes in weight or waist circumference were reported in some meta-analyses, but they were mainly driven by the energy content of the dietary intervention. In more detail, when the dietary intervention is performed without limitation in the energy intake, enhanced dairy consumption may lead to increased energy intake and weight gain [40,44,45]. Conversely, dairy products may have modest effects in facilitating weight loss in the context of an energy-restricted diet [40,41,44,45] and there is some suggestion that yogurt may be protective against long-term weight gain [72].

The potential mechanisms through which dairy products may affect body weight include a reduction in appetite due to a higher intake of proteins, particularly casein, which has a good satiating and thermogenic effect [60,74,75]. Furthermore, the increase in calcium intake may be beneficial for weight loss because it can reduce fatty acids absorption and lipogenesis and stimulate lipolysis, by decreasing plasma calcitriol [1,25 (OH)2 D] levels and parathyroid hormone or calciotropic hormones synthesis [76].

A marginal or null effect of total or specific dairy foods has been reported on fasting glucose and glycated hemoglobin in population without diabetes. Dairy products contain variable amounts of sugar, which may explain the marginal and non-clinically relevant increase in fasting glucose reported by O'Connor et al. [48]; however, no information about the sugar content is provided. Furthermore, it not possible to estimate the confounding effect on plasma glucose of other foods consumed in association with dairy products. Conversely, there are some indications that fermented dairy products added with probiotics might exert beneficial effects on fasting glucose and glycated hemoglobin. This is in line with the protective effect of yogurt consumption on the risk of developing type 2 diabetes reported by some authors [60,77].

The evidence regarding insulin sensitivity and subclinical inflammation is scant and overall does not support a significant impact of this food group.

The effects of dairy on blood pressure have also been shown to be substantially neutral; notwithstanding, several studies have provided evidence for biologically plausible mechanisms through which dairy foods might lower blood pressure (i.e., calcium, vitamin D, potassium, and phosphorous content, bioactive small peptides effect, and probiotic activity) [78]. Moreover, habitual dairy intake is associated with lower blood pressure levels in cross-sectional observational studies. Additionally, in the large randomized DASH trial dairy intake was associated with lower blood pressure; however, the dietary intervention included multiple dietary components besides low-fat dairy food (i.e., reduced total and saturated fat intake and increased consumption of vegetables and fruit), and therefore it is not possible in this study to estimate the effects of each single intervention [79].

The effects of dairy foods on plasma lipids have been extensively studied. The well-established knowledge that saturated fats in the diet increase serum cholesterol, which in turn leads to an increased risk of CVD, was the basis on which over time, guidelines for CVD prevention in the general adult population focused on the control of plasma LDL cholesterol, to be achieved, among others, by limiting dairy foods consumption and by substituting whole fat with low-fat products [80]. Indeed, the evidence reviewed here does not support an adverse effect of moderate dairy food consumption (either low or full fat) on plasma lipids, and the literature on this topic is constantly growing. Notably, despite dairy fat consists of around 60% SFAs, they represent a mixture of various subtypes, with different effects on the lipid profile. For instance, besides long-chain SFAs with LDL-raising effect as lauric acid, myristic acid, palmitic acid and stearic acid, dairy fatty acids include (9.8%) medium-chain saturated fatty acids (MCSFAs) (between 6 to 12 carbons, i.e. 6:0 to 12:0), (31.9%) odd-chain SFAs (OCSFAs) (15:0, 17:0), (25%) monounsaturated and (2.3%) polyunsaturated fatty acids (18:1n-9, 18:2n-6 and 18:3n-3), branched-chain saturated fatty acids, and trace amounts of natural (ruminant) trans fats (i.e., trans-palmitoleic acid, 156–158 trans-16:1n-7) [81,82].

Accordingly, a meta-analysis of five randomized controlled trials from Denmark, Norway, and Australia supports heterogeneous effects of dairy on lipid profiles depending on the type of food consumed [52]. Indeed, in this paper, hard cheese compared to butter lowers total cholesterol by 5% and LDL cholesterol by approximately 6.5% despite a similar ratio of polyunsaturated/saturated fatty acids (P/S ratio); this indicates that the different effects of cheese and butter on plasma lipid levels in this meta-analysis cannot be due to the relative amounts of SFAs and PUFAs in the two diets. More probably, the different responses should be attributed to other dairy components or to the specific processing methods utilized for butter and cheese production.

In this regard, it is notable that dairy products enriched with probiotics have a clear hypocholesterolemic effect, since they can modify the gut flora by promoting bacteria strains able to ferment dietary fiber (non-digestible carbohydrates) from seeds, wholegrains, legumes, and vegetables, which are not metabolized or absorbed while passing through the upper gastrointestinal tract. Fiber fermentation leads to the production of short-chain fatty acids (SCFAs) that exert local and systemic effects, such as the inhibition of the hepatic cholesterol synthesis and the stimulation of liver cholesterol uptake [83].

Furthermore, SCFAs, especially butyrate, can lower the plasma levels of pro-inflammatory markers (high-sensitivity C-reactive protein, TNF-α and IL-6) by acting at the gene expression level or by the activation of the MAPK pathway [84].

In summary, by combining data on CV events and risk factors, this study provides coherent evidence for a neutral effect of a moderate consumption of total dairy food on CVD irrespective of fat content. A beneficial effect of some specific items (i.e., fermented products and products added with probiotics), largely mediated through their effect on major CV risk factors—mainly lipids and subclinical inflammation—also emerges.

5. Conclusions

This study highlights the complexity of the relationship between different dairy foods and cardiovascular diseases as well as their risk factors. Altogether, the results indicate that the association of dairy intake with cardiovascular risk is largely driven by the food type (i.e., cheese, yogurt, milk). These findings may inform dietary recommendations for CVD prevention by allowing healthy people with normal plasma cholesterol levels a more liberal consumption of up to 200 g/day of total dairy foods (including milk, cheese, and yogurt), irrespective of being full or low fat. Within this amount of consumption, fermented dairy should be preferred (i.e., one generous serving/day of yogurt or three small servings/week of cheese).

Further studies should be undertaken to clarify the mechanisms of the beneficial impact of probiotics and other components of dairy foods such as whey protein and calcium on the cardiovascular risk-factor profile.

Supplementary Materials: The following are available online at https://www.mdpi.com/article/10.3390/nu14040831/s1: Supplemental Methods S1: Search strategy. Supplemental Figure S1: PRISMA flow diagram. Table S1: Summary of meta-analyses of randomized controlled trials on the effect of dairy products intake on body weight and waist circumference. Table S2: Summary of meta-analyses of randomized controlled trials on the effect of dairy products intake on fasting glucose, glycated hemoglobin and insulin resistance. Table S3: Summary of meta-analyses of randomized controlled trials on the effect of dairy products intake on blood pressure. Table S4: Summary of meta-analyses of randomized controlled trials on the effect of dairy products intake on plasma lipids. Table S5: Summary of meta-analyses of randomized controlled trials on the effect of dairy products intake on markers of subclinical inflammation.

Author Contributions: Conceptualization, G.R. and O.V.; methodology, M.V., G.R. and O.V.; data curation, M.V.; writing—original draft preparation, A.G., I.C. and M.V.; writing—review and editing, G.R. and O.V.; supervision, G.R. and O.V.; project administration, G.R. and O.V.; funding acquisition, G.R. and O.V. All authors have read and agreed to the published version of the manuscript.

Funding: This research was partially funded by (1) a research grant from the "Barilla Center for Food and Nutrition Foundation (BCFN)" within the framework of a project aimed at an evidence-based reformulation of the Food Pyramid for the prevention of cardiovascular disease; (2) Ministero della Salute Progetto di Ricerca Finalizzata RF-2016-02364513.

Institutional Review Board Statement: Not applicable.

Informed Consent Statement: Not applicable.

Data Availability Statement: Not applicable.

Conflicts of Interest: G.R. is member of the scientific advisory board of the "BCFN Foundation" and of the Barilla Health and Well-being advisory board, of the Nutrition Foundation of Italy and of "Istituto Nutrizionale Carapelli". The funders had no role in the design of the study; in the collection, analyses, or interpretation of data; in the writing of the manuscript, or in the decision to publish the results.

References

1. Roth, G.A.; Abate, D.; Abate, K.H.; Abay, S.M.; Abbafati, C.; Abbasi, N.; Abbastabar, H.; Abd-Allah, F.; Abdela, J.; Abdelalim, A.; et al. Global, Regional, and National Age-Sex-Specific Mortality for 282 Causes of Death in 195 Countries and Territories, 1980–2017: A Systematic Analysis for the Global Burden of Disease Study 2017. *Lancet* **2018**, *392*, 1736–1788. [CrossRef]
2. Kyu, H.H.; Abate, D.; Abate, K.H.; Abay, S.M.; Abbafati, C.; Abbasi, N.; Abbastabar, H.; Abd-Allah, F.; Abdela, J.; Abdelalim, A.; et al. Global, Regional, and National Disability-Adjusted Life-Years (DALYs) for 359 Diseases and Injuries and Healthy Life Expectancy (HALE) for 195 Countries and Territories, 1990–2017: A Systematic Analysis for the Global Burden of Disease Study 2017. *Lancet* **2018**, *392*, 1859–1922. [CrossRef]
3. Forouhi, N.G.; Krauss, R.M.; Taubes, G.; Willett, W. Dietary Fat and Cardiometabolic Health: Evidence, Controversies, and Consensus for Guidance. *BMJ* **2018**, *361*, k2139. [CrossRef]
4. Keys, A.; Anderson, J.T.; Grande, F. Prediction of Serum-Cholesterol Responses of Man to Changes in Fats in the Diet. *Lancet* **1957**, *273*, 959–966. [CrossRef]
5. Hegsted, D.M.; McGandy, R.B.; Myers, M.L.; Stare, F.J. Quantitative Effects of Dietary Fat on Serum Cholesterol in Man. *Am. J. Clin. Nutr.* **1965**, *17*, 281–295. [CrossRef]
6. Ference, B.A.; Ginsberg, H.N.; Graham, I.; Ray, K.K.; Packard, C.J.; Bruckert, E.; Hegele, R.A.; Krauss, R.M.; Raal, F.J.; Schunkert, H.; et al. Low-Density Lipoproteins Cause Atherosclerotic Cardiovascular Disease. 1. Evidence from Genetic, Epidemiologic, and Clinical Studies. A Consensus Statement from the European Atherosclerosis Society Consensus Panel. *Eur. Heart J.* **2017**, *38*, 2459–2472. [CrossRef]
7. Visseren, F.L.J.; Mach, F.; Smulders, Y.M.; Carballo, D.; Koskinas, K.C.; Bäck, M.; Benetos, A.; Biffi, A.; Boavida, J.-M.; Capodanno, D.; et al. 2021 ESC Guidelines on Cardiovascular Disease Prevention in Clinical Practice. *Eur. Heart J.* **2021**, *42*, 3227–3337. [CrossRef]
8. Van Horn, L.; Carson, J.A.; Appel, L.J.; Burke, L.E.; Economos, C.; Karmally, W.; Lancaster, K.; Lichtenstein, A.H.; Johnson, R.K.; Thomas, R.J.; et al. Recommended Dietary Pattern to Achieve Adherence to the American Heart Association/American College of Cardiology (AHA/ACC) Guidelines: A Scientific Statement from the American Heart Association. *Circulation* **2016**, *134*, e505–e529. [CrossRef] [PubMed]
9. Fardet, A.; Rock, E. From a Reductionist to a Holistic Approach in Preventive Nutrition to Define New and More Ethical Paradigms. *Healthcare* **2015**, *3*, 1054–1063. [CrossRef] [PubMed]
10. Nestel, P.J.; Beilin, L.J.; Clifton, P.M.; Watts, G.F.; Mori, T.A. Practical Guidance for Food Consumption to Prevent Cardiovascular Disease. *Heart Lung Circ.* **2021**, *30*, 163–179. [CrossRef] [PubMed]
11. Kwok, C.S.; Gulati, M.; Michos, E.D.; Potts, J.; Wu, P.; Watson, L.; Loke, Y.K.; Mallen, C.; Mamas, M.A. Dietary Components and Risk of Cardiovascular Disease and All-Cause Mortality: A Review of Evidence from Meta-Analyses. *Eur. J. Prev. Cardiol.* **2019**, *26*, 1415–1429. [CrossRef] [PubMed]
12. Mozaffarian, D. Dietary and Policy Priorities for Cardiovascular Disease, Diabetes, and Obesity: A Comprehensive Review. *Circulation* **2016**, *133*, 187–225. [CrossRef]
13. Riccardi, G.; Giosuè, A.; Calabrese, I.; Vaccaro, O. Dietary Recommendations for Prevention of Atherosclerosis. *Cardiovasc. Res.* **2021**, cvab173. [CrossRef]
14. Riccardi, G.; Vitale, M.; Vaccaro, O. Are Europeans Moving towards Dietary Habits More Suitable for Reducing Cardiovascular Disease Risk? *Nutr. Metab. Cardiovasc. Dis.* **2020**, *30*, 1857–1860. [CrossRef] [PubMed]
15. Eilander, A.; Harika, R.K.; Zock, P.L. Intake and Sources of Dietary Fatty Acids in Europe: Are Current Population Intakes of Fats Aligned with Dietary Recommendations? *Eur. J. Lipid Sci. Technol. EJLST* **2015**, *117*, 1370–1377. [CrossRef]
16. Moher, D.; Liberati, A.; Tetzlaff, J.; Altman, D.G.; Altman, D.; Antes, G.; Atkins, D.; Barbour, V.; Barrowman, N.; Berlin, J.A.; et al. Preferred Reporting Items for Systematic Reviews and Meta-Analyses: The PRISMA Statement. *Ann. Intern. Med.* **2009**, *151*, 264–269. [CrossRef] [PubMed]

17. O'Sullivan, T.A.; Hafekost, K.; Mitrou, F.; Lawrence, D. Food Sources of Saturated Fat and the Association with Mortality: A Meta-Analysis. *Am. J. Public Health* **2013**, *103*, e31–e42. [CrossRef]
18. Eleftheriou, D.; Benetou, V.; Trichopoulou, A.; La Vecchia, C.; Bamia, C. Mediterranean Diet and Its Components in Relation to All-Cause Mortality: Meta-Analysis. *Br. J. Nutr.* **2018**, *120*, 1081–1097. [CrossRef]
19. Guo, J.; Astrup, A.; Lovegrove, J.A.; Gijsbers, L.; Givens, D.I.; Soedamah-Muthu, S.S. Milk and Dairy Consumption and Risk of Cardiovascular Diseases and All-Cause Mortality: Dose-Response Meta-Analysis of Prospective Cohort Studies. *Eur. J. Epidemiol.* **2017**, *32*, 269–287. [CrossRef]
20. Schwingshackl, L.; Schwedhelm, C.; Hoffmann, G.; Lampousi, A.M.; Knüppel, S.; Iqbal, K.; Bechthold, A.; Schlesinger, S.; Boeing, H. Food Groups and Risk of All-Cause Mortality: A Systematic Review and Meta-Analysis of Prospective Studies. *Am. J. Clin. Nutr.* **2017**, *105*, 1462–1473. [CrossRef]
21. Qin, L.-Q.; Xu, J.-Y.; Han, S.-F.; Zhang, Z.-L.; Zhao, Y.-Y.; Szeto, I.M. Dairy Consumption and Risk of Cardiovascular Disease: An Updated Meta-Analysis of Prospective Cohort Studies. *Asia Pac. J. Clin. Nutr.* **2015**, *24*, 90–100. [CrossRef] [PubMed]
22. Alexander, D.D.; Bylsma, L.C.; Vargas, A.J.; Cohen, S.S.; Doucette, A.; Mohamed, M.; Irvin, S.R.; Miller, P.E.; Watson, H.; Fryzek, J.P. Dairy Consumption and CVD: A Systematic Review and Meta-Analysis. *Br. J. Nutr.* **2016**, *115*, 737–750. [CrossRef]
23. Soedamah-Muthu, S.S.; Ding, E.L.; Al-Delaimy, W.K.; Hu, F.B.; Engberink, M.F.; Willett, W.C.; Geleijnse, J.M. Milk and Dairy Consumption and Incidence of Cardiovascular Diseases and All-Cause Mortality: Dose-Response Meta-Analysis of Prospective Cohort Studies. *Am. J. Clin. Nutr.* **2011**, *93*, 158–171. [CrossRef] [PubMed]
24. Soedamah-Muthu, S.S.; de Goede, J. Dairy Consumption and Cardiometabolic Diseases: Systematic Review and Updated Meta-Analyses of Prospective Cohort Studies. *Curr. Nutr. Rep.* **2018**, *7*, 171–182. [CrossRef] [PubMed]
25. Bechthold, A.; Boeing, H.; Schwedhelm, C.; Hoffmann, G.; Knüppel, S.; Iqbal, K.; De Henauw, S.; Michels, N.; Devleesschauwer, B.; Schlesinger, S.; et al. Food Groups and Risk of Coronary Heart Disease, Stroke and Heart Failure: A Systematic Review and Dose-Response Meta-Analysis of Prospective Studies. *Crit. Rev. Food Sci. Nutr.* **2019**, *59*, 1071–1090. [CrossRef] [PubMed]
26. Mazidi, M.; Mikhailidis, D.P.; Sattar, N.; Howard, G.; Graham, I.; Banach, M. Consumption of Dairy Product and Its Association with Total and Cause Specific Mortality—A Population-Based Cohort Study and Meta-Analysis. *Clin. Nutr.* **2019**, *38*, 2833–2845. [CrossRef]
27. Mullie, P.; Pizot, C.; Autier, P. Daily Milk Consumption and All-Cause Mortality, Coronary Heart Disease and Stroke: A Systematic Review and Meta-Analysis of Observational Cohort Studies. *BMC Public Health* **2016**, *16*, 1236. [CrossRef]
28. Jakobsen, M.U.; Trolle, E.; Outzen, M.; Mejborn, H.; Grønberg, M.G.; Lyndgaard, C.B.; Stockmarr, A.; Venø, S.K.; Bysted, A. Intake of Dairy Products and Associations with Major Atherosclerotic Cardiovascular Diseases: A Systematic Review and Meta-Analysis of Cohort Studies. *Sci. Rep.* **2021**, *11*, 1303. [CrossRef]
29. Zhang, K.; Chen, X.; Zhang, L.; Deng, Z. Fermented Dairy Foods Intake and Risk of Cardiovascular Diseases: A Meta-Analysis of Cohort Studies. *Crit. Rev. Food Sci. Nutr.* **2020**, *60*, 1189–1194. [CrossRef] [PubMed]
30. Tong, X.; Chen, G.-C.; Zhang, Z.; Wei, Y.-L.; Xu, J.-Y.; Qin, L.-Q. Cheese Consumption and Risk of All-Cause Mortality: A Meta-Analysis of Prospective Studies. *Nutrients* **2017**, *9*, 63. [CrossRef] [PubMed]
31. Chen, G.-C.; Wang, Y.; Tong, X.; Szeto, I.M.Y.; Smit, G.; Li, Z.-N.; Qin, L.-Q. Cheese Consumption and Risk of Cardiovascular Disease: A Meta-Analysis of Prospective Studies. *Eur. J. Nutr.* **2017**, *56*, 2565–2575. [CrossRef] [PubMed]
32. Gao, X.; Jia, H.-Y.; Chen, G.-C.; Li, C.-Y.; Hao, M. Yogurt Intake Reduces All-Cause and Cardiovascular Disease Mortality: A Meta-Analysis of Eight Prospective Cohort Studies. *Chin. J. Integr. Med.* **2020**, *26*, 462–468. [CrossRef] [PubMed]
33. Wu, L.; Sun, D. Consumption of Yogurt and the Incident Risk of Cardiovascular Disease: A Meta-Analysis of Nine Cohort Studies. *Nutrients* **2017**, *9*, 315. [CrossRef]
34. Um, C.Y.; Prizment, A.; Hong, C.-P.; Lazovich, D.; Bostick, R.M. Associations of Calcium and Dairy Product Intakes with All-Cause, All-Cancer, Colorectal Cancer and CHD Mortality among Older Women in the Iowa Women's Health Study. *Br. J. Nutr.* **2019**, *121*, 1188–1200. [CrossRef]
35. Pala, V.; Sieri, S.; Chiodini, P.; Masala, G.; Palli, D.; Mattiello, A.; Panico, S.; Tumino, R.; Frasca, G.; Fasanelli, F.; et al. Associations of Dairy Product Consumption with Mortality in the European Prospective Investigation into Cancer and Nutrition (EPIC)-Italy Cohort. *Am. J. Clin. Nutr.* **2019**, *110*, 1220–1230. [CrossRef] [PubMed]
36. Ding, M.; Li, J.; Qi, L.; Ellervik, C.; Zhang, X.; Manson, J.E.; Stampfer, M.; Chavarro, J.E.; Rexrode, K.M.; Kraft, P.; et al. Associations of Dairy Intake with Risk of Mortality in Women and Men: Three Prospective Cohort Studies. *BMJ* **2019**, *367*, l6204. [CrossRef]
37. Koskinen, T.T.; Virtanen, H.E.K.; Voutilainen, S.; Tuomainen, T.-P.; Mursu, J.; Virtanen, J.K. Intake of Fermented and Non-Fermented Dairy Products and Risk of Incident CHD: The Kuopio Ischaemic Heart Disease Risk Factor Study. *Br. J. Nutr.* **2018**, *120*, 1288–1297. [CrossRef]
38. Johansson, I.; Esberg, A.; Nilsson, L.M.; Jansson, J.-H.; Wennberg, P.; Winkvist, A. Dairy Product Intake and Cardiometabolic Diseases in Northern Sweden: A 33-Year Prospective Cohort Study. *Nutrients* **2019**, *11*, 284. [CrossRef]
39. Sochol, K.M.; Johns, T.S.; Buttar, R.S.; Randhawa, L.; Sanchez, E.; Gal, M.; Lestrade, K.; Merzkani, M.; Abramowitz, M.K.; Mossavar-Rahmani, Y.; et al. The Effects of Dairy Intake on Insulin Resistance: A Systematic Review and Meta-Analysis of Randomized Clinical Trials. *Nutrients* **2019**, *11*, 2237. [CrossRef]
40. Geng, T.; Qi, L.; Huang, T. Effects of Dairy Products Consumption on Body Weight and Body Composition Among Adults: An Updated Meta-Analysis of 37 Randomized Control Trials. *Mol. Nutr. Food Res.* **2018**, *62*, 1700410. [CrossRef]

41. Stonehouse, W.; Wycherley, T.; Luscombe-Marsh, N.; Taylor, P.; Brinkworth, G.; Riley, M. Dairy Intake Enhances Body Weight and Composition Changes during Energy Restriction in 18-50-Year-Old Adults-A Meta-Analysis of Randomized Controlled Trials. *Nutrients* **2016**, *8*, 394. [CrossRef]
42. Booth, A.O.; Huggins, C.E.; Wattanapenpaiboon, N.; Nowson, C.A. Effect of Increasing Dietary Calcium through Supplements and Dairy Food on Body Weight and Body Composition: A Meta-Analysis of Randomised Controlled Trials. *Br. J. Nutr.* **2015**, *114*, 1013–1025. [CrossRef]
43. Benatar, J.R.; Sidhu, K.; Stewart, R.A.H. Effects of High and Low Fat Dairy Food on Cardio-Metabolic Risk Factors: A Meta-Analysis of Randomized Studies. *PLoS ONE* **2013**, *8*, e76480. [CrossRef]
44. Abargouei, A.S.; Janghorbani, M.; Salehi-Marzijarani, M.; Esmaillzadeh, A. Effect of Dairy Consumption on Weight and Body Composition in Adults: A Systematic Review and Meta-Analysis of Randomized Controlled Clinical Trials. *Int. J. Obes.* **2012**, *36*, 1485–1493. [CrossRef]
45. Chen, M.; Pan, A.; Malik, V.S.; Hu, F.B. Effects of Dairy Intake on Body Weight and Fat: A Meta-Analysis of Randomized Controlled Trials. *Am. J. Clin. Nutr.* **2012**, *96*, 735–747. [CrossRef]
46. Companys, J.; Pedret, A.; Valls, R.M.; Solà, R.; Pascual, V. Fermented Dairy Foods Rich in Probiotics and Cardiometabolic Risk Factors: A Narrative Review from Prospective Cohort Studies. *Crit. Rev. Food Sci. Nutr.* **2021**, *61*, 1966–1975. [CrossRef] [PubMed]
47. Dixon, A.; Robertson, K.; Yung, A.; Que, M.; Randall, H.; Wellalagodage, D.; Cox, T.; Robertson, D.; Chi, C.; Sun, J. Efficacy of Probiotics in Patients of Cardiovascular Disease Risk: A Systematic Review and Meta-Analysis. *Curr. Hypertens. Rep.* **2020**, *22*, 74. [CrossRef]
48. O'Connor, S.; Turcotte, A.-F.; Gagnon, C.; Rudkowska, I. Increased Dairy Product Intake Modifies Plasma Glucose Concentrations and Glycated Hemoglobin: A Systematic Review and Meta-Analysis of Randomized Controlled Trials. *Adv. Nutr.* **2019**, *10*, 262–279. [CrossRef]
49. Usinger, L.; Reimer, C.; Ibsen, H. Fermented Milk for Hypertension. *Cochrane Database Syst. Rev.* **2012**, CD008118. [CrossRef] [PubMed]
50. Dong, J.-Y.; Szeto, I.M.Y.; Makinen, K.; Gao, Q.; Wang, J.; Qin, L.-Q.; Zhao, Y. Effect of Probiotic Fermented Milk on Blood Pressure: A Meta-Analysis of Randomised Controlled Trials. *Br. J. Nutr.* **2013**, *110*, 1188–1194. [CrossRef] [PubMed]
51. Schwingshackl, L.; Hoffmann, G.; Iqbal, K.; Schwedhelm, C.; Boeing, H. Food Groups and Intermediate Disease Markers: A Systematic Review and Network Meta-Analysis of Randomized Trials. *Am. J. Clin. Nutr.* **2018**, *108*, 576–586. [CrossRef] [PubMed]
52. De Goede, J.; Geleijnse, J.M.; Ding, E.L.; Soedamah-Muthu, S.S. Effect of Cheese Consumption on Blood Lipids: A Systematic Review and Meta-Analysis of Randomized Controlled Trials. *Nutr. Rev.* **2015**, *73*, 259–275. [CrossRef]
53. Ziaei, R.; Ghavami, A.; Khalesi, S.; Ghiasvand, R.; Mokari Yamchi, A. The Effect of Probiotic Fermented Milk Products on Blood Lipid Concentrations: A Systematic Review and Meta-Analysis of Randomized Controlled Trials. *Nutr. Metab. Cardiovasc. Dis. NMCD* **2021**, *31*, 997–1015. [CrossRef] [PubMed]
54. Pourrajab, B.; Fatahi, S.; Dehnad, A.; Kord Varkaneh, H.; Shidfar, F. The Impact of Probiotic Yogurt Consumption on Lipid Profiles in Subjects with Mild to Moderate Hypercholesterolemia: A Systematic Review and Meta-Analysis of Randomized Controlled Trials. *Nutr. Metab. Cardiovasc. Dis. NMCD* **2020**, *30*, 11–22. [CrossRef] [PubMed]
55. Sun, J.; Buys, N. Effects of Probiotics Consumption on Lowering Lipids and CVD Risk Factors: A Systematic Review and Meta-Analysis of Randomized Controlled Trials. *Ann. Med.* **2015**, *47*, 430–440. [CrossRef] [PubMed]
56. Cho, Y.A.; Kim, J. Effect of Probiotics on Blood Lipid Concentrations: A Meta-Analysis of Randomized Controlled Trials. *Medicine* **2015**, *94*, e1714. [CrossRef]
57. Shimizu, M.; Hashiguchi, M.; Shiga, T.; Tamura, H.; Mochizuki, M. Meta-Analysis: Effects of Probiotic Supplementation on Lipid Profiles in Normal to Mildly Hypercholesterolemic Individuals. *PLoS ONE* **2015**, *10*, e0139795. [CrossRef] [PubMed]
58. Moosavian, S.P.; Rahimlou, M.; Saneei, P.; Esmaillzadeh, A. Effects of Dairy Products Consumption on Inflammatory Biomarkers among Adults: A Systematic Review and Meta-Analysis of Randomized Controlled Trials. *Nutr. Metab. Cardiovasc. Dis. NMCD* **2020**, *30*, 872–888. [CrossRef]
59. Van Aerde, M.A.; Soedamah-Muthu, S.S.; Geleijnse, J.M.; Snijder, M.B.; Nijpels, G.; Stehouwer, C.D.A.; Dekker, J.M. Dairy Intake in Relation to Cardiovascular Disease Mortality and All-Cause Mortality: The Hoorn Study. *Eur. J. Nutr.* **2013**, *52*, 609–616. [CrossRef]
60. Astrup, A.; Geiker, N.R.W.; Magkos, F. Effects of Full-Fat and Fermented Dairy Products on Cardiometabolic Disease: Food Is More Than the Sum of Its Parts. *Adv. Nutr.* **2019**, *10*, 924S–930S. [CrossRef] [PubMed]
61. Vitale, M.; Masulli, M.; Calabrese, I.; Rivellese, A.A.; Bonora, E.; Signorini, S.; Perriello, G.; Squatrito, S.; Buzzetti, R.; Sartore, G.; et al. Impact of a Mediterranean Dietary Pattern and Its Components on Cardiovascular Risk Factors, Glucose Control, and Body Weight in People with Type 2 Diabetes: A Real-Life Study. *Nutrients* **2018**, *10*, 1067. [CrossRef]
62. Astrup, A.; Rice Bradley, B.H.; Brenna, J.T.; Delplanque, B.; Ferry, M.; Torres-Gonzalez, M. Regular-Fat Dairy and Human Health: A Synopsis of Symposia Presented in Europe and North America (2014–2015). *Nutrients* **2016**, *8*, 463. [CrossRef] [PubMed]
63. Yu, E.; Hu, F.B. Dairy Products, Dairy Fatty Acids, and the Prevention of Cardiometabolic Disease: A Review of Recent Evidence. *Curr. Atheroscler. Rep.* **2018**, *20*, 24. [CrossRef] [PubMed]
64. Bhupathi, V.; Mazariegos, M.; Cruz Rodriguez, J.B.; Deoker, A. Dairy Intake and Risk of Cardiovascular Disease. *Curr. Cardiol. Rep.* **2020**, *22*, 11. [CrossRef] [PubMed]
65. Kimura, N.; Keys, A. Coronary Heart Disease in Seven Countries. X. Rural Southern Japan. *Circulation* **1993**, *88*, 2524–2531.

66. Simons, L.A. Interrelations of Lipids and Lipoproteins with Coronary Artery Disease Mortality in 19 Countries. *Am. J. Cardiol.* **1986**, *57*, 5G–10G. [CrossRef]
67. Renaud, S.; de Lorgeril, M. Dietary Lipids and Their Relation to Ischaemic Heart Disease: From Epidemiology to Prevention. *J. Intern. Med. Suppl.* **1989**, *731*, 39–46. [CrossRef] [PubMed]
68. Artaud-Wild, S.M.; Connor, S.L.; Sexton, G.; Connor, W.E. Differences in Coronary Mortality Can Be Explained by Differences in Cholesterol and Saturated Fat Intakes in 40 Countries but Not in France and Finland. A Paradox. *Circulation* **1993**, *88*, 2771–2779. [CrossRef]
69. Mulet-Cabero, A.-I.; Wilde, P.J. Role of Calcium on Lipid Digestion and Serum Lipids: A Review. *Crit. Rev. Food Sci. Nutr.* **2021**, 1–14. [CrossRef]
70. Sutanto, H.; Lyon, A.; Lumens, J.; Schotten, U.; Dobrev, D.; Heijman, J. Cardiomyocyte Calcium Handling in Health and Disease: Insights from in Vitro and in Silico Studies. *Prog. Biophys. Mol. Biol.* **2020**, *157*, 54–75. [CrossRef]
71. Anderson, J.J.B.; Kruszka, B.; Delaney, J.A.C.; He, K.; Burke, G.L.; Alonso, A.; Bild, D.E.; Budoff, M.; Michos, E.D. Calcium Intake from Diet and Supplements and the Risk of Coronary Artery Calcification and Its Progression among Older Adults: 10-Year Follow-up of the Multi-Ethnic Study of Atherosclerosis (MESA). *J. Am. Heart Assoc.* **2016**, *5*, e003815. [CrossRef] [PubMed]
72. Mozaffarian, D.; Wu, J.H.Y. Flavonoids, Dairy Foods, and Cardiovascular and Metabolic Health: A Review of Emerging Biologic Pathways. *Circ. Res.* **2018**, *122*, 369–384. [CrossRef] [PubMed]
73. Pereira, D.I.A.; Gibson, G.R. Effects of Consumption of Probiotics and Prebiotics on Serum Lipid Levels in Humans. *Crit. Rev. Biochem. Mol. Biol.* **2002**, *37*, 259–281. [CrossRef] [PubMed]
74. Dove, E.R.; Hodgson, J.M.; Puddey, I.B.; Beilin, L.J.; Lee, Y.P.; Mori, T.A. Skim Milk Compared with a Fruit Drink Acutely Reduces Appetite and Energy Intake in Overweight Men and Women. *Am. J. Clin. Nutr.* **2009**, *90*, 70–75. [CrossRef] [PubMed]
75. Westerterp-Plantenga, M.S.; Nieuwenhuizen, A.; Tomé, D.; Soenen, S.; Westerterp, K.R. Dietary Protein, Weight Loss, and Weight Maintenance. *Annu. Rev. Nutr.* **2009**, *29*, 21–41. [CrossRef] [PubMed]
76. Zemel, M.B.; Richards, J.; Milstead, A.; Campbell, P. Effects of Calcium and Dairy on Body Composition and Weight Loss in African-American Adults. *Obes. Res.* **2005**, *13*, 1218–1225. [CrossRef] [PubMed]
77. Gijsbers, L.; Ding, E.L.; Malik, V.S.; de Goede, J.; Geleijnse, J.M.; Soedamah-Muthu, S.S. Consumption of Dairy Foods and Diabetes Incidence: A Dose-Response Meta-Analysis of Observational Studies. *Am. J. Clin. Nutr.* **2016**, *103*, 1111–1124. [CrossRef]
78. McGrane, M.M.; Essery, E.; Obbagy, J.; Lyon, J.; Macneil, P.; Spahn, J.; Van Horn, L. Dairy Consumption, Blood Pressure, and Risk of Hypertension: An Evidence-Based Review of Recent Literature. *Curr. Cardiovasc. Risk Rep.* **2011**, *5*, 287–298. [CrossRef]
79. Appel, L.J.; Moore, T.J.; Obarzanek, E.; Vollmer, W.M.; Svetkey, L.P.; Sacks, F.M.; Bray, G.A.; Vogt, T.M.; Cutler, J.A.; Windhauser, M.M.; et al. A Clinical Trial of the Effects of Dietary Patterns on Blood Pressure. DASH Collaborative Research Group. *N. Engl. J. Med.* **1997**, *336*, 1117–1124. [CrossRef] [PubMed]
80. Drouin-Chartier, J.-P.; Côté, J.A.; Labonté, M.-È.; Brassard, D.; Tessier-Grenier, M.; Desroches, S.; Couture, P.; Lamarche, B. Comprehensive Review of the Impact of Dairy Foods and Dairy Fat on Cardiometabolic Risk. *Adv. Nutr.* **2016**, *7*, 1041–1051. [CrossRef]
81. Månsson, H.L. Fatty Acids in Bovine Milk Fat. *Food Nutr. Res.* **2008**, *52*, 1821. [CrossRef] [PubMed]
82. Ran-Ressler, R.R.; Bae, S.; Lawrence, P.; Wang, D.H.; Brenna, J.T. Branched-Chain Fatty Acid Content of Foods and Estimated Intake in the USA. *Br. J. Nutr.* **2014**, *112*, 565–572. [CrossRef] [PubMed]
83. Bozzetto, L.; Costabile, G.; Della Pepa, G.; Ciciola, P.; Vetrani, C.; Vitale, M.; Rivellese, A.A.; Annuzzi, G. Dietary Fibre as a Unifying Remedy for the Whole Spectrum of Obesity-Associated Cardiovascular Risk. *Nutrients* **2018**, *10*, 943. [CrossRef]
84. Yao, Y.; Cai, X.; Fei, W.; Ye, Y.; Zhao, M.; Zheng, C. The Role of Short-Chain Fatty Acids in Immunity, Inflammation and Metabolism. *Crit. Rev. Food Sci. Nutr.* **2022**, *62*, 1–12. [CrossRef] [PubMed]

nutrients

MDPI

Article

Relationship between Serum 25-Hydroxyvitamin D Level and Risk of Recurrent Stroke

Guowei Li [1,2,*], Likang Li [1], Jonathan D. Adachi [3], Ruoting Wang [1], Zebing Ye [4], Xintong Liu [5], Lehana Thabane [2,6] and Gregory Y. H. Lip [7,8]

1 Center for Clinical Epidemiology and Methodology (CCEM), Guangdong Second Provincial General Hospital, Guangzhou 510317, China; lilikangccem@hotmail.com (L.L.); wangruoting1996@163.com (R.W.)
2 Department of Health Research Methods, Evidence, and Impact (HEI), McMaster University, Hamilton, ON L8S 4L8, Canada; thabanl@mcmaster.ca
3 Department of Medicine, McMaster University, Hamilton, ON L8S 4L8, Canada; jd.adachi@sympatico.ca
4 Department of Cardiology, Guangdong Second Provincial General Hospital, Guangzhou 510317, China; tgccem@hotmail.com
5 Department of Neurology, Guangdong Second Provincial General Hospital, Guangzhou 510317, China; jzhzang@163.com
6 Centre for Evaluation of Medicines, St. Joseph's Healthcare, Hamilton, ON L8N 4A6, Canada
7 Liverpool Centre for Cardiovascular Science, University of Liverpool, Liverpool L69 3BX, UK; gregory.lip@liverpool.ac.uk
8 Aalborg Thrombosis Research Unit, Department of Clinical Medicine, Aalborg University, 9000 Aalborg, Denmark
* Correspondence: lig28@mcmaster.ca; Tel.: 86-020-3264-0264; Fax: 86-020-8916-9025

Abstract: Evidence for the association between vitamin D and risk of recurrent stroke remains sparse and limited. We aimed to assess the relationship between serum circulating 25-hydroxyvitamin D (25(OH)D) level and risk of recurrent stroke in patients with a stroke history, and to identify the optimal 25(OH)D level in relation to lowest recurrent stroke risk. Data from the nationwide prospective United Kingdom Biobank were used for analyses. Primary outcome was time to first stroke recurrence requiring a hospital visit during follow-up. We used Cox proportional hazards regression model with restricted cubic splines to explore 25(OH)D level in relation to recurrent stroke. The dose-response relationship between 25(OH)D and recurrent stroke risk was also estimated, taking the level of 10 nmol/L as reference. A total of 6824 participants (mean age: 60.6 years, 40.8% females) with a baseline stroke were included for analyses. There were 388 (5.7%) recurrent stroke events documented during a mean follow-up of 7.6 years. Using Cox proportional hazards regression model with restricted cubic splines, a quasi J-shaped relationship between 25(OH)D and risk of recurrent stroke was found, where the lowest recurrent stroke risk lay at the 25(OH)D level of approximate 60 nmol/L. When compared with 10 nmol/L, a 25(OH)D level of 60 nmol/L was related with a 48% reduction in the recurrent stroke risk (hazard ratio = 0.52, 95% confidence interval: 0.33–0.83). Based on data from a large-scale prospective cohort, we found a quasi J-shaped relationship between 25(OH)D and risk of recurrent stroke in patients with a stroke history. Given a lack of exploring the cause–effect relationship in this observational study, more high-quality evidence is needed to further clarify the vitamin D status in relation to recurrent stroke risk.

Keywords: vitamin D; recurrent stroke; 25-hydroxyvitamin; stroke prevention

Citation: Li, G.; Li, L.; Adachi, J.D.; Wang, R.; Ye, Z.; Liu, X.; Thabane, L.; Lip, G.Y.H. Relationship between Serum 25-Hydroxyvitamin D Level and Risk of Recurrent Stroke. *Nutrients* **2022**, *14*, 1908. https://doi.org/10.3390/nu14091908

Academic Editor: Sareen Gropper

Received: 15 April 2022
Accepted: 30 April 2022
Published: 2 May 2022

1. Introduction

Stroke is a serious public health concern, especially with the aging population, ranking as the second leading cause of death worldwide [1]. A previous stroke is substantially related with increased risk of subsequent stroke, with a recurrent rate ranging from 5% to 40% within five years [2–4]. Hence, there has been a move towards a more holistic and integrated care approach to stroke management, including attention to lifestyle and comorbidities [5].

Recently, vitamin D has been widely found to associate with risk of stroke, based on data from observational studies. Several meta-analyses consistently reported that low level of circulating 25-hydroxyvitamin D (25(OH)D) was significantly related with the onset of stroke [6–9]. Nevertheless, evidence for the association between vitamin D and risk of *recurrent* stroke remains sparse and limited. Some studies reported that among patients with a previous stroke, those with the first quartile of 25(OD)D level had a significantly highest risk of recurrent stroke [10,11]. However, their small sample sizes and short-term follow-up precluded further investigation of adequate vitamin D levels in relation to stroke recurrence. Furthermore, the optimal level of 25(OH)D associated with the lowest risk of recurrent stroke remains largely unknown, especially given the non-linear relationship between 25(OH)D level and stroke risk.

Therefore, in this study, we aimed to assess the association between serum 25(OH)D level and risk of recurrent stroke in patients with a prior stroke history, and second, to identify the optimal 25(OH)D level in relation to lowest recurrent stroke risk. Data from the nationwide prospective United Kingdom (UK) Biobank were used for analyses.

2. Methods

2.1. Participants and Setting

Descriptions about the UK Biobank have been published in the literature and on the website (www.ukbiobank.ac.uk, accessed on 10 December 2021) [12]. In brief, the UK Biobank is a nationwide prospective cohort study that recruited >500,000 community dwellers between year 2006 and 2010, with the goals of improving diagnosis, prevention, treatment, and prognosis of diseases for the middle-aged and older adults. The study collected data from participant self-reports, interview with trained staff and physical measures, and used multiple data sources for linkage. The UK Biobank was approved by the Northwest Multicenter Research Ethics committee (11/NW/0382). The Guangdong Second General Provincial Hospital Research Ethics Committee approved the current analysis (2022-KY-KZ-119-01). All participants provided written informed consent.

For our current study, we limited eligible participants to those with a history of stroke at baseline. Information on history of stroke was identified from patient self-reports, the ICD-10 and ICD-9 code at baseline. The patient selection process is displayed in Supplemental Figure S1 for this study.

2.2. Outcome Measures

Our primary outcome was time to first stroke recurrence requiring a hospital visit during follow-up. The ICD-10 codes were used to determine the recurrent stroke events and their corresponding survival time (Supplemental Table S1 shows the codes used for stroke identification). Secondary outcomes included ischemic and hemorrhagic stroke. Patients were followed up to stroke recurrence, 31 March 2017, or death, whichever came first.

2.3. Serum 25(OH)D Levels and Other Independent Variables

Serum 25(OH)D level (in nmol/L) was measured from the non-fasted blood sample drawn at the time of study enrollment, using the Liaison XL 25(OH)D assay.

Data on other independent variables at baseline included age, sex, smoking and drinking, ethnicity, education, body mass index (BMI), physical activity, atrial fibrillation, hypertension, heart failure, hypercholesterolemia, and diabetes. We also collected information on intake of statins, non-steroidal anti-inflammatory drugs (NSAIDs), anticoagulants, antihypertensive and antidiabetic medications, and mineral and vitamin supplementation. To enhance the under-recognition of data on comorbidities and medication intake at baseline, we used the information from patient's self-reports, interviews with trained staff regarding medications and treatment that patients received, and ICD codes. We documented the existence of a variable if the patient had a positive response to any of the aforementioned data fields.

3. Statistical Analysis

Continuous and categorical variables were depicted with mean (standard deviation, SD) and frequency (percentage), respectively. Comparisons of baseline information between patients with and without recurrent stroke were performed by t-test for continuous variables and chi-square test for categorical variables. We used the Kaplan–Meier method to graph failure curve for recurrent stroke. Kernel density estimation was used to estimate the probability density of participants' serum 25(OH)D level.

To illustrate the relationship between 25(OH)D level and risk of recurrent stroke, we conducted a Cox proportional hazards regression model that was adjusted for the baseline characteristics as listed in Table 1 including age, sex, BMI, smoking and drinking, physical activity, comorbidities, medications, and supplementation. Results were presented with hazard ratios (HRs) and corresponding 95% confidence intervals (CIs). To explore the optimal of 25(OH)D level, we employed the restricted cubic splines with knots at 5th, 35th, 65th, and 95th percentiles to smooth the non-linear association between 25(OH)D level and recurrent stroke risk. We also showed the dose-response relationship between 25(OH)D level and recurrent stroke risk to estimate the HRs for pre-determined levels at 20, 30, 40, 50, 60, 70, and 80 nmol/L, taking the 25(OH)D of 10 nmol/L as reference. Similar analyses were also conducted for secondary outcomes of ischemic and hemorrhagic stroke.

Table 1. Patients' baseline characteristics and comparisons between patients with and without recurrent stroke.

Characteristics	Total Participants (n = 6824)	Stroke Recurrence during Follow-Up		
		Yes (n = 388)	No (n = 6436)	*p*-Value
Age: mean (SD), years	60.6 (6.9)	61.9 (6.4)	60.6 (6.9)	<0.01
Female sex: n (%)	2783 (40.8)	136 (35.1)	2647 (41.1)	0.02
BMI: mean (SD), kg/m^2	28.8 (5.1)	28.5 (4.9)	28.9 (5.1)	0.13
Smoking: n (%)				
Never	2747 (40.6)	140 (36.4)	2607 (40.9)	0.05
Former	2915 (43.1)	167 (43.4)	2748 (43.1)	
Current	1097 (16.2)	78 (20.3)	1019 (16.0)	
Alcohol drinking: n (%)				
Never	446 (6.6)	28 (7.3)	418 (6.5)	0.83
Former	588 (8.7)	32 (8.3)	556 (8.7)	
Current	5761 (84.8)	326 (84.5)	5435 (84.8)	
White ethnicity: n (%)	6465 (95.3)	367 (94.8)	6098 (95.3)	0.68
With college or university degree: n (%)	567 (8.5)	31 (8.2)	536 (8.5)	0.83
Physical activity (≥600 MET min per week): n (%)	3792 (73.1)	196 (66.9)	3596 (73.5)	0.01
Comorbidity: n (%)				
Atrial fibrillation	535 (7.8)	39 (10.1)	496 (7.7)	0.10
Hypertension	4222 (61.9)	269 (69.3)	3953 (61.4)	<0.01
Hypercholesterolemia	2996 (43.9)	188 (48.5)	2808 (43.6)	0.06
Diabetes	1003 (14.7)	92 (23.7)	911 (14.2)	<0.01
Heart failure	206 (3.0)	17 (4.4)	189 (2.9)	0.11

Table 1. *Cont.*

Characteristics	Total Participants (n = 6824)	Stroke Recurrence during Follow-Up		
		Yes (n = 388)	No (n = 6436)	*p*-Value
Medication and supplementation intake: n (%)				
NSAIDs	694 (10.2)	38 (9.8)	656 (10.2)	0.80
Antihypertensive drugs	4075 (59.8)	259 (66.8)	3816 (59.4)	<0.01
Antidiabetic drugs	763 (11.2)	76 (19.6)	687 (10.7)	<0.01
Statins	4462 (65.4)	274 (70.6)	4188 (65.1)	0.03
Anticoagulants	631 (9.2)	51 (13.1)	580 (9.0)	<0.01
Vitamins	1934 (28.7)	110 (28.6)	1824 (28.7)	0.99
Minerals and other dietary supplementation	2685 (39.6)	144 (37.5)	2541 (39.7)	0.39
Serum 25(OH)D: mean (SD), nmol/L	46.5 (22.4)	45.6 (25.9)	46.5 (22.1)	0.48

SD = standard deviation; BMI = body mass index; NSAIDs = non-steroidal anti-inflammatory drugs.

We performed two *post hoc* sensitivity analyses to assess the robustness of the main findings. A competing risk analysis using the Fine and Gray model was conducted for the relationship between 25(OH)D levels and risk of recurrent stroke, by treating all-cause death as competing events for recurrent stroke. Recognizing that the 25(OH)D levels may have oscillation by months and the length between the onset of stroke and blood sampling date may be different, we performed another sensitivity analysis in the multivariable Cox model after further adjusted for these two variables (the month for 25(OH)D measures, and the length between the stroke onset and blood sampling date).

Two pre-defined subgroups were conducted by sex (males and females) and age (<65 and ≥65 years) to explore whether there existed potential effect modifications. Furthermore, to directly compare with results from previous studies, we estimated the associations between different quartiles of 25(OH)D level and recurrent stroke risk, with the first quartile as reference value. Likewise, we assessed different 25(OH)D levels in relation to recurrent stroke risk using the recognized cut-off points for vitamin D sufficiency (>50 nmol/L), insufficiency (25–50), and deficiency (<25), taking vitamin D deficiency as reference.

All tests were two-sided, and the significance level was set as 0.05. Stata version 17 (StataCorp, College Station, TX, USA) and R version 3.5.1 (R Foundation for Statistical Computing, Vienna, Austria) were used for analyses.

4. Results

There were 6824 participants (mean age: 60.6 years, 40.8% females) with a baseline stroke included in this analysis. There were approximately 40% and 7% of participants who never smoked and consumed alcohol, respectively. The majority of participants was physically active (73%) and with hypertension (62%). Other information on comorbidities, medication, and supplementation intake was also presented in Table 1. The mean 25(OH)D level was 46.5 nmol/L.

During a mean follow-up of 7.6 (SD = 1.8) years, there were 388 (5.7%) recurrent stroke events documented including 250 ischemic, 87 hemorrhagic, and 51 unspecified stroke (Supplemental Figure S2 displays the Kaplan–Meier failure curve for recurrent stroke). Table 1 displays comparisons of baseline characteristics between participants with and without recurrent stroke. Participants with recurrent stroke were older, and less likely to be females and physically active when compared with controls (all *p*-values < 0.05). Significantly higher percentages of hypertension and diabetes were found in patients with recurrent stroke. Participants in the recurrent stroke group were more likely to consume statins, anticoagulants, antihypertensive, and antidiabetic medications. Supplemental

Figure S3 shows the probability density function of serum 25(OH)D level stratified by participants with and without recurrent stroke. A lower 25(OH)D level was found in patients with recurrent stroke (45.6 vs. 46.5); however, the difference was not statistically significant (p-value = 0.48; Table 1).

A quasi J-shaped relationship between 25(OH)D and risk of recurrent stroke was found in this study (Figure 1), where the lowest recurrent stroke risk lay at the 25(OH)D level of 58.2 nmol/L. As Table 2 presents, when taking the 25(OH)D of 10 nmol/L as reference, a 25(OH)D level of 60 nmol/L was related with a 48% reduction in the recurrent stroke risk (HR = 0.52, 95% CI: 0.33–0.83). A total of 658 (9.6%) deaths occurred before recurrent stroke; therefore, these deaths were the competing events for stroke recurrence. The competing risk analysis by treating deaths as competing events yielded similar findings to the main results, with the potential lowest recurrent stroke risk at the 25(OH)D level of approximate 60 nmol/L (subhazards ratio = 0.66, 95% CI: 0.40–1.12; Supplemental Table S2). Another sensitivity analysis by further adjusted for the month for 25(OH)D measures, the length between the onset of stroke and blood sampling date, and these two variables, showed similar results to the main findings, with the smallest HR of 0.51, 0.51, and 0.50 correspondingly at the 25(OH)D level of 60 nmol/L (Supplemental Table S2).

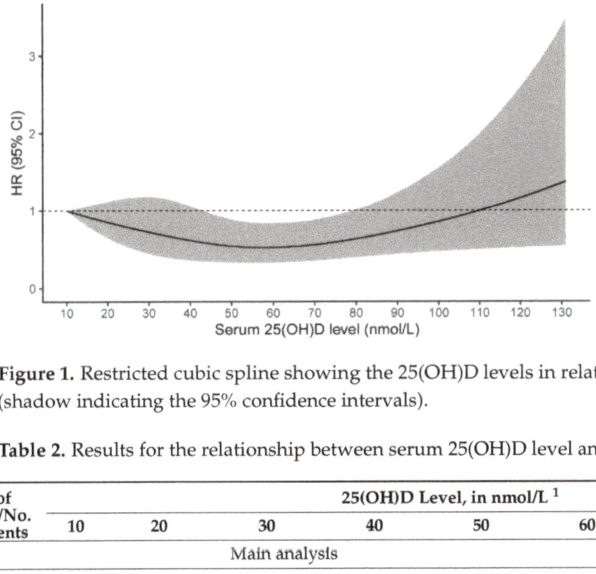

Figure 1. Restricted cubic spline showing the 25(OH)D levels in relation to risk of recurrent stroke (shadow indicating the 95% confidence intervals).

Table 2. Results for the relationship between serum 25(OH)D level and recurrent stroke risk.

Outcome/Analysis		**No. of Events/No. of Patients**	**25(OH)D Level, in nmol/L** [1]							
			10	**20**	**30**	**40**	**50**	**60**	**70**	**80**
Main analysis										
Primary outcome	**Total recurrent stroke**	388/6824	Ref	0.85 (0.65–1.12)	0.72 (0.44–1.18)	0.62 (0.36–1.06)	0.54 (0.33–0.89)	0.52 (0.33–0.84)	0.56 (0.35–0.88)	0.63 (0.39–1.01)
Secondary outcome	**Ischemic stroke**	250/6824	Ref	0.91 (0.64–1.28)	0.82 (0.44–1.54)	0.75 (0.37–1.51)	0.70 (0.37–1.32)	0.69 (0.38–1.26)	0.72 (0.40–1.31)	0.79 (0.43–1.44)
	Hemorrhagic stroke	87/6824	Ref	0.64 (0.36–1.13)	0.44 (0.16–1.21)	0.38 (0.12–1.16)	0.40 (0.15–1.09)	0.42 (0.16–1.06)	0.40 (0.16–1.02)	0.38 (0.14–1.02)
Subgroup analysis										
By sex	**Males**	252/4041	Ref	1.01 (0.73–1.41)	0.99 (0.54–1.80)	0.88 (0.45–1.72)	0.74 (0.40–1.37)	0.70 (0.39–1.25)	0.76 (0.43–1.35)	0.91 (0.51–1.63)
	Females	136/2783	Ref	0.54 (0.33–0.88)	0.33 (0.14–0.78)	0.29 (0.11–0.72)	0.32 (0.14–0.74)	0.34 (0.16–0.74)	0.31 (0.14–0.67)	0.25 (0.11–0.60)
By age	**<65 years**	205/4355	Ref	0.85 (0.58–1.24)	0.71 (0.36–1.37)	0.56 (0.28–1.12)	0.45 (0.24–0.85)	0.43 (0.23–0.79)	0.47 (0.25–0.86)	0.56 (0.30–1.04)
	≥65 years	183/2469	Ref	0.86 (0.58–1.28)	0.75 (0.36–1.57)	0.68 (0.28–1.64)	0.66 (0.29–1.48)	0.66 (0.31–1.39)	0.68 (0.33–1.41)	0.72 (0.34–1.52)

Ref = reference; [1] Results shown as hazard ratios (95% confidence intervals) from the models that used restricted cubic splines and were adjusted for age, sex, BMI, smoking and drinking, physical activity, comorbidities, medications, and supplementation.

Analyses for secondary outcomes and by subgroup yielded in general similar results to the main findings, with the lowest smallest recurrent stroke risk at 25(OH)D of about 60 nmol/L (HRs ranging from 0.43 to 0.70); however, the potentially lowest risks of hemorrhagic stroke (HR = 0.38) and for females (HR = 0.29) were observed at approximately 40 nmol/L (Table 2; Supplemental Figures S4–S9).

Table 3 shows results from additional analyses for the relationship between 25 (OH)D and recurrent stroke risk. When compared with the first quartile, the third quartile of 25(OH)D level (ranging from 43.5 to 61.3 nmol/L) was found to significantly associate with a 32% reduction in recurrent stroke risk (HR = 0.68, 95% CI: 0.48–0.96). Participants with insufficient (HR = 0.60) or sufficient 25(OH)D levels (HR = 0.59) had a significantly and similarly reduced risk of recurrent stroke, when taking 25(OH)D deficiency as reference.

Table 3. Result from additional analyses for the relationship between 25 (OH)D and recurrent stroke risk.

Serum 25(OH)D Level	Recurrent Stroke		
	No. of Events/No. of Patients	**HR (95% CI) [1]**	***p*-Value**
	Defined by quartile [2]		
1st quartile	117/1719	Ref	-
2nd quartile	92/1707	0.77 (0.56–1.07)	0.12
3rd quartile	86/1693	0.68 (0.48–0.96)	0.03
4th quartile	93/1705	0.77 (0.55–1.08)	0.13
	Defined by status		
Deficiency (<25 nmol/L)	97/1239	Ref	-
Insufficiency (25–50 nmol/L)	149/2906	0.60 (0.44–0.81)	< 0.01
Sufficiency (>50 nmol/L)	142/2679	0.59 (0.43–0.82)	< 0.01

HR = hazard ratio; CI = confidence interval; Ref = reference; [1] Results from the models that used restricted cubic splines and were adjusted for age, sex, BMI, smoking and drinking, physical activity, comorbidities, medications, and supplementation. [2] The cut-off points to define quartiles were 28.9, 43.5, and 61.3 nmol/L.

5. Discussion

In this nationwide prospective cohort study, we explored what was the optimal vitamin D level in relation to risk of recurrent stroke in patients with a prior stroke history. There was a quasi J-shaped relationship between 25(OH)D and risk of recurrent stroke observed, with the lowest risk found at 25(OH)D level of approximate 60 nmol/L. When compared with 10 nmol/L, a 25(OH)D level of 60 nmol/L was significantly associated with a 48% reduction in recurrent stroke risk.

The majority of literature had consistently found that low 25(OH)D levels were linearly or non-linearly associated with risk of *first-ever* stroke, although the relationship remained controversial. In this study, we found the optimal 25(OH)D level lay at approximate 60 nmol/L regarding the risk of *recurrent* stroke, which was in line with the quasi J-shaped association found from a recent meta-analysis that reported the 25(OH)D level of 50 nmol/L was related with the lowest stroke risk [8]. Even though with a different inflection, our study again confirmed that either a low or high level of 25(OH)D was related with elevated risk of recurrent stroke. Our different optimal 25(OH)D value from the published meta-analysis may be due to different population (with stroke history versus free from stroke) and outcome (stroke recurrence versus onset of stroke), data sources (individual patient data versus published summary data) and study settings (single country versus multi-country). Nevertheless, our results may provide some evidence about the vitamin D status in relation to risk of recurrent stroke.

The role of 25(OH)D level in stroke recurrence was largely remained uninvestigated. Vitamin D status may be associated with stroke size and disease severity, which could subsequently impact the propensity towards stroke recurrence [13,14]. Unfortunately, there were no data on NIHSS (National Institutes of Health Stroke Scale) to evaluate the stroke severity and size between patients with and without recurrent stroke. However, our additional analyses showed similar results to previous studies that further adjusted for NIHSS (Table 3) [10,11]. Vitamin D deficiency had been linked with cardiovascular risk including hypertension, diabetes mellitus, and arterial stiffness, thereby contributing to enhanced stroke risk [15–17]. Moreover, vitamin D may own neuroprotective effects, while vitamin D deficiency could promote inflammation and vascular remodelling to increase the risk of stroke [18]. Indeed, in this study, when compared with low level of 25(OH)D or vitamin D deficiency, a decreased risk of recurrent stroke was consistently found. Moreover, high dosages of vitamin D administration in animal experiments could result in widespread arterial calcification, especially when with co-existing diabetes, atherosclerosis, and kidney disease [19]. This also in part supported the elevated risk of recurrent stroke with high 25(OH)D levels (Figure 1). Nevertheless, given the non-randomization design, whether the observed 25(OH)D level was a surrogate for healthy lifestyle and/or frailty status, and whether there was residual biases and confounding effect existed in this study, remained uncertain. Therefore, our findings regarding the relationship between 25(OH)D level and recurrent stroke risk should be interpreted with caution, requiring further high-quality evidence and ideally from randomized controlled trials for clarification and verification.

There have been four studies evaluating the association between 25(OH)D level and risk of recurrent stroke in the literature, three from China [10,11,20] and one from the US [21]. All these studies categorized 25(OH)D level as either four quartiles [10,11] or binary for analyses [20,21]. These studies failed to comprehensively assess vitamin D status in relation to recurrent stroke. Importantly, they did not consider the potential association between high 25(OH)D levels and increased recurrent stroke risk, which would mislead the audience about the incremental beneficial effect of high vitamin D status. Their various cut-off points for 25(OH)D categorization used for analyses could also make their findings difficult to interpret, as it may not be appropriate to use quartiles or dichotomization for primary analyses regarding the J-shaped relationship given the substantial heterogeneity in the defined subgroups. For instance, Huang et al. used the fourth quartile (>22.8 ng/mL, i.e., >57 nmol/L) as a reference group [11], while our results demonstrated a slight decrease followed by continuous increase in recurrent stroke risk with elevated 25(OH)D levels in the group of >57 nmol/L (Figure 1). Moreover, their small sample sizes (ranging from 220 to 946) prohibited further attempts at exploring stroke subtypes and subgroup effects.

By contrast, our study used data from a large-scale prospective cohort with a follow-up of approximate eight years for analyses. The relationship between 25(OH)D and recurrent stroke risk was depicted by graphs from restricted cubic splines in combination with estimates of multiple individual 25(OH)D points, generating a detailed non-linear association between 25(OH)D in relation to recurrent stroke. Furthermore, we performed analyses for stroke subtypes and stratified by age and sex to test the potential effect modifications. Of note, the relationship for hemorrhagic stroke and females seemed to demonstrate a different pattern from the main analysis (Supplemental Figures S5 and S7). Part of the interpretation may be due to the relatively small sample size of recurrent stroke events in females and for hemorrhagic stroke (Table 2). Nonetheless, these exploratory results for different subgroups required further adequately powered and well-designed studies for further investigation and clarification.

6. Strength and Limitations

This study used data from a nationwide cohort to comprehensively assess serum 25(OH)D level in relation to recurrent stroke, with results shown as illustrations and estimates from specific 25(OH)D values. Rigorous methodology and robust analyses also supported the study findings. The high risk of stroke recurrence and its substantial impact on mortality remained a severe public health concern [22,23]. While there was an evidence gap in the existing guidelines regarding 25(OH)D level in relation to recurrent stroke, our findings might highlight the importance of adequate vitamin D status in relation to recurrent stroke risk, although we had limited data to explore the causal mechanisms.

Several limitations exited in this study. First, we could not fully preclude confounding effects especially of those unmeasured variables in this observational study, which may compromise the validity and strength of our results [24]. For instance, due to lack of data on NIHSS, whether and to what extent the relationship between vitamin D status and recurrent stroke risk could be influenced by stroke severity and size, remained unknown. Likewise, the relationship between 25(OH)D and recurrent stroke risk may be driven by some unmeasured factors associated with lifestyle and frailty, which would impair our observed findings. Some baseline comorbidities (hypertension, diabetes, and heart failure) and medications (NSAIDs, antihypertensive, and antidiabetic drugs) were significantly associated with serum 25(OH)D levels. Even though we adjusted for all the comorbidities and medications in the multivariable model, unquantified moderator and confounding effects would remain. It was reported that the use of Liaison assay would systematically underestimate the values of 25(OH)D, which may lead to study populations being misclassified as vitamin D deficiency [25]. Therefore, our results should be interpreted with caution, especially regarding the absolute 25(OH)D values and the inflection points found from the quasi J-shaped relationship curves. We only had data on baseline 25(OH)D measures; thus, no analysis for the change in vitamin D status in relation to recurrent stroke could be performed. It would be a worthwhile endeavor to further explore the change in vitamin D status and its potential usefulness for stroke prognosis and risk evaluation. Furthermore, it was uncertain about whether the splines for hemorrhagic stroke and in females were because of either insufficient statistical power or the true absence of a shaped relationship with 25(OH)D. Collectively, our findings from an observational study were primarily hypothesis generating with an exploratory nature, which warranted further exploration to clarify the vitamin D status in relation to risk of recurrent stroke.

7. Conclusions

Based on data from a large-scale prospective cohort, we found a quasi J-shaped relationship between 25(OH)D and risk of recurrent stroke in patients with a stroke history, which might provide some insights into the vitamin D status for recurrent stroke prevention. Given a lack of exploring the cause–effect relationship in this observational study, more high-quality evidence is needed to further clarify the vitamin D status in relation to recurrent stroke risk.

Supplementary Materials: The following supporting information can be downloaded at: https://www.mdpi.com/article/10.3390/nu14091908/s1, Table S1: Codes used for ascertainment of stroke events at baseline and during follow-up; Table S2. Sensitivity analysis results for the relationship between 25(OH)D and risk of recurrent ischemic stroke; Figure S1: Flow diagram showing the selection of participants in this study; Figure S2: Kaplan–Meier failure curve for incident recurrent stroke; Figure S3: Kernel density estimate for probability density function of 25(OH)D stratified by participants with and without recurrent stroke; Figure S4: Restricted cubic splines showing 25(OH)D in relation to recurrent ischemic stroke with the lowest risk laying at 56.3 nmol/L; Figure S5: Restricted cubic splines showing 25(OH)D in relation to recurrent hemorrhagic stroke with the potentially lowest risk laying at 41.2 nmol/L; Figure S6: Restricted cubic splines showing 25(OH)D in relation to recurrent stroke in males with the lowest risk laying at 58.7 nmol/L; Figure S7: Restricted cubic splines showing 25(OH)D in relation to recurrent stroke in females with the potentially lowest risk laying at 38.7 nmol/L; Figure S8: Restricted cubic splines showing 25(OH)D in relation to recurrent stroke in patients <65 years with the lowest risk laying at 58.4 nmol/L; Figure S9: Restricted cubic splines showing 25(OH)D in relation to recurrent stroke in patients $\geq$65 years with the lowest risk laying at 53.3 nmol/L.

Author Contributions: G.L., L.L., Z.Y. and X.L. contributed equally to this study. G.L., L.L., Z.Y., X.L. and G.Y.H.L.: conceived and designed the study. G.L., L.L., Z.Y. and X.L.: obtained data, performed analyses and interpretation, and drafted the manuscript. G.L., J.D.A., R.W., L.T. and G.Y.H.L.: provided professional and statistical support, and made critical revisions. G.L. and G.Y.H.L. acted as the guarantors of this work. All authors have read and agreed to the published version of the manuscript.

Funding: This work was supported by the Science Foundation of Guangdong Second Provincial General Hospital (grant number: YY2018-002, recipient: GL) and the Medical Scientific Research Foundation of Guangdong Province of China (grant number: A2020453, recipient: GL).

Institutional Review Board Statement: The UK Biobank study was approved by the North West Multicenter Research Ethics Committee (11/NW/0382). The Guangdong Second General Provincial Hospital Research Ethics Committee approved the current analysis (2022-KY-KZ-119-01).

Informed Consent Statement: All participants provided written consent before enrolment.

Data Availability Statement: The data can be available on application to the UK Biobank (www.ukbiobank.ac.uk/). Data described for the analyses and in the manuscript will be made available upon request.

Acknowledgments: We would like to thank the participants and staff of the UK Biobank study for their valuable contributions. This research has been conducted using the UK Biobank Resource under Application Number 63844.

Conflicts of Interest: G.Y.H.L. has served as a consultant for Novartis, Bayer/Janssen, Biotronik, BMS/Pfizer, Medtronic, Boehringer Ingelheim, Verseon, and Daiichi-Sankyo and as a speaker for Medtronic, Bayer, BMS/Pfizer, Boehringer Ingelheim, and Daiichi-Sankyo. No fees have been received directly or personally. All other authors have declared no conflicts of interest.

References

1. Virani, S.S.; Alonso, A.; Aparicio, H.J.; Benjamin, E.J.; Bittencourt, M.S.; Callaway, C.W.; Carson, A.P.; Chamberlain, A.M.; Cheng, S.; Delling, F.N.; et al. Heart Disease and Stroke Statistics—2021 Update: A Report From the American Heart Association. *Circulation* **2021**, *143*, e254–e743. [CrossRef] [PubMed]
2. Khanevski, A.N.; Bjerkreim, A.T.; Novotny, V.; Næss, H.; Thomassen, L.; Logallo, N.; Kvistad, C.E.; The NOR-STROKE Study Group. Recurrent ischemic stroke: Incidence, predictors, and impact on mortality. *Acta Neurol. Scand.* **2019**, *140*, 3–8. [CrossRef] [PubMed]
3. Flach, C.; Muruet, W.; Wolfe, C.D.; Bhalla, A.; Douiri, A. Risk and Secondary Prevention of Stroke Recurrence: A Population-Base Cohort Study. *Stroke* **2020**, *51*, 2435–2444. [CrossRef] [PubMed]
4. Donkor, E.S. Stroke in the21stCentury: A Snapshot of the Burden, Epidemiology, and Quality of Life. *Stroke Res. Treat.* **2018**, *2018*, 3238165. [CrossRef]
5. Lip, G.Y.H.; Ntaios, G. "Novel Clinical Concepts in Thrombosis": Integrated Care for Stroke Management—Easy as ABC. *Thromb. Haemost.* **2021**, *122*, 316–319. [CrossRef]
6. Zhou, R.; Wang, M.; Huang, H.; Li, W.; Hu, Y.; Wu, T. Lower Vitamin D Status Is Associated with an Increased Risk of Ischemic Stroke: A Systematic Review and Meta-Analysis. *Nutrition* **2018**, *10*, 277. [CrossRef] [PubMed]
7. Jani, R.; Mhaskar, K.; Tsiampalis, T.; Kassaw, N.A.; González, M.; Ángel, M.; Panagiotakos, D.B. Circulating 25-hydroxy-vitamin D and the risk of cardiovascular diseases. Systematic review and meta-analysis of prospective cohort studies. *Nutr. Metab. Cardiovasc. Dis.* **2021**, *31*, 3282–3304. [CrossRef]
8. Shi, H.; Chen, H.; Zhang, Y.; Li, J.; Fu, K.; Xue, W.; Teng, W.; Tian, L. 25-Hydroxyvitamin D level, vitamin D intake, and risk of stroke: A dose-response meta-analysis. *Clin. Nutr.* **2020**, *39*, 2025–2034. [CrossRef]
9. Su, C.; Jin, B.; Xia, H.; Zhao, K. Association between Vitamin D and Risk of Stroke: A PRISMA-Compliant Systematic Review and Meta-Analysis. *Eur. Neurol.* **2021**, *84*, 399–408. [CrossRef]
10. Qiu, H.; Wang, M.; Mi, D.; Zhao, J.; Tu, W.; Liu, Q. Vitamin D status and the risk of recurrent stroke and mortality in ischemic stroke patients: Data from a 24-month follow-up study in China. *J. Nutr. Health Aging* **2017**, *21*, 766–771. [CrossRef]
11. Huang, H.; Zheng, T.; Wang, S.; Wei, L.; Wang, Q.; Sun, Z. Serum 25-hydroxyvitamin D predicts early recurrent stroke in ischemic stroke patients. *Nutr. Metab. Cardiovasc. Dis.* **2016**, *26*, 908–914. [CrossRef]
12. Sudlow, C.; Gallacher, J.; Allen, N.; Beral, V.; Burton, P.; Danesh, J.; Downey, P.; Elliott, P.; Green, J.; Landray, M.; et al. UK Biobank: An Open Access Resource for Identifying the Causes of a Wide Range of Complex Diseases of Middle and Old Age. *PLoS Med.* **2015**, *12*, e1001779. [CrossRef] [PubMed]
13. Siniscalchi, A.; Lochner, P.; Anticoli, S.; Chirchiglia, D.; De Sarro, G.; Gallelli, L. What is the Current Role for Vitamin D and the Risk of Stroke? *Curr. Neurovascular Res.* **2019**, *16*, 178–183. [CrossRef] [PubMed]
14. Yarlagadda, K.; Ma, N.; Doré, S. Vitamin D and Stroke: Effects on Incidence, Severity, and Outcome and the Potential Benefits of Supplementation. *Front. Neurol.* **2020**, *11*, 384. [CrossRef]
15. Michos, E.D.; Carson, K.A.; Schneider, A.L.C.; Lutsey, P.L.; Xing, L.; Sharrett, A.R.; Alonso, A.; Coker, L.H.; Gross, M.; Post, W.; et al. Vitamin D and Subclinical Cerebrovascular Disease: The Atherosclerosis Risk in Communities brain magnetic resonance imaging study. *JAMA Neurol.* **2014**, *71*, 863. [CrossRef]
16. Berridge, M.J. Vitamin D deficiency and diabetes. *Biochem. J.* **2017**, *474*, 1321–1332. [CrossRef]

17. Kheiri, B.; Abdalla, A.; Osman, M.; Ahmed, S.; Hassan, M.; Bachuwa, G. Vitamin D deficiency and risk of cardiovascular diseases: A narrative review. *Clin. Hypertens.* **2018**, *24*, 9. [CrossRef] [PubMed]
18. Makariou, S.; Michel, P.; Tzoufi, M.; Challa, A.; Milionis, H. Vitamin D and Stroke: Promise for Prevention and Better Outcome. *Curr. Vasc. Pharmacol.* **2014**, *12*, 117–124. [CrossRef]
19. Drüeke, T.B.; Massy, Z.A. Role of vitamin D in vascular calcification: Bad guy or good guy? *Nephrol. Dial. Transplant.* **2012**, *27*, 1704–1707. [CrossRef]
20. Ji, W.; Zhou, H.; Wang, S.; Cheng, L.; Fang, Y. Low serum levels of 25-hydroxyvitamin D are associated with stroke recurrence and poor functional outcomes in patients with ischemic stroke. *J. Nutr. Health Aging* **2016**, *21*, 892–896. [CrossRef]
21. Welles, C.C.; Whooley, M.A.; Karumanchi, S.A.; Hod, T.; Thadhani, R.; Berg, A.H.; Ix, J.H.; Mukamal, K.J. Vitamin D deficiency and cardiovascular events in patients with coronary heart disease: Data from the Heart and Soul Study. *Am. J. Epidemiol.* **2014**, *179*, 1279–1287. [CrossRef] [PubMed]
22. Kolmos, M.; Christoffersen, L.; Kruuse, C. Recurrent Ischemic Stroke—A Systematic Review and Meta-Analysis. *J. Stroke Cerebrovasc. Dis.* **2021**, *30*, 105935. [CrossRef] [PubMed]
23. Talelli, P.; Greenwood, R.J. Review: Recurrent stroke: Where do we stand with the secondary prevention of noncardioembolic ischaemic strokes? *Ther. Adv. Cardiovasc. Dis.* **2008**, *2*, 387–405. [CrossRef] [PubMed]
24. Fewell, Z.; Smith, G.D.; Sterne, J.A.C. The Impact of Residual and Unmeasured Confounding in Epidemiologic Studies: A Simulation Study. *Am. J. Epidemiol.* **2007**, *166*, 646–655. [CrossRef]
25. Fraser, W.D.; Tang, J.C.Y.; Dutton, J.J.; Schoenmakers, I. Vitamin D Measurement, the Debates Continue, New Analytes Have Emerged, Developments Have Variable Outcomes. *Calcif. Tissue Res.* **2019**, *106*, 3–13. [CrossRef]

nutrients MDPI

Review

Arachidonic Acid as Mechanotransducer of Renin Cell Baroreceptor

Undurti N. Das

UND Life Sciences, 2221 NW 5th St., Battle Ground, WA 98604, USA; undurti@lipidworld.com

Abstract: For normal maintenance of blood pressure and blood volume a well-balanced renin-angiotensin-aldosterone system (RAS) is necessary. For this purpose, renin is secreted as the situation demands by the juxtaglomerular cells (also called as granular cells) that are in the walls of the afferent arterioles. Juxtaglomerular cells can sense minute changes in the blood pressure and blood volume and accordingly synthesize, store, and secrete appropriate amounts of renin. Thus, when the blood pressure and blood volume are decreased JGA cells synthesize and secrete higher amounts of renin and when the blood pressure and blood volume is increased the synthesis and secretion of renin is decreased such that homeostasis is restored. To decipher this important function, JGA cells (renin cells) need to sense and transmit the extracellular physical forces to their chromatin to control renin gene expression for appropriate renin synthesis. The changes in perfusion pressure are sensed by Integrin β1 that is transmitted to the renin cell's nucleus via lamin A/C that produces changes in the architecture of the chromatin. This results in an alteration (either increase or decrease) in renin gene expression. Cell membrane is situated in an unique location since all stimuli need to be transmitted to the cell nucleus and messages from the DNA to the cell external environment can be conveyed only through it. This implies that cell membrane structure and integrity is essential for all cellular functions. Cell membrane is composed to proteins and lipids. The lipid components of the cell membrane regulate its (cell membrane) fluidity and the way the messages are transmitted between the cell and its environment. Of all the lipids present in the membrane, arachidonic acid (AA) forms an important constituent. In response to pressure and other stimuli, cellular and nuclear shape changes occur that render nucleus to act as an elastic mechanotransducer that produces not only changes in cell shape but also in its dynamic behavior. Cell shape changes in response to external pressure(s) result(s) in the activation of cPLA2 (cytosolic phospholipase 2)-AA pathway that stretches to recruit myosin II which produces actin-myosin cytoskeleton contractility. Released AA can undergo peroxidation and peroxidized AA binds to DNA to regulate the expression of several genes. Alterations in the perfusion pressure in the afferent arterioles produces parallel changes in the renin cell membrane leading to changes in renin release. AA and its metabolic products regulate not only the release of renin but also changes in the vanilloid type 1 (TRPV1) expression in renal sensory nerves. Thus, AA and its metabolites function as intermediate/mediator molecules in transducing changes in perfusion and mechanical pressures that involves nuclear mechanotransduction mechanism. This mechanotransducer function of AA has relevance to the synthesis and release of insulin, neurotransmitters, and other soluble mediators release by specialized and non-specialized cells. Thus, AA plays a critical role in diseases such as diabetes mellitus, hypertension, atherosclerosis, coronary heart disease, sepsis, lupus, rheumatoid arthritis, and cancer.

Keywords: renin; arachidonic acid; juxtaglomerular cells; granular cells; nucleus; lamin; actin-myosin; afferent

Citation: Das, U.N. Arachidonic Acid as Mechanotransducer of Renin Cell Baroreceptor. *Nutrients* **2022**, *14*, 749. https://doi.org/10.3390/nu14040749

Academic Editor: Marica Bakovic

Received: 26 December 2021
Accepted: 2 February 2022
Published: 10 February 2022

1. Introduction

Maintenance of normal blood pressure and blood volume is essential for health. Renin-angiotensin-aldosterone system (RAS) pays a crucial role in this aspect. Renin is a crucial part of the RAS. Juxtaglomerular (JG) cells also called as granular cells (renin cells) are the seat of renin synthesis and they store and secrete renin as the situation demands (see Figure 1). JG cells are in the walls of the afferent arterioles of the glomerulus and are considered as the specialized smooth muscle cells. In view of the tight packing of the cells of the juxtaglomerular apparatus especially of the granular cells between the endothelial cells, myocytes of the afferent arteriole and other cells (see Figure 1) even slight changes in the blood volume and pressure within the afferent arteriole are sensed by the renin cells (granular cells). Thus, the renin expressing cells are uniquely located to sense and respond to changes in blood pressure and the extracellular fluid such that they can either enhance or decrease the synthesis and release of renin to restore homeostasis. Recent studies revealed that the baroreceptor function of the renin cells can be attributed to its nuclear mechanotransducer ability to sense and transmit the extracellular physical forces to its chromatin to regulate renin gene expression and its secretion. Perfusion pressure changes sensed by the integrin β1 are transmitted to the renin cell's nucleus via lamin A/C that results in changes in the architecture of the chromatin and corresponding alterations in the expression of the renin gene [1]. It is likely that the changes in the perfusion pressure sensed by the renin cells and the renin gene via a nuclear mechanotransduction mechanism is regulated or controlled by certain intermediate molecules. I propose that such an intermediate molecule(s) could be arachidonic acid (AA) and its metabolites that have modulatory influence on the secretion and action of RAS.

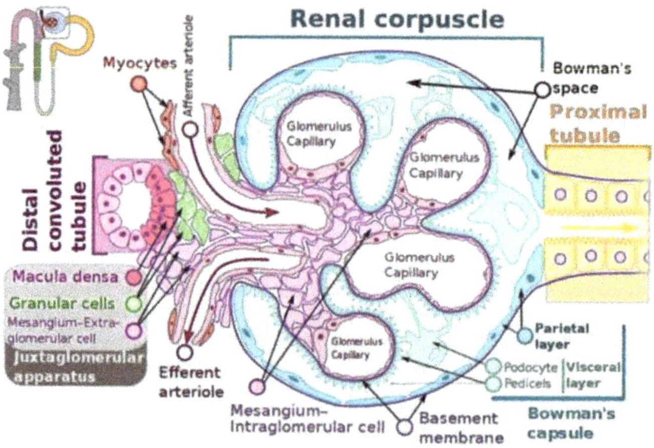

Figure 1. Cross section of the JGA and the glomerulus and their relationship to the afferent and efferent arterioles. Note the close relationship between JGA and the afferent and efferent arterioles. The JGA monitors the composition of the fluid in the distal convoluted tubule and adjusts the glomerular filtration rate accordingly. The picture shows the glomerulus and the surrounding structures. JGA = Juxtaglomerular apparatus.

2. JGA (Juxta Glomerular Apparatus) Functions as a Baroreceptor

Baroreceptors are mechanoreceptor sensory neurons capable of responding to changes in the stretch of the blood vessel. Thus, a change in the pressure of blood vessel triggers an alteration in the action potential generation rates that is conveyed to the solitary nucleus in the medulla oblongata via autonomic reflexes to influence the cardiac output and vascular resistance [2].

Baroreceptors are of two types: high-pressure arterial baroreceptors and low-pressure baroreceptors (also known as cardiopulmonary or volume receptors) which are located within the carotid sinuses and the aortic arch (high pressure baroreceptors) and within the atria, ventricles, and pulmonary vasculature (low-pressure baroreceptor) respectively.

The main function of the baroreceptors is to maintain systemic blood pressure. Increase in blood pressure stretches the baroreceptors resulting in an increase in the vagal tone nerve and inhibition of the sympathetic outflow that results in restoration of blood pressure to normal. In contrast, fall in blood pressure decreases signal output that disinhibits the sympathetic control sites and decreases parasympathetic activity resulting in an increase in blood pressure. Stimulation of the low-pressure baroreceptors enhances salt and water retention in addition to influencing intake of salt and water.

The information carried from the baroreceptors is integrated in the medulla oblongata that communicates with the heart muscle, the cardiac pacemaker and the arterioles and veins of the body. Any reduction in the sympathetic innervation of the kidneys causes vasodilation and increased blood flow to the kidneys resulting in an increase in urine production. On the other hand, whenever blood pressure falls, the sympathetic nerves stimulate renin release that ultimately results in the production of angiotensin II, which, in turn, enhances aldosterone that increases retention of Na^+ and water. It is interesting that changes in afferent arterial pressure alters glomerular filtration rate.

It is noteworthy that kidneys regulate the rate of blood flow over a wide range of blood pressures. Despite significant changes in the blood pressure, the glomerular filtration rate changes very little due to two internal autoregulatory mechanisms: *the myogenic mechanism* and the *tubuloglomerular feedback mechanism* that operate independent of the outside influence.

3. Arteriole Myogenic Mechanism

When blood pressure increases, smooth muscle cells located in the arteriole wall get stretched resulting in its contraction to resist the increase in the blood pressure. As a result, very little change in the flow occurs. In contrast, when blood pressure is decreased, the smooth muscle cells relax leading to a decrease in the resistance that allows a continued even flow of blood.

4. Tubuloglomerular Feedback

JGA is involved in the tubuloglomerular feedback mechanism (Figure 1) that utilizes adenosine triphosphate (ATP), adenosine, and nitric oxide (NO) that are capable of either contract or relax the afferent arteriolar smooth muscle cells. The macula dense cells that are in intimate contact with the afferent and efferent arterioles of the glomerulus, respond to changes in the fluid flow rate and Na^+ concentration. Activated macula densa cells release ATP and adenosine that stimulate the myogenic juxtaglomerular cells of the afferent arteriole to slow blood flow and reduce glomerular filtration rate. On the other hand, decrease in glomerular filtration rate results in less Na^+ excretion that results in decreased ATP and adenosine production leading to the afferent arteriolar dilatation and increase in the glomerular filtratation rate. NO relaxes the afferent arteriole whereas ATP and adenosine constrict it. Thus, a close interaction exists between NO and adenosine and ATP on the glomerular filtratation rate.

The distal convoluted tubule that is in direct contact with the arterioles forms a part of the JGA known as the macula densa, which can monitor the fluid composition that is flowing through the distal convoluted tubule. Thus, any changes in the concentration of Na^+ and the rate of fluid movement in the tubule macula densa leads to releases ATP and adenosine that regulate the glomerular filtration rate.

The regulation of renin release is done by the macula densa cells of the JGA of the afferent arteriole (Figures 1 and 2). Renin, in turn, modulates the formation of angiotensin II, a stimulator of the release of aldosterone from the adrenal cortex. Aldosterone enhances Na^+ reabsorption by the kidney and along with which water retention occurs leading to an increase in blood pressure.

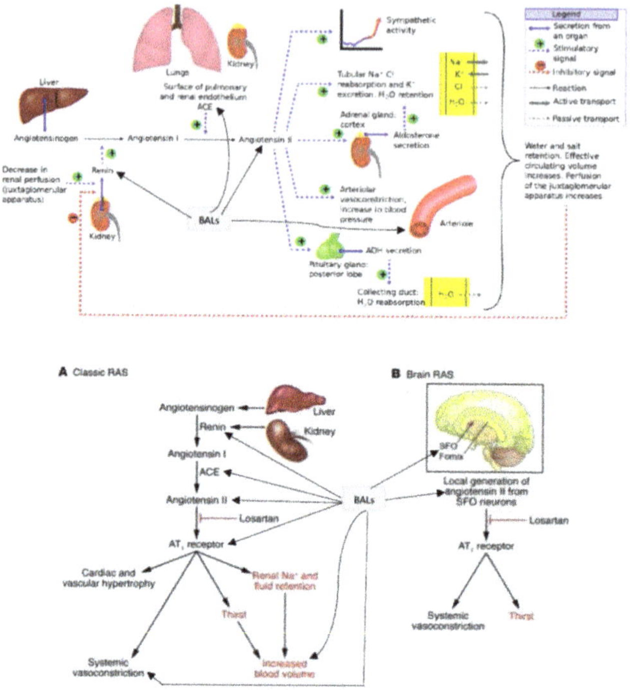

Figure 2. Scheme showing the release of renin from JGA, formation of angiotensin-II and regulation of fluid and electrolyte balance and blood pressure. (**A**) Classic RAS. Angiotensinogen from the liver is converted to angiotensin II by renin and angiotensin-converting enzyme (ACE). Angiotensin II activates AT1 receptors in target tissues producing changes in blood volume and blood pressure. (**B**) Brain RAS. Production of angiotensin II in the SFO affects thirst and salt appetite.IL-6 = Interleukin-6 TNF = Tumor necrosis factor RAS = Renin-angiotensin-aldosterone system AT receptor =Angiotensin receptor ACE = Angiotensin converting enzyme SFO = Subfornical organ.

5. Mechanotransduction from Cell Membrane to the Nucleus

Cell membrane integrity is essential for optimal cell response to various stimuli (both external and internal) since, all stimuli need to be conveyed to the genome through the cell membrane and vice versa [3]. This applies to the cells of the JGA and their response to various stimuli that regulate renin secretion. It has been reported that the nucleus serves as an elastic mechanotransducer of cellular shape deformation and controls its (cell) dynamic behavior [4,5]. Changes in the cell shape in response to external pressures induce inner nuclear membrane unfolding that, in turn, activates cPLA2 (cytosolic phospholipase 2) leading to myosin II recruitment resulting in actin-myosin cytoskeleton contractility (see Figures 3 and 4). These events produce alterations in the perfusion pressure in the afferent arterioles with corresponding changes in the renin cell membrane and parallel changes in renin release [2]. Thus, renin cells that are near the afferent arteriolar endothelial cells/myocytes/mesangium glomerular cells sense changes in the afferent arterioles and convey the same to the renin cell membrane. This results in the activation of cPLA2 (and possibly other phospholipases).

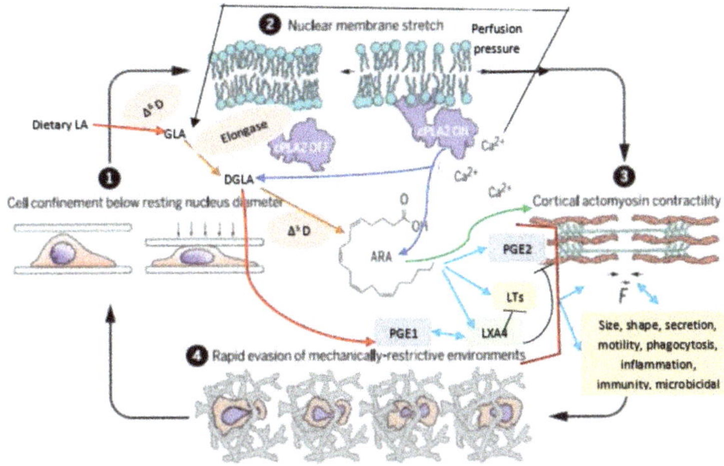

Figure 3. Scheme showing the mechanotransducer function of nucleus in response to changes in pressure and stretch. Pressure or stretch stimuli {1–2} increase nuclear membrane tension. This results in calcium release, cPLA2 activation and AA release {3} (other unsaturated fatty acids may also be released from the cell membrane) that results in actomyosin force generation (AA may also act on other cytoskeletal structures). AA metabolites regulate inflammation, immune response, and possess antimicrobial actions. Modified from Ref. [4].

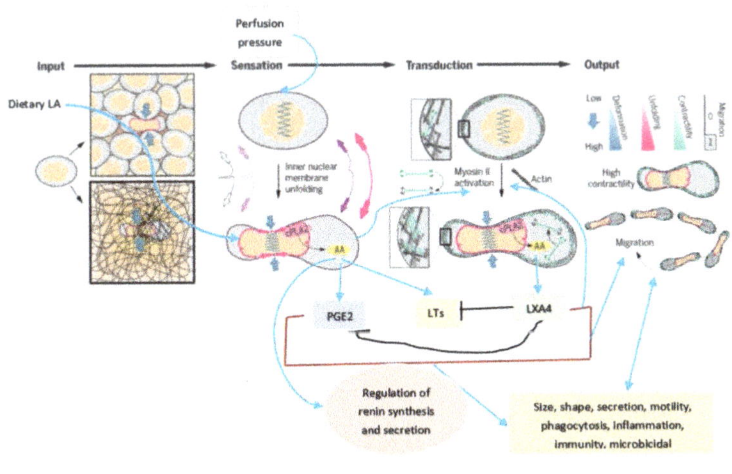

Figure 4. Scheme showing how nucleus can act as an elastic mechanotransducer in response to changes in the pressure and stretch stimuli (modified from Ref. [5]).

Cell shape changes occur in response to pressure and stretch stimuli. This causes the nuclear membrane unfolding and activates cPLA2 resulting in release of AA (and other unsaturated fatty acids) from the cell membrane. Released AA induces changes in the actin cytoskeleton behavior and other cytoskeleton structures (including lamin, integrin β1). In response to changes in the mechanical forces (pressure and stretch) the cells (especially myocytes, macular densa, granular cells, mesangium-extra-glomerular cells, afferent and efferent endothelial cells) respond and interact with each other and transduce mechanical signaling through focal adhesion contacts and F-actin and other cytoskeleton structures (both extra and intracellular) that are transmitted to the nucleus through the nuclear pores and nuclear translocation of mechanosensing-dependent transcription factors (YAP-TAZ and MRTF- transcription factors). This leads to chromatin activation with consequent changes in the expression of renin gene. If tumor cells, leukocytes, or macrophages, etc., are exposed to pressure and stretch stimuli, cells migrate or metastasize. AA (and other unsaturated fatty acids) released lead to the formation of its metabolites that produce changes in the cell size, shape, alter motility, phagocytosis, and influence inflammation, immune response and show microbicidal action (see Figure 5).

Perfusion pressure changes and/or direct mechanical stimuli produce significant changes in renin gene expression and its phenotype. This has been attributed to the baroreceptor(s) that reside in renin cells that are sensed by the integrin β1 at the renin cell membrane. The stimuli that are sensed by the baroreceptors of the renin cell are transduced to the nuclear membrane and chromatin by lamin A/C that changes the renin gene expression and renin bioavailability such that homeostasis is restored [1,2]. These results imply that changes in the cell membrane dynamics are transmitted to the nucleus for changes in the respective genes. This transmission of messages from the cell membrane to the nucleus is performed by the cytoskeleton system that function as the second messenger of the mechanotransduction process. Perfusion pressure changes are conveyed to the renin cell membrane such that alterations in cPLA2 activity occurs in afferent arteriolar endothelial cells/myocytes/mesangium glomerular cells and renin cells.

6. AA Functions as a Mechanotransducer

Dietary essential fatty acids (EFAs) cis-linoleic acid (LA, 18:2 n-6) and alpha-linolenic acid (ALA, 18:3 n-3) are essential for humans to survive. EFAs are converted to their long-chain metabolites gamma-linolenic acid (GLA, 18:3 n-6), dihomo-GLA (DGLA, 20:3 n-6) and arachidonic acid (AA, 20:4 n-6) derived from LA and eicosapentaenoic acid (EPA, 20:5 n-3) and docosahexaenoic acid (DHA, 22:6 n-3) from ALA. All these fatty acids form an important constituent of cell membrane and are in their (cell membrane) phospholipid fraction. PUFAs by virtue of their unsaturation can enhance the fluidity of the cell membranes (including nuclear and mitochondrial membranes). Thus, it is envisaged that increased presence of PUFAs (especially AA, EPA and DHA) enhances the fluidity of the cell membrane whereas cholesterol and other saturated fatty acids render membrane more rigid. The importance of membrane fluidity lies in the fact that increased cell membrane fluidity increases the number of receptors and their affinity to their respective proteins/growth factors and hormones (reviewed in [3]). This is especially relevant to insulin action and development of insulin resistance. The cell membrane fluidity is also an important factor to enable cells such as leukocytes, macrophages, platelets, T and B cells and tumor cells to adhere to the extracellular matrix and other tissues and enable them to pass through capillaries. In the event the cell membrane of is sufficiently fluid, they will be able to pass through the capillaries some of which are much smaller in diameter compared to various blood cells and prevent thrombosis whereas, in general, tumor cells are sufficiently rigid that is responsible for their ability to produce vascular thrombosis.

The ability of cells to respond to various pressures and stretch stimuli (including shear stress of blood flow) results in cell membrane and nucleus to act as elastic mechanotransducer of cellular shape deformation and control their (cell) dynamic behavior including change in their shape, motility, ability to secrete chemicals needed to produce their respective actions, phagocytosis (endocytosis and exocytosis), inflammation, immune response, and other functions. In this context, the ability of AA (and DGLA, EPA and DHA) to form precursors to various biologically active metabolites assumes significance. The observation that nucleus acts as an elastic mechanotransducer of cellular shape deformation that results in the activation of cPLA2 and consequent release of AA (and possibly, DGLA, EPA and DHA; [4,5]) implies that the released fatty acids could be utilized for the formation of various metabolites from these released fatty acids (see Figure 5). Thus, it is envisaged that the released AA (and DGLA, EPA, and DHA) is converted into various metabolites that have several biological actions depending on the necessity. In this context, it is noteworthy that various metabolites formed from AA (prostaglandins, leukotrienes, thromboxanes, lipoxins; and those formed from EPA and DHA including resolvins, protectins and maresins) have very short half-life (few seconds) but potent biological actions that ensures much-needed local actions that are short lived but sufficiently strong to meet the local needs.

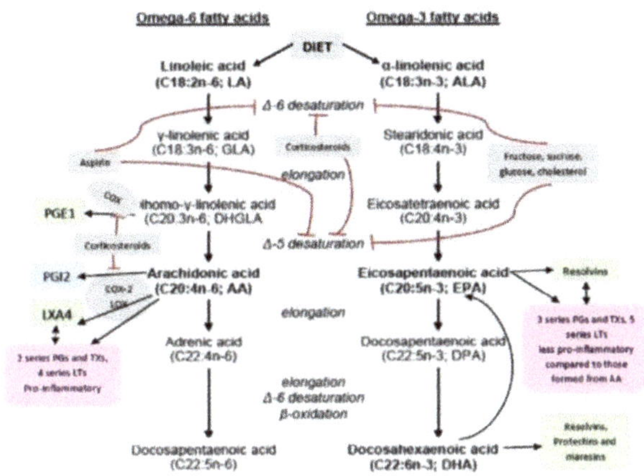

(A)

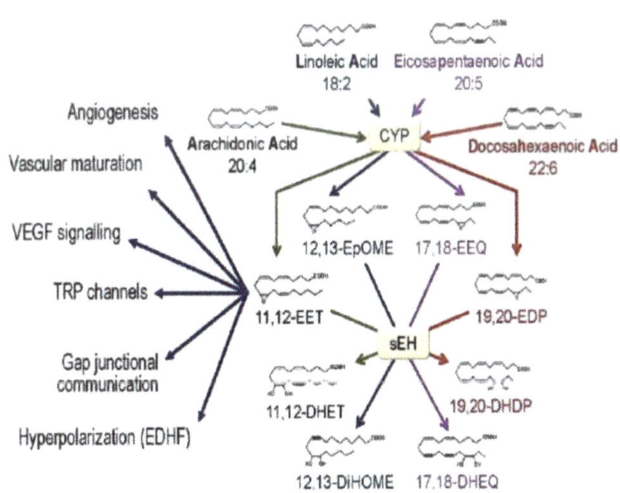

(B)

Figure 5. *Cont.*

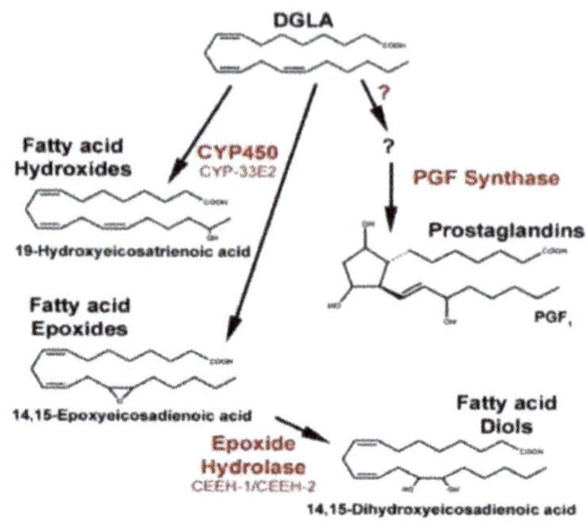

(C)

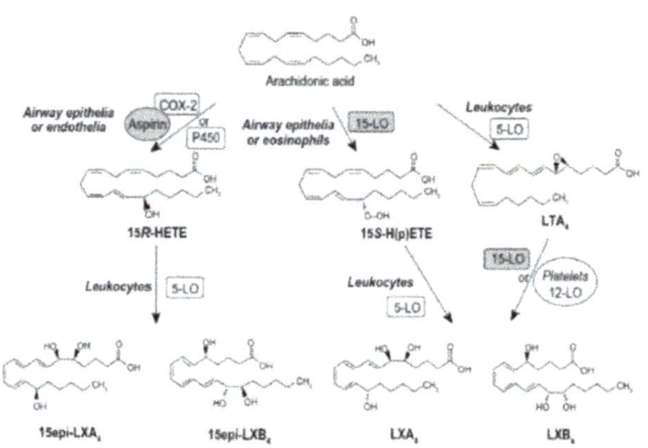

(D)

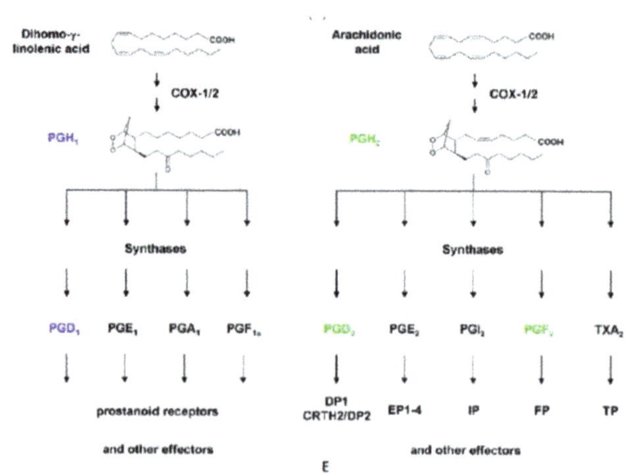

E

Figure 5. *Cont.*

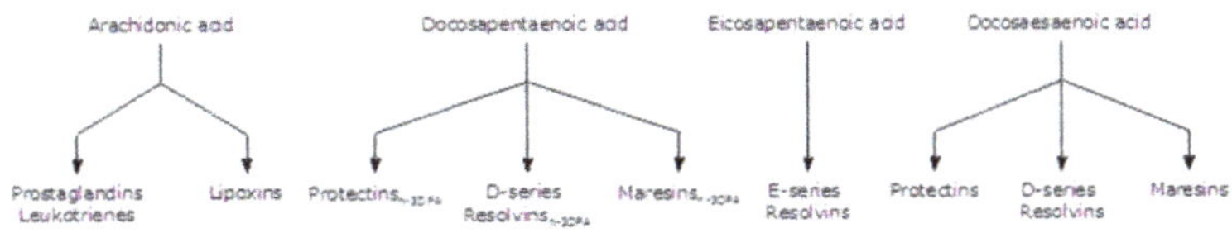

F

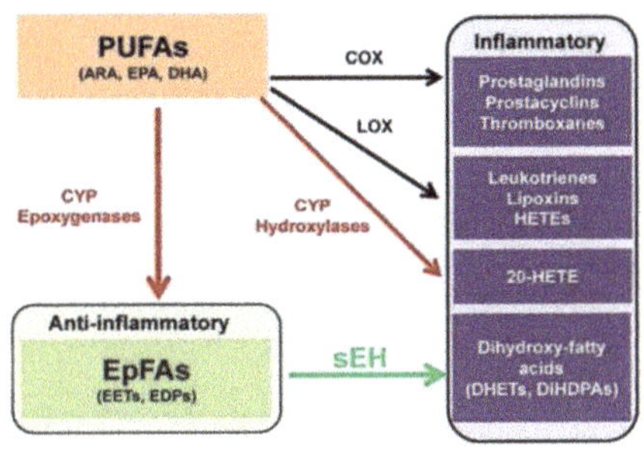

G

Figure 5. Scheme showing the metabolism of EFAs, and the various products formed from GLA, DGLA, AA, EPA, DPA and DHA. CYP = Cytochrome P450 enzymes sHE = Soluble epoxy hydrolase. (**A**) Metabolism of EFAs LA and ALA. (**B**) Metabolism of LA and EPA by cytochrome P450 enzymes and their actions. (**C**) Metabolism of DGLA and the products from the same. (**D**) Metabolism of AA and formation of LXA4 from the same. (**E**) A comparison of the metabolism of DGLA and AA and various products formed from these fatty acids. (**F**) A comparison of the metabolites formed from AA, DPA, EPA and DHA. (**G**) Metabolism of PUFAs and products formed from the same by the action of various enzymes.

7. AA and Its Metabolites Regulate Gene(s) Expression and Renin Release and Action

AA and its products regulate the release of renin at least, in part by producing changes in the transient receptor potential vanilloid type 1 (TRPV1) expressed in renal sensory nerves [6–10]. This implies that AA and its metabolites function as mediators that bring changes in renin gene expression in response to alterations in perfusion pressure involving nuclear mechanotransduction mechanism.

Several studies revealed that PGE2, PGI2, PGE synthase and PGE2 receptors EP2 and EP4 regulate renin secretion by the macula densa [11,12]. All four PGE2 receptor subtypes (EP1, EP2, EP3, EP4) control renal vascular tone, EP1 and EP3 receptors increasing, and EP2 and EP4 receptors, decreasing it [13]. PGE2 produces both renal vascular dilatation at low concentrations (1 nmol/) and vasoconstriction at higher concentrations (100 nmol/) [13].

This dual action of PGE2 (formed from AA) depends on the expression of PGE2 receptors, their number, and the concentration of PGE2 formed locally that, in turn, depends on AA released from the afferent arteriolar endothelial cells/myocytes/mesangium glomerular cells. The presence of EP1, EP2, EP3 and EP4 receptors in the vessels as well as in the thick ascending limb and collecting duct, glomerulus and collecting duct and the ability of PGE2 to modulate vascular tone and epithelial transport suggests that it regulates renin release, blood pressure and blood volume in response to various stimuli [14].

Akin to the actions of PGE2 on vascular tissue (both as a vasodilator and vasoconstrictor), it also has both pro- and anti-inflammatory actions. The pro-inflammatory actions of PGE2 can be ascribed to its ability to polarize macrophages by acting on mesenchymal stem cells (MSCs) binding to its receptors EP2 and EP4 depending on its local concentration [15–22]. Despite the belief that PGE2 is pro-inflammatory in nature, several studies suggested that PGE2 can also trigger anti-inflammatory events [17–25]. The pro- and anti-inflammatory actions of PGE2 depend on its ability to bind to different types of PGE2 receptors (EPs). At low concentrations, PGE2 binds to the high affinity EP4 receptor to enhance IL-23 production, whereas high PGE2 concentrations bind to EP2 receptor to suppress IL-23 release [19,20] that explains its dual pro and anti-inflammatory actions [26–29]. In addition, once PGE2 concentrations reach a peak, it triggers the generation of LXA4 and suppresses LTB4 production by inducing 5- and 15-lipoxygenases that triggers the pro-inflammatory status to be switched over to anti-inflammatory pathway (see Figure 6, [22,30]). These observations [15–31] emphasize the critical role of PGE2 in inflammation, vasoconstriction, vasodilatation and in the regulation of renin release from the JGA. This switchover of the metabolism of AA from PGE2 to LXA4 depends on the source of AA. There is a biphasic release of AA from the cell membrane lipid pool. The first pulse of AA release due to the activation of iPLA2 is utilized to form predominantly PGE2. On the other hand, the second pulse of AA release due to cPLA2 and/or sPLA2 activation is directed to form LXA4. Thus, factors regulating the activation of different forms of PLA2 are critical in the regulation of inflammation, vasoconstriction or vasodilatation and release or inhibition of renin release. It is likely that PGE1 derived from DGLA, the precursor of AA, also has actions like PGE2 in triggering the formation of LXA4 (see Figure 6) though PGE1 may be much less potent compared to PGE2. EPA and DHA also regulate inflammation, vascular tissue response and renin release. PGE3 and LTs (formed from EPA) and resolvins, protectins and maresins (derived from EPA and DHA) regulate inflammation, vascular tone, and renin release like PGE2, LTE4 and LXA4 (derived from AA) except that they are much less potent compared to those derived from AA.

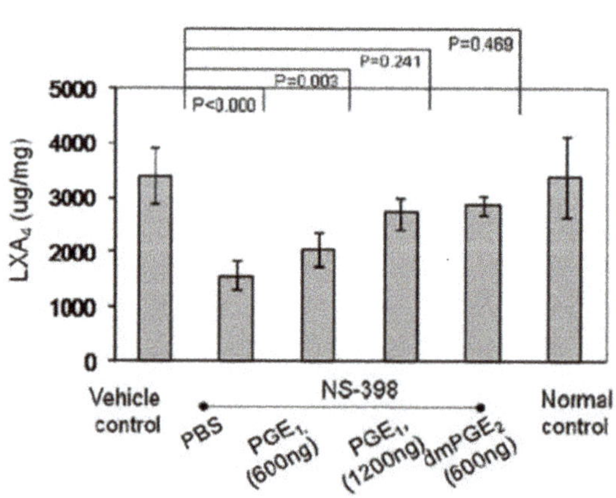

Figure 6. Regulation of LXA4 generation by PGE1 and PGE2.In experimental animals induced to develop collagen-induced arthritis, PGE1 and PGE2 enhance LXA4 formation PGE2 > PGE1). This data is taken from Ref. [31]. Normal control refers to the control foot pad of the opposite side.

The critical role of AA in various cellular processes including inflammation, vascular tone and renin release is supported by the observation that its peroxidized products bind to DNA and regulate gene(s) expression(s) [32,33]. It is likely that AA that is released in response to the changes in pressure and stretch get converted into its various metabolites (PGs, LTs, TXs, and LXA4) to regulate renin release. The possibility that AA by itself can regulate renin release cannot be discounted.

8. AA and Its Metabolites in Renin Synthesis, Secretion, and Action

Renin is an aspartyl protease which limits the activity of renin-angiotensin-aldosterone system (RAS). Renin-expressing cells in the kidney synthesize active renin from prorenin, which is stored in the cells to release it on demand. Both cAMP and Ca^{2+} pathways, cAMP-binding protein (CREB), and PPAR-γ (peroxisome proliferator-activated receptor-γ) exert a positive effect on renin gene transcription. In contrast to this, transcription factor NF-κB (nuclear factor-κB) interferes with the CREB-binding site and thus, inhibits the pro-inflammatory cytokines action on renin gene expression [34,35]. This may explain the negative feed-back regulation exerted by angiotensin-II/aldosterone on renin gene expression (and thus reduce renin secretion and levels). In contrast to this, AA, PGE1, PGE2, LXA4, resolvins, protectins and maresins may enhance renin gene expression by virtue of their inhibitory action on NF-κB, IL-6 and TNF-α, whereas AA metabolites, 12(s)-HPETE and 12-HETE, which are pro-inflammatory molecules inhibit renin gene expression [3,6,8,9,11,29]. These results emphasize as to how AA can both enhance and reduce renin synthesis and release and regulate blood volume and blood pressure. Based on these results, it is envisaged that AA release that occurs in response to changes in the pressure and stretch stimuli regulates renin release. Whenever there is a decrease in pressure and stretch stimuli as a result reduced blood volume and blood pressure, AA that is released is preferentially converted to PGE1, PGE2, LXA4, resolvins, protectins and maresins that enhance renin synthesis whereas an increase in pressure and stretch stimuli that occurs secondary to an increase blood volume and blood pressure the released AA is converted to HEPETE and HETE that inhibit renin synthesis and release. Thus, AA (like PGE2 as discussed above) can have both stimulatory and inhibitory action on renin synthesis and release.

9. Renin Secretion at the Cellular Level

Dysregulation of renal renin synthesis and secretion can cause significant alterations in blood volume, blood pressure, and as a result significant damage to various organs could occur. Hence, understanding the control of renal renin synthesis and renal renin secretion is essential. In general, experiments are performed in experimental animals to study the control of renin at the organ and cellular level and in vitro systems. Although such studies do give valuable information about renin synthesis and secretion, one need to know that there could occur substantial differences between experimental animals and humans pertaining to quantitative differences in the number of renin-producing cells, the rate of intracellular renin processing, and the rate of renin secretion. In the adult kidney, renin is synthesized by JGA granular cells located close to the renal afferent arterioles at the entrance into the glomerular capillary network (see Figure 1). The transcription rate of the renin gene is an essential event that determines the production rate of renin.

Among the factors that regulate renin gene expression include cAMP and Ca^{2+}pathways, cAMP-binding protein (CREB), and PPAR-γ (peroxisome proliferator-activated receptor-γ) that have a positive effect on renin gene transcription. On the other hand, NF-κB, IL-6, and TNF-α reduce renin gene expression and its secretion (see Figure 2, [34,35]. In addition, AA, and several of its metabolites (similarly EPA and DHA and their metabolites) also regulate renin gene expression [3,6,8,9,11,29]. In this context, it is noteworthy that AA (and EPA and DHA) can interact with syntaxin to regulate exocytosis and thus, control renin secretion.

Cyclic AMP stimulates renin secretion (by activating protein kinase A) [36]. Thus, catecholamines that activate cAMP formation (via β1-receptors) and PGI2 and PGE2 that inhibit cAMP degradation and nitric oxide (via cyclic GMP (cGMP)) and pharmacological inhibitors of cAMP-phosphodiesterases such as theophylline stimulate renin secretion [37]. In contrast to this, maneuvers that increase the cytosolic calcium concentration in renin-secreting cells inhibit renin release [38]. Thus, angiotensin-II and endothelin inhibit renin release [39,40]. Similarly, increased perfusion pressure of the afferent arterioles increases the calcium concentration in JGA cells and thus, inhibits renin release [41]. cGMP generation by ANP in renin-secreting cells also inhibits renin secretion [42].

Several neurotransmitters and neuropeptides stimulate renin secretion by enhancing cAMP pathway, whereas neuropeptide Y suppresses renin secretion by decreasing cAMP formation. Norepinephrine binds to β1-receptors on renin-secreting cells and thus, stimulates renin secretion. Renal vasoconstriction precedes sympathetic output to the kidneys [43–52]. The tubular macula densa cells and the preglomerular endothelial cells produce NO and prostaglandins (PGs) [53]. NO, PGI2 and PGE2 stimulate renin secretion [54,55].

10. AA Interacts with Syntaxin to Regulate Exocytosis/Renin Secretion

Renin is stored in granules in juxtaglomerular (JG) cells. SNAREs (soluble N-ethylmal eimide-sensitive factor attachment proteins), VAMP2 (vesicle associated membrane protein 2) and VAMP3 have a role in cAMP-stimulated exocytosis as is seen in other endocrine cells. VAMP2 and VAMP3 mRNA are expressed in JG cells and VAMP2 (but not VAMP3) is co-localized with renin-containing granules. Silencing VAMP2 blocks cAMP-induced renin release (by ~50%) whereas silencing VAMP3 had no effect [56]. This soluble NSF (N-ethylmaleimide-sensitive factor) attachment protein receptor (SNARE) protein VAMP2 mediated cAMP-stimulated renin release and exocytosis. Furthermore, JG cells with renin-containing granules co express and colocalize the isoform SNAP23 indicating that the SNARE protein SNAP23 is involved in cAMP-stimulated renin release, implying that renin release is a SNARE-dependent process [56–59].

Synaptic vesicle exocytosis is essential for neurotransmitter release at nerve terminals and renin by renin (JGA) cells. The vesicular transport machinery consists of four key components: (i) v-SNARE protein, which is a vesicle membrane protein, (ii) t-SNARE, the target membrane protein, (iii) membrane fusion *N*-ethylmaleimide-sensitive fusion cytosolic protein (NSF), and (iv) SNAPs that are adaptors for NSF termed (soluble NSF attachment proteins) (Figure 7). The v- and t-SNAREs are complementary to each other that are needed for vesicle docking. The assembled v- and t-SNARE acts as a receptor for the SNAPs that can incorporate the fusion protein, NSF. The docking and fusion particle containing all the four basic parts, thus formed, is called the SNARE complex. The energy needed for the vesicle fusion is derived from the hydrolysis of ATP by NSF, which is an ATPase [58]. Renin granule exocytosis needs SNAREs for the release of renin from juxtaglomerular cells [56,57,60]. Recent studies revealed that there is a critical role for AA and DHA in exocytosis of neurotransmitters in which the fusion of neurotransmitter-containing vesicles with the neuronal plasma membrane occurs. AA and DHA potentiate secretion from neurosecretory cells. AA increases soluble NSF attachment protein receptor complex and SNARE complex formation [61].

Syntaxin is one of the members of the SNARE family of proteins that mediates membrane fusion, an event needed for intracellular membrane trafficking [62]. Syntaxins diffuse in the plasma membrane to form clusters that have an important role in vesicle secretion process [63]. Exocytosis of secretory vesicles needs syntaxin to form clusters, which depends on the changes in the lipid bilayer of the cell membrane, that causes the aggregation of membrane proteins [64]. All SNARE proteins target membranes during fusion that is needed for exocytosis and is called the SNARE core complex, which is formed from SNARE proteins on the secretory vesicles (vesicle-associated membrane protein or VAMP) and a soluble SNARE protein (synaptosome-associated protein or SNAP). Syntaxins and VAMPs are anchored to the membranes (see Figure 7). Syntaxin 1 is needed for exocytosis of neurotransmitters. Membrane extension and neurotransmitter release needs fusion of intracellular vesicles with the plasma membrane involving SNARE protein assembly, membrane fusion, and its disassembly [65–69]. There are about 15 members of the syntaxin family. Syntaxins in conjunction with the cytoplasmic NSF and SNAP proteins mediate vesicle fusion and thus, are needed for exocytosis and endocytosis. All mammalian syntaxins, except for syntaxin 11, are transmembrane proteins and endosomal syntaxins is involved in specialized processes such as neurite outgrowth and myelin sheath formation [70].

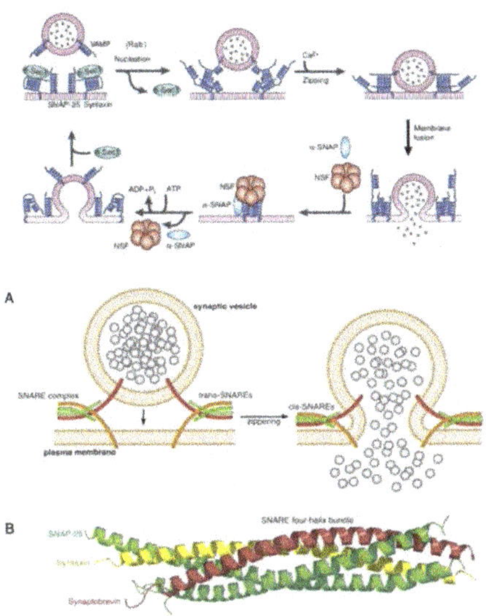

Figure 7. Schemes showing the mechanism of synaptic vesicle exocytosis. For details see text. (**A**) Scheme showing how SNAP-25 and Syntaxin participates in membrane fusion and exocytosis. (**B**) Scheme showing how SNAP-25, syntaxin and synaptobrevin interact with each other.

It is evident from the preceding discussion that membrane-associated SNAREs, syntaxin 1, SNAP25 (synaptosomal-associated protein 25 kDa) and synaptobrevin are needed for membrane fusion [71]. Fusion of intracellular vesicular storage material with the cytosolic surface of the plasma membrane involving SNARE proteins is essential in membrane extension and exocytosis of neurotransmitters. The pairing of SNARE with syntaxin-1 and SNAP-25 on the intracellular surface of the plasma membrane is needed for neurotransmitter exocytosis. Phospholipases (PLs) that release AA are highly enriched in nerve growth cones and are involved in neurite outgrowth. Syntaxin 3 (STX3) that is needed for neurite growth is activated by AA and DHA suggesting that these fatty acids are essential for membrane expansion at the growth cones, and exocytosis of neurotransmitters [61,71–76]. Thus, AA and DHA are critical in the exocytosis of various neurotransmitters.

In a similar fashion, AA may modulate renin release from JGA. JG cells express VAMP2 and VAMP3 mRNA especially, VAMP2 that is explicitly co-localized with JGA granules that contain renin. VAMP2 plays a role in cAMP mediated renin exocytosis [56]. The expression and colocalization of SNAP23 in renin-containing granules of JG cells implies that SNAP23 is involved in renin release and thus, renin release is a SNARE-dependent process [56–59]. SNARE, syntaxin-1 and SNAP-25 interact with each other and are needed for exocytosis. Since both AA and DHA (AA > DHA) activate syntaxin, it is proposed that these fatty acids regulate renin release. Furthermore, several AA metabolites can either increase or decrease renin release [6–14], depending on the metabolite formed from AA. For instance, 12(s)-HPETE and 12-HETE and 12- and 15-LOX metabolites inhibit renin release whereas PGE2 and PGI2 increase renin release depending on their concentration and binding of PGE2 to EP1-EP4 receptors. Thus, AA can either increase or decrease renin release depending on the metabolite formed and its concentration and binding to specific receptors [6–14]. These results agree with the results that nucleus serves as a sensor and responds to changes in pressure and stretch {1–2 in the Figure 3} that results in the release of calcium, cPLA2 activation and release of AA [3] (and other unsaturated fatty acids from the membrane) that ultimately causes actomyosin force generation (AA can directly modulate cytoskeletal structures). This is akin to changes in the blood flow and vasomotor tone of the afferent arteriole of the glomerulus that is sensed by the cells of the JGA leading to changes in their nuclear membrane tension which results in the released AA (and of EPA and DHA) and their subsequent conversion to its products such as prostaglandins, thromboxanes, leukotrienes, lipoxins, and resolvins from EPA and DHA and protectins and maresins from DHA that can either increase or decrease blood flow and produce respective changes in the vasomotor tone. These changes ultimately result either an increase or decrease in renin release as the situation demands. Based on these evidences, it is envisaged that AA serves as a mechanotransducer of renin cell baroreceptor and thus, modulate renal blood flow, body fluid volume and blood pressure (see Figures 1–4 and 8).

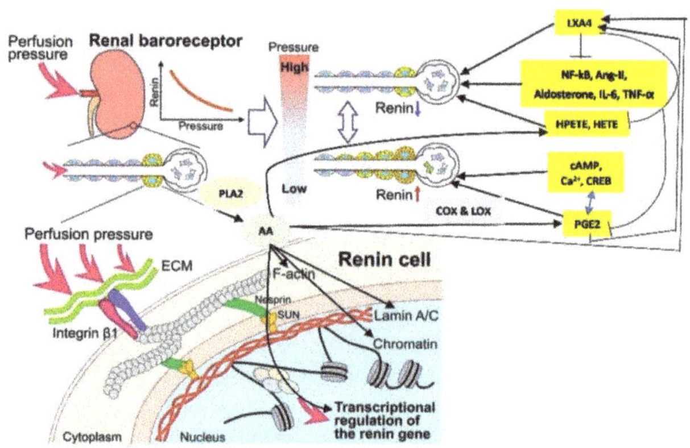

Figure 8. Scheme showing how AA can function as a mechanotransducer in response to changes in the perfusion pressure in the afferent arteriole and transmit renal baroreceptor actions. ECM = Extracellular matrix AA = Arachidonic acid. PGE2. +Prostaglandin E2 LXA4 = Lipoxin A4 NF-kB = Nuclear factor-kappa B IL-6 = Interleukin-6 TNF = Tumor necrosis factor HPETE = Hydroperoxyeicosatetraenoic acid HETE = Hydroxyeicosatetraenoic acid CREB = cAMP-binding protein. AA by itself or by its various metabolites can act on F-actin, Lamin A/C and chromatin to regulate renin gene expression and thus, regulate renin synthesis and exocytosis. Metabolites of AA such as PGE2, LXA4, HPETE and HETE can either enhance or decrease renin formation in response to changes in perfusion pressure (see Figure 8). PGE2 can either increase or decrease renin synthesis depending on its concentration and binding to its various receptors. PGE2, when it reaches its optimum level, triggers the synthesis of LXA4 from AA and LXA4, in turn, inhibits PGE2 formation. NF-kB, angiotensin-II, aldosterone, IL-6 and TNF-α stimulate PLA2 to induce the release of AA from cell membrane lipid pool and enhance the formation of PGE2 and other pro-inflammatory eicosanoids. LXA4 inhibits the expression of Nf-kB, and inhibits the synthesis of IL-6, TNF-α. HPETE and HETE are pro-inflammatory molecules formed from AA and can inhibit renin synthesis. This positive and negative feedback regulation among AA, PGE2, LXA4, NF-kB, IL-6, TNF-α, angiotensin-II and cAMP ensure that renin synthesis and exocytosis occur in an optimum way depending on the perfusion pressure, blood volume and blood pressure (see Figure 8).

11. Conclusions and Therapeutic Implications

Exocytosis is the fusion of secretory vesicles with the plasma membrane that is essential for the discharge of vesicle content into the extracellular space. This is associated with incorporation of new proteins and lipids into the plasma membrane. Exocytosis can be seen in many cells {called as (constitutive exocytosis) or it can occur in specialized cells such as neurons, endocrine and exocrine cells (this includes insulin secretion by pancreatic β cells, renin by renin cells of JGA), called as regulated exocytosis}. Constitutive exocytosis is needed to secrete digestive proenzymes, peptide hormones, immunoglobulins, chylomicra lipoproteins by non-neuronal cells. In majority of the instances, exocytosis is triggered by an increase in the cytosolic free Ca^{2+} concentration. In neurons and endocrine cells, secretory vesicles fuse with the plasma membrane in response to stimulation, and their secretion is linked to synapsins (in neurons) or actin (in endocrine cells). Several Ca^{2+}-binding proteins and synaptotagmin are essential for exocytosis at synapses. GTP-binding proteins are also involved in exocytosis. It is evident from the preceding discussion that SNARE that vesicle SNAREs (synaptobrevin and homologues) bind to target SNAREs (syntaxin/SNAP-25 and homologues), whereupon SNAPs and NSF bind to elicit membrane fusion and subsequent exocytosis [77]. Spatial control of exocytosis is needed for any biological processes such as neurotransmitter secretion, immune surveillance, cell migration and wound healing, and for development of cell polarity and cell growth (see Figure 9). In neurons, exocytosis of secretory granules and vesicles is confined to the synaptic cleft [78]. During immune surveillance, contact between an antigen-presenting cell (APC) and a cytotoxic T cell is essential and leads to formation of the immunological synapse that results in vesicle transport of soluble agents and membrane proteins within this zone of cell–cell contact that is needed to secrete cytokines, perforin, etc., to regulate immune response and induce apoptosis of tumor cells [79]. Proper cell migration and proliferation is needed for wound healing wherein exocytosis is coupled to local changes in plasma membrane and cytoskeleton dynamics [80]. During cell division, localized exocytosis occurs to mediate ingression of the cleavage furrow [81,82]. Development of cell polarity directs exocytosis at specialized sites of membrane growth [83,84].

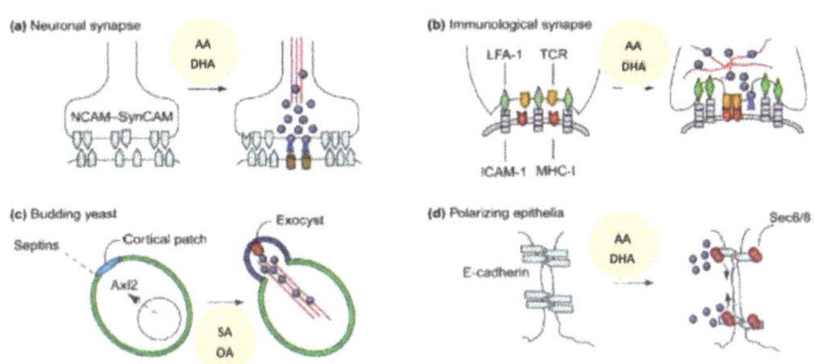

Figure 9. Molecular cues and membrane domains of localized exocytosis (modified from Ref. [83]). SA = Stearic acid. OA = Oleic acid. (**a**) Role of AA and DHA in neurite growth and neurotransmitters exocytosis {similar events lead to insulin and renin secretion}. (**b**) Under normal physiological conditions wherein robust immune surveillance exists, T-cell receptors (TCRs, orange) and LFA-1 cell adhesion molecules interact with MHC I and ICAM-1 molecules to form the immunological synapse. (**c**) In budding yeast, both extrinsic and intrinsic cues lead to cell–cell adhesion and is confined to E-cadherin. The exocyst is restricted to the tight junction as cells become more polarized. (**d**) SNARE proteins are localized at the site of delivery and fusion of intracellular vesicles because of local changes in intramembrane and membrane skeleton dynamics.

In this context, it is important to note that AA and DHA and their metabolites participate and regulate neurite growth and neurotransmitters exocytosis, insulin and renin secretion, immune surveillance, cell migration, wound healing and cell growth and multiplication [6–14,61,71–76,85–99]. These results coupled with the observation that alterations in the pressure or stretch leads to an alteration in the nuclear membrane tension that triggers calcium release, cPLA2 activation and release of AA [4,5] implies the role of AA to function as a mechanotransducer. These results suggest that AA can function as a mechanotransducer. Released AA induces changes in the actin cytoskeleton behavior and other cytoskeleton structures (including lamin, integrin β1). This ultimately results in changes in the expression of genes and their products. The released AA can be metabolized to form a wide variety of its products based on the activities of COX, LOX and p450 enzymes that have potent biological actions (see Figure 5). These metabolites of AA have multiple actions that impinge on almost all types of cells and tissues and several physiological processes [100,101]. It is noteworthy that AA is an important component of all cell membranes. Hence, it is likely that all types of cells in the body that are exposed to pressure or stretch and/or have constitutive or regulated exocytotic function need AA for their physiological function. The pressure or stretch can be exerted by blood flow, changes in blood volume, blood viscosity and could be brought about by internal or external stimuli. It is likely that insulin secretion by pancreatic β cells, renin secretion by renin cells of JGA and vascular endothelial cells that are constantly exposed to shear stress of blood volume and changes in blood pressure are the most important cells/tissues that need constant supply of AA to bring about their respective physiological actions. Thus, it is suggested that secretion of insulin by β cells and renin by renin cells of JGA and vasodilator and vasoconstrictor molecules production by endothelial cells is regulated by their AA content and metabolism. This is supported by the reports that pancreatic β cell secretion of insulin and renin production are regulated by AA and its metabolites [6–14,86–88]. Vascular endothelial cells produce several vasoactive substances that regulate vasomotor tone and blood pressure [102–104]. Of all, AA metabolites PGI2, PGE2, LTs, TXs, LXA4, and dihydroxy fatty acids are critical (see Figure 5). The balance between vasoconstrictors and vasodilators produced from AA (and EPA and DHA) seem to have both local and systemic actions in the regulation of blood pressure. In this context, AA seems to be important since it forms the precursor to both vasoconstrictor (LTs, TXs) and vasodilator (PGI2, PGE2, and LXA4) metabolites. Furthermore, AA can regulate nitric oxide (NO) generation that is necessary to maintain vascular endothelial integrity, tissue perfusion and blood pressure [105–108]. The crosstalk between PGE2 (a vasodilator, platelet aggregator and pro-inflammatory molecule) and LXA4 (a vasodilator, platelet anti-aggregator and anti-inflammatory molecule) seems to play a critical role in insulin and renin secretion (see Figures 6 and 9, [29–31]).

Since AA regulates exocytosis, it is imperative that cells (especially β, renin cells of JGA and vascular endothelial cells) need a constant supply of adequate amounts of AA. Since the availability of AA from diet is limited, cells/tissues depend on endogenous production of AA from dietary LA. Conversion of LA to AA needs desaturases and elongases enzymes (see Figure 5A), whereas the conversion of AA to their respective metabolites needs COX-1, COX-2, and 5-, 12-, and 15-LOX enzymes (see Figure 5). Based on the local conditions, AA and DHA are utilized to form their respective metabolites that determine the cellular process (such as exocytosis of insulin, renin, neurite growth, neurotransmitter release, immune response, cell growth, cell motility, wound healing, cancer cell growth or death or metastasis, etc.). Based on the available evidence, it is proposed that AA not only functions as a mechanotransducer to convey messages from the membrane to the nucleus but also appears to regulate cytoskeleton organization, cell shape, size, mitosis (meiosis) and gene expression [80,109–115].

It is evident from the current evidence that lipids are the prime constituents of the fusing membranes and play a critical role in exocytosis. In addition to fatty acids (AA and DHA) there appears to be a role for other lipids such as phosphatidylinositol-4,5-bisphosphate [PtdIns(4,5)P2], cholesterol, phosphatidic acid (PA), phosphatidylserine, phosphoinositides {such as PtdIns(4,5)P2, PtdIns(3)P, PIP2} also play a role in exocytosis [116]. It is noteworthy that AA forms an important constituent of these lipids (both structurally and functionally). Thus, the role of AA (and possibly, DHA and other unsaturated fatty acids) in exocytosis, immune surveillance, cell migration, wound healing, development of cell polarity and cell growth is definite. This is supported by the observation that plasma concentrations of AA in the phospholipid fraction are low in diabetes mellitus, hypertension, coronary heart disease, sepsis, pneumonia, rheumatoid arthritis, lupus (in RA and lupus decrease of GLA, the precursor of AA is also seen) (see Tables 1 and 2, [117,118] diseases in which exocytosis, immune surveillance, cell migration, wound healing, cell growth are critical to recover. Atherosclerosis free aortae have abundant essential fatty acids (LA and ALA and AA is the metabolite of LA) whereas atherosclerotic lesions are deficient in EFAs. This suggests that vascular endothelial cell integrity and function is dependent on the availability of EFAs [119–122].

Table 1. The percentage of distribution of fatty acids from plasma phospholipid fraction in patients with hypertension (HTN), coronary heart disease (CHD), type 2 diabetes mellitus, and diabetic nephropathy that are common with advanced age.

Fatty Acid	Control	HTN	CHD	Type 2 DM	Diabetic Nephropathy
16:0	25.9 ± 3.0	29.3 ± 2.7 *	27.8 ± 3.5	26.6 ± 5.2	26.8 ± 2.7
18:0	20.9 ± 3.6	23.2 ± 4.9 *	18:0 ± 10.7	14.6 ± 4.1	11.6 ± 3.6 *
18:1 n-9	13.0 ± 2.3	12.1 ± 1.5	11.5 ± 3.1	12.0 ± 2.6	14.5 ± 3.1
18:2 n-6 (LA)	18.6 ± 3.1	14.5 ± 3.1 *	17.8 ± 5.0	13.9 ± 5.3	15.1 ± 3.1
18:3 n-6 (GLA)	0.14 ± 0.1	0.4 ± 0.3 *	0.1 ± 0.1 *	0.2 ± 0.3	0.1 ± 0.2
20:3 n-6 (DGLA)	3.4 ± 1.0	3.1 ± 0.9	2.7 ± 1.1	1.7 ± 1.0 *	2.0 ± 0.8 *
20:4 n-6 (AA)	9.4 ± 1.8	7.8 ± 2.0 *	7.0 ± 2.1 *	4.6 ± 1.8 *	6.6 ± 2.6 *
22:5 n-6	0.7 ± 0.4	0.4 ± 0.4 *	1.0 ± 0.9	2.1 ± 0.6 *	1.3 ± 0.5 *
18:3 n-6/18:2 n-6	0.008	0.026	0.005	0.017	0.008
20:4 n-6/18:2 n-6	0.51	0.54	0.39	0.33	0.43
20:4 n-6/20:3 -6	2.8	2.53	2,59	2.8	3.3
18:3 n-3 (ALA)	0.2 ± 0.1	0.4 ± 0.2 *	0.3 ± 0.5	0.1 ± 0.2 *	0.1 ± 0.1 *
20:5 n-3 (EPA)	0.4 ± 0.4	0.6 ± 0.6	0.1 ± 0.2 *	0.3 ± 0.3	0.2 ± 0.3
22:5 n-3	0.5 ± 0.2	0.4 ± 0.5	0.3 ± 0.3 *	1.6 ± 1.3	1.7 ± 1.1
22:6 n-3 (DHA)	1.4 ± 0.5	1.2 ± 0.6	0.8 ± 0.4 *	0.5 ± 0.4 *	0.5 ± 0.3 *
20:5 n-3/18:3 n-3	1.8	1.39	0.41	3.2	4.0

All values are expressed as mean ± SD. * $p < 0.05$ compared to control. This data is taken from Refs [3,118].

Table 2. Fatty acid analysis of the plasma PL (phospholipid) fraction in patients with pneumonia, Figure 3 and [117].

	Control (n = 10)	Pneumonia (n = 12)	Septicemia (n = 14)	RA (n = 12)	SLE (Lupus) (n = 5)
16:0	24.8 ± 3.4	32.5 ± 3.6	26.95 ± 4.1	30.2 ± 3.0	32.0 ± 3.75
18:0	23.3 ± 4.1	21.4 ± 7.1	24.58 ± 6.0	19.0 ±6.1	14.6 ± 5.82
18:1 n-9	13.1 ± 2.3	15.6 ± 3.2	16.5 ± 3.3 *	14.8 ± 2.1	16.0 ± 2.78
18:2 n-6	17.7 ± 3.1	14.2 ± 0.3 *	16.3 ± 2.4	17.5 ± 2.7	20.8 ± 2.2
18:3 n-6	0.13 ± 0.09	0.13 ± 0.08	0.04 ± 0.05 *	0.02 ± 0.04 **	0.01 ± 0.01 **
20:3 n-6	3.2 ± 0.79	1.5 ± 0.4 *	0.46 ± 0.54 *	2.5 ± 0.58	2.12 ± 0.52
20:4 n-6	8.8± 2.0	5.1 ± 0.4 *	5.8 ± 1.6 *	9.5 ± 2.2	8.93 ± 2.0
22:4 n-6	0.42 ± 0.23	0.8 ± 0.9	0.34 ± 0.28	0.26 ± 0.37 **	0.18 ± 0.18 **
22:5 n-6	0.73 ± 0.55	0.45 ± 0.63	1.5 ± 1.02 *	0.6 ± 0.7	0.8 ± 1.0
18:3 n-3	0.27 ± 0.12	0.09 ± 0.04 *	0.16 ± 0.11 *	0.12 ± 0.16 *	0.1 ± 0.1 *
20:5 n-3	0.25 ± 0.26	0.23 ± 0.24	0.01 ± 0.01 *	0.05 ± 0.14 **	0.04 ± 0.04 **
22:6 n-3	1.43 ± 0.43	0.54 ± 0.43 *	1.2 ± 1.14	0.62 ± 0.56 *	0.88 ± 0.75 *

** $p < 0.001$ compared to control; * $p < 0.05$ compared to control.

It is noteworthy that tumor cells are deficient in AA (see Table 3) and selectively undergo apoptosis (compared to normal cells) on supplementation with this fatty acid (reviewed in [3,123]. These results suggest that supplementation of adequate amounts of AA could be employed to selectively eliminate tumor cells. Similar approach can be implemented for the prevention and management of diabetes mellitus, hypertension, coronary heart disease, sepsis, pneumonia, rheumatoid arthritis, lupus, and other diseases. It is interesting that AA supplementation, when its deficiency exists, leads to no change or decrease in the formation of PGE2 but augments formation of anti-inflammatory, vasodilator, platelet anti-aggregator, and anti-cancer molecule LXA4 [124–126]. Although in the present discussion, emphasis has been on AA, it is likely that other unsaturated fatty acids such as LA, GLA, DGLA, ALA, EPA, DPA and DHA may show mechanotransducer function as well. Both our in vitro and in vivo studies revealed that of all the saturated and unsaturated fatty acids correct is the best in elaborating its cytoprotective, anti-inflammatory and anti-diabetic actions. These beneficial actions of AA are due to enhanced formation of LXA4, a potent anti-inflammatory and anti-cancer molecule. GLA, DGLA, EPA and DHA also enhanced LXA4 formation, but the amount formed is significantly less [33]. The ability of GLA, DGLA, EPA and DHA to enhance LXA4 formation could be attributed to their ability to displace AA from the cell membrane lipid pool. It was reported that resolvins and protectins enhance LXA4 formation and thus, mediate some, if not all, of their anti-inflammatory actions [6,82,127,128]. Our studies revealed that LXA4 is more potent than resolvins and protectins in its anti-inflammatory, cytoprotective and anti-cancer actions [129–131]). These results suggest that AA and LXA4 are more important and potent compared to other unsaturated fatty acids (such as GLA, DGLA, EPA and DHA) and their metabolites (resolvins, protectins and maresins) (reviewed in [3]). This may explain as to why studies performed with EPA and DHA did not result in dramatic response in the prevention of inflammation, diabetes mellitus and cancer due to deficiency of AA in these diseases and failure to its (AA) supplement in adequate amounts. Hence, it is suggested that AA need to be given along with EPA and DHA to obtain optimal results [119]. Similarly, methods need to be developed such that AA is selectively delivered to the JGA to regulate renin release and function. AA can be given orally, intravenously for long periods without significant side effects [3,132–134]. In view of these evidence, it is imperative that potential clinical use of a simple nutrient AA in several diseases need to be explored.

Table 3. Content of fatty acids in normal liver, hepatoma cells, and in microsomal suspensions from normal liver and Yoshida hepatoma cells. All values of mean ± S. E. This data is form Ref. [3].

Measurement (Fatty Acid)	Normal Intact Liver	Intact Yoshida Cells	Normal Liver Microsomes	Yoshida Microsomes
16:0	18.5 ± 0.2	18.7 ± 2.0	18.9 ± 1.1	18.5 ±0.5
18:0	17.5 ± 0.5	13.3 ± 1.1	22.0 ± 3.0	13.7 ± 0.2
18:1, n-9 (oleic acid)	12.1 ± 1.0	21.5 ± 0.8	8.6 ± 1.0	18.1 ± 0.3
20:4 (AA)	16.7 ± 2.4	8.7 ± 0.7	19.1 ± 2.4	9.6 ± 0.8
22:5	-	2.9 ± 0.1	-	2.4 ± 0.3
22:6 (DHA)	6.3 ± 0.2	5.2 ± 0.6	6.1 ± 0.3	5.3 ± 0.4

Funding: No external funding was received for this work.

Institutional Review Board Statement: Not applicable.

Informed Consent Statement: Not applicable.

Data Availability Statement: All the data has been given in the manuscript/references.

Conflicts of Interest: The author declare no conflict of interest.

Abbreviations

AA	Arachidonic acid
EPA	Eicosapentaenoic acid
DHA	Docosahexaenoic acid
LXA4	Lipoxin A4
RAS	Renin-angiotensin aldosterone system
And-II	Angiotensin-II
JGA	Juxtaglomerular apparatus
PLA2	Phospholipase A2
TRPV1	Transient receptor potential vanilloid type
1 ATP	Adenosine triphosphate
LA	Cis-linoleic acid
ALA	Alpha-linolenic acid
GLA	Gamma-linolenic acid
COX-2	Cyclooxygenase-2
PGE2	Prostaglandin E2
PGs	Prostaglandins

LTs	Leukotrienes
TXs	Thromboxanes
mPGES1 and PGES2	Microsomal PGE synthases 1 and 2 EP2 and EP4 = PGE2 receptors
LOX	Lipoxygenase
MSCs	Mesenchymal stem cells
CREB	cAMP-binding protein
PPAR-γ	Peroxisome proliferator-activated receptor
γ cAMP	Cyclic adenosine monophosphate
HPETE	Hydroperoxyeicosatetraenoic acid
HETE	Hydroxyeicosatetraenoic acid
PGI2	Prostacyclin
cGMP	Cyclic guanosine monophosphate
NO	Nitric oxide
SNAREs	Soluble N-ethylmaleimide-sensitive factor attachment proteins
NSF	N-ethylmaleimide-sensitive factor
SNAP	Synaptosome-associated protein
APC	Antigen-presenting cell
VAMP	Vesicle-associated membrane protein
STX3	Syntaxin 3
PUFAs	Polyunsaturated fatty acids
EFAs	Essential fatty acids
NF-kB	Nuclear factor-kappa B
IkB	Inhibitory kappa B
IL-6	Interleukin-6
TNF	Tumor necrosis factor

References

1. Watanabe, H.; Belyea, B.C.; Paxton, R.L.; Li, M.; Dzamba, B.J.; DeSimone, D.W.; Gomez, R.A.; Sequeira-Lopez, M.L.S. Renin Cell Baroreceptor, a Nuclear Mechanotransducer Central for Homeostasis. *Circ. Res.* **2021**, *129*, 262–276. [CrossRef] [PubMed]
2. Heesch, C.M. Reflexes that Control Cardiovascular Function. *Am. J. Physiol. Content* **1999**, *277*, S234. [CrossRef] [PubMed]
3. Das, U. "Cell Membrane Theory of Senescence" and the Role of Bioactive Lipids in Aging, and Aging Associated Diseases and their Therapeutic Implications. *Biomolecules* **2021**, *11*, 241. [CrossRef]
4. Lomakin, A.J.; Cattin, C.J.; Cuvelier, D.; Alraies, Z.; Molina, M.; Nader, G.P.F.; Srivastava, N.; Sáez, P.J.; Garcia-Arcos, J.M.; Zhitnyak, I.Y.; et al. The Nucleus Acts as a Ruler Tailoring Cell Responses to Spatial Constraints. *Science* **2020**, *370*, eaba2894. [CrossRef]
5. Venturini, V.; Pezzano, F.; Castro, F.C.; Häkkinen, H.-M.; Jiménez-Delgado, S.; Colomer-Rosell, M.; Marro, M.; Tolosa-Ramon, Q.; Paz-López, S.; Valverde, M.A.; et al. The Nucleus Measures Shape Changes for Cellular Proprioception to Control Dynamic Cell Behavior. *Science* **2020**, *370*, eaba2644. [CrossRef] [PubMed]
6. Xie, C.; Wang, D.H. Inhibition of Renin Release by Arachidonic Acid Metabolites, 12(s)-HPETE and 12-HETE: Role of TRPV1 Channels. *Endocrinology* **2011**, *152*, 3811–3819. [CrossRef]
7. Nasrallah, R.; Hébert, R.L. Prostacyclin Signaling in the Kidney: Implications for Health and Disease. *Am. J. Physiol. Physiol.* **2005**, *289*, F235–F246. [CrossRef] [PubMed]
8. Antonipillai, I.; Nadler, J.L.; Robin, E.C.; Horton, R. The Inhibitory Role of 12- and 15-Lipoxygenase Products on Renin Release. *Hypertension* **1987**, *10*, 61–66. [CrossRef] [PubMed]
9. Gerkens, J.; Williams, A.; Branch, R. Effect of Precursors of the 1, 2, and 3 Series Prostaglandins on Renin Release and Renal Blood Flow in the Dog. *Prostaglandins* **1981**, *22*, 513–520. [CrossRef]
10. Weber, P.C. Renal Prostaglandins in the Control of Renin1. *Contr. Nephrol.* **1978**, *12*, 92–105. [CrossRef]
11. Facemire, C.S.; Nguyen, M.; Jania, L.; Beierwaltes, W.H.; Kim, H.-S.; Koller, B.H.; Coffman, T.M. A Major Role for the EP4 Receptor in Regulation of Renin. *Am. J. Physiol. Renal Physiol.* **2011**, *301*, F1035–F1041. [CrossRef] [PubMed]
12. Matzdorf, C.; Kurtz, A.; Höcherl, K. COX-2 Activity Determines the Level of Renin Expression but Is Dispensable for Acute Upregulation of Renin Expression in Rat Kidneys. *Am. J. Physiol. Renal Physiol.* **2007**, *292*, F1782–F1790. [CrossRef] [PubMed]
13. Schweda, F.; Klar, J.; Narumiya, S.; Nüsing, R.M.; Kurtz, A. Stimulation of Renin Release by Prostaglandin E2 Is Mediated by EP2 and EP4 Receptors in Mouse Kidneys. *Am. J. Physiol. Renal Physiol.* **2004**, *287*, F427–F433. [CrossRef] [PubMed]
14. Breyer, M.D.; Breyer, R. Prostaglandin E Receptors and the Kidney. *Am. J. Physiol. Renal Physiol.* **2000**, *279*, F12–F23. [CrossRef] [PubMed]
15. Vasandan, A.B.; Jahnavi, S.; Shashank, C.; Prasad, P.; Kumar, A.; Prasanna, S.J. Human Mesenchymal Stem Cells Program Macrophage Plasticity by Altering their Metabolic Status via a PGE2-Dependent Mechanism. *Sci. Rep.* **2016**, *6*, 38308. [CrossRef] [PubMed]

16. Rozenberg, A.; Rezk, A.; Boivin, M.-N.; Darlington, P.J.; Nyirenda, M.; Li, R.; Jalili, F.; Winer, R.; Artsy, E.A.; Uccelli, A.; et al. Human Mesenchymal Stem Cells Impact Th17 and Th1 Responses Through a Prostaglandin E2 and Myeloid-Dependent Mechanism. *Stem Cells Transl. Med.* **2016**, *5*, 1506–1514. [CrossRef] [PubMed]
17. Németh, K.; Leelahavanichkul, A.; Yuen, P.S.; Mayer, B.; Parmelee, A.; Doi, K.; Robey, P.G.; Leelahavanichkul, K.; Koller, B.H.; Brown, J.M.; et al. Bone Marrow Stromal Cells Attenuate Sepsis via Prostaglandin E2–Dependent Reprogramming of Host Macrophages to Increase their Interleukin-10 Production. *Nat. Med.* **2008**, *15*, 42–49. [CrossRef] [PubMed]
18. Ylöstalo, J.H.; Bartosh, T.J.; Coble, K.; Prockop, D.J. Human Mesenchymal Stem/Stromal Cells Cultured as Spheroids are Self-activated to Produce Prostaglandin E2 that Directs Stimulated Macrophages into an Anti-inflammatory Phenotype. *Stem Cells* **2012**, *30*, 2283–2296. [CrossRef] [PubMed]
19. Poloso, N.J.; Urquhart, P.; Nicolaou, A.; Wang, J.; Woodward, D.F. PGE2 Differentially Regulates Monocyte-Derived Dendritic Cell Cytokine Responses Depending on Receptor Usage (EP2/EP4). *Mol. Immunol.* **2013**, *54*, 284–295. [CrossRef]
20. Kalim, K.W.; Groettrup, M. Prostaglandin E2 inhibits IL-23 and IL-12 Production by Human Monocytes through Down-Regulation of their Common p40 Subunit. *Mol. Immunol.* **2013**, *53*, 274–282. [CrossRef]
21. Loynes, C.A.; Lee, J.A.; Robertson, A.L.; Steel, M.J.; Ellett, F.; Feng, Y.; Levy, B.D.; Whyte, M.K.; Renshaw, S.A. PGE 2 Production at Sites of Tissue Injury Promotes an Anti-Inflammatory Neutrophil Phenotype and Determines the Outcome of Inflammation Resolution in Vivo. *Sci. Adv.* **2018**, *4*, eaar8320. [CrossRef] [PubMed]
22. Chan, M.M.-Y.; Moore, A.R. Resolution of Inflammation in Murine Autoimmune Arthritis Is Disrupted by Cyclooxygenase-2 Inhibition and Restored by Prostaglandin E2-Mediated Lipoxin A4 Production. *J. Immunol.* **2010**, *184*, 6418–6426. [CrossRef] [PubMed]
23. Park, H.J.; Kim, J.; Saima, F.T.; Rhee, K.-J.; Hwang, S.; Kim, M.Y.; Baik, S.K.; Eom, Y.W.; Kim, H.-S. Adipose-Derived Stem Cells Ameliorate Colitis by Suppression of Inflammasome Formation and Regulation of M1-Macrophage Population through Prostaglandin E2. *Biochem. Biophys. Res. Commun.* **2018**, *498*, 988–995. [CrossRef]
24. An, J.-H.; Song, W.-J.; Li, Q.; Kim, S.-M.; Yang, J.-I.; Ryu, M.-O.; Nam, A.R.; Bhang, D.H.; Jung, Y.-C.; Youn, H.-Y. Prostaglandin E2 Secreted from Feline Adipose Tissue-Derived Mesenchymal Stem Cells Alleviate DSS-Induced Colitis by Increasing Regulatory T Cells in Mice. *BMC Veter.-Res.* **2018**, *14*, 354. [CrossRef] [PubMed]
25. Terraza-Aguirre, C.; Campos-Mora, M.; Elizondo-Vega, R.; Contreras-López, R.; Luz-Crawford, P.; Jorgensen, C.; Djouad, F. Mechanisms behind the Immunoregulatory Dialogue between Mesenchymal Stem Cells and Th17 Cells. *Cells* **2020**, *9*, 1660. [CrossRef]
26. Chiossone, L.; Conte, R.; Spaggiari, G.M.; Serra, M.; Romei, C.; Bellora, F.; Becchetti, F.; Andaloro, A.; Moretta, L.; Bottino, C. Mesenchymal Stromal Cells Induce Peculiar Alternatively Activated Macrophages Capable of Dampening Both Innate and Adaptive Immune Responses. *Stem Cells* **2016**, *34*, 1909–1921. [CrossRef]
27. Liu, X.; Qu, X.; Chen, Y.; Liao, L.; Cheng, K.; Shao, C.; Zenke, M.; Keating, A.; Zhao, R.C.H. Mesenchymal Stem/Stromal Cells Induce the Generation of Novel IL-10–Dependent Regulatory Dendritic Cells by SOCS3 Activation. *J. Immunol.* **2012**, *189*, 1182–1192. [CrossRef]
28. Yamane, H.; Sugimoto, Y.; Tanaka, S.; Ichikawa, A. Prostaglandin E(2) Receptors, EP2 and EP4, Differentially Modulate TNF-Alpha and IL-6 Production Induced by Lipopolysaccharide in Mouse Peritoneal Neutrophils. *Biochem. Biophys. Res. Commun.* **2000**, *278*, 224–228. [CrossRef]
29. Das, U.N. Bioactive Lipids in Age-Related Disorders. *Rev. New Drug Targets Age-Related Disord.* **2020**, *1260*, 33–83. [CrossRef]
30. Das, U.N. Essential Fatty Acids, Lipid Peroxidation and Apoptosis. *Prostaglandins Leukot Essen Fat. Acids* **1999**, *61*, 157–164. [CrossRef]
31. Gilroy, D.; Colvillenash, P.R.; Willis, D.K.; Chivers, J.; Paulclark, M.J.; Willoughby, D. Inducible Cyclooxygenase May Have Anti-Inflammatory Properties. *Nat. Med.* **1999**, *5*, 698–701. [CrossRef] [PubMed]
32. Gundala, N.K.V.; Naidu, V.G.M.; Das, U.N. Arachidonic Acid and LipoxinA4 Attenuate Streptozotocin-Induced Cytotoxicity to RIN5 F Cells in Vitro and Type 1 and Type 2 Diabetes Mellitus in Vivo. *Nutrition* **2017**, *35*, 61–80. [CrossRef]
33. Todorov, V.T.; Völkl, S.; Müller, M.; Bohla, A.; Klar, J.; Kunz-Schughart, L.A.; Hehlgans, T.; Kurtz, A. Tumor Necrosis Factor-Alpha Activates NFkappaB to Inhibit Renin Transcription by Targeting cAMP-Responsive Element. *J. Biol. Chem.* **2004**, *279*, 1458–1467. [CrossRef] [PubMed]
34. Todorov, V.T.; Völkl, S.; Friedrich, J.; Kunz-Schughart, L.A.; Hehlgans, T.; Vermeulen, L.; Haegeman, G.; Schmitz, M.L.; Kurtz, A. Role of CREB1 and NF {κ}B-p65 in the Down-Regulation of Renin Gene Expression by Tumor Necrosis Factor {α}. *J. Biol. Chem.* **2005**, *280*, 24356–24362. [CrossRef] [PubMed]
35. Friis, U.G.; Jensen, B.L.; Sethi, S.; Andreasen, D.; Hansen, P.B.; Skøtt, O. Control of Renin Secretion from Rat Juxtaglomerular Cells by cAMP-Specific Phosphodiesterases. *Circ Res.* **2002**, *90*, 996–1003. [CrossRef] [PubMed]
36. Kurtz, A. Renin Release: Sites, Mechanisms, and Control. *Annu. Rev Physiol.* **2011**, *73*, 377–399. [CrossRef] [PubMed]
37. Beierwaltes, W.H. The Role of Calcium in the Regulation of Renin Secretion. *Am. J. Physiol. Physiol.* **2010**, *298*, F1–F11. [CrossRef]
38. Kurtz, A.; Penner, R. Angiotensin II Induces Oscillations of Intracellular Calcium and Blocks Anomalous Inward Rectifying Potassium Current in Mouse Renal Juxtaglomerular Cells. *Proc. Natl. Acad. Sci. USA* **1989**, *86*, 3423–3427. [CrossRef] [PubMed]
39. Ritthaler, T.; Scholz, H.; Ackermann, M.; Riegger, G.; Kurtz, A.; Krämer, B.K. Effects of Endothelins on Renin Secretion from Isolated Mouse Renal Juxtaglomerular Cells. *Am. J. Physiol. Content* **1995**, *268*, F39–F45. [CrossRef] [PubMed]

40. Lai, E.Y.; Wang, Y.; Persson, A.E.G.; Manning, R.D., Jr.; Liu, R. Pressure Induces Intracellular Calcium Changes in Juxtaglomerular Cells in Perfused Afferent Arterioles. *Hypertens. Res.* **2011**, *34*, 942–948. [CrossRef]
41. Kurtz, A.; Della Bruna, R.; Pfeilschifter, J.; Taugner, R.; Bauer, C. Atrial Natriuretic Peptide Inhibits Renin Release from Juxtaglomerular Cells by a cGMP-Mediated Process. *Proc. Natl. Acad. Sci. USA* **1986**, *83*, 4769–4773. [CrossRef]
42. Vandongen, R.; Peart, W.S.; Boyd, G.W. Adrenergic Stimulation of Renin Secretion in the Isolated Perfused Rat Kidney. *Circ. Res.* **1973**, *32*, 290–296. [CrossRef] [PubMed]
43. Imbs, J.-L.; Schmidt, M.; Schwartz, J. Effect of Dopamine on Renin Secretion in the Anesthetized Dog. *Eur. J. Pharmacol.* **1975**, *33*, 151–157. [CrossRef]
44. Holdaas, H.; Langaard, O.; Eide, I.; Kiil, F. Conditions for Enhancement of Renin Release by Isoproterenol, Dopamine, and Glucagon. *Am. J. Physiol. Content* **1982**, *242*, F267–F273. [CrossRef] [PubMed]
45. Kurtz, A.; Muff, R.; Born, W.; Lundberg, J.M.; I Millberg, B.; Gnädinger, M.P.; E Uehlinger, D.; Weidmann, P.; Hökfelt, T.; A Fischer, J. Calcitonin Gene-Related Peptide Is a Stimulator of Renin Secretion. *J. Clin. Investig.* **1988**, *82*, 538–543. [CrossRef] [PubMed]
46. Porter, J.P.; Ganong, W.F. Vasoactive Intestinal Peptide and Renin Secretion. *Ann. N. Y. Acad. Sci.* **1988**, *527*, 465–477. [CrossRef] [PubMed]
47. Hautmann, M.; Friis, U.G.; Desch, M.; Todorov, V.; Castrop, H.; Segerer, F.; Otto, C.; Schutz, G.; Schweda, F. Pituitary Adenylate Cyclase–Activating Polypeptide Stimulates Renin Secretion via Activation of PAC1 Receptors. *J. Am. Soc. Nephrol.* **2007**, *18*, 1150–1156. [CrossRef] [PubMed]
48. Hackenthal, E.; Aktories, K.; Jakobs, K.H.; E Lang, R. Neuropeptide Y Inhibits Renin Release by a Pertussis Toxin-Sensitive Mechanism. *Am. J. Physiol. Content* **1987**, *252*, F543–F550. [CrossRef] [PubMed]
49. Bischoff, A.; Michel, M.C. Renal effects of neuropeptide Y. *Pflügers Archiv* **1998**, *435*, 443–453. [CrossRef]
50. Boivin, V.; Jahns, R.; Gambaryan, S.; Ness, W.; Boege, F.; Lohse, M.J. Immunofluorescent Imaging of Beta 1- and Beta 2-Adrenergic Receptors in Rat Kidney. *Kidney Int.* **2001**, *59*, 515–531. [CrossRef] [PubMed]
51. Konstantinos, P.T.; Dimitriadis, K. Radiance-Htn Trio: How the Saga of Renal Denervation Revisits Hypertensiontherapy. *Cardiovasc. Res.* **2021**, *117*, e141–e143.
52. Castrop, H.; Höcherl, K.; Kurtz, A.; Schweda, F.; Todorov, V.; Wagner, C. Physiology of Kidney Renin. *Physiol. Rev.* **2010**, *90*, 607–673. [CrossRef] [PubMed]
53. Jensen, B.L.; Schmid, C.; Kurtz, A. Prostaglandins Stimulate Renin Secretion and Renin mRNA in Mouse Renal Juxtaglomerular Cells. *Am. J. Physiol. Physiol.* **1996**, *271*, F659–F669. [CrossRef] [PubMed]
54. Kurtz, A.; Götz, K.-H.; Hamann, M.; Wagner, C. Stimulation of Renin Secretion by Nitric Oxide Is Mediated by Phosphodiesterase 3. *Proc. Natl. Acad. Sci. USA* **1998**, *95*, 4743–4747. [CrossRef] [PubMed]
55. Mendez, M.; Gross, K.W.; Glenn, S.T.; Garvin, J.L.; Carretero, O.A. Vesicle-Associated Membrane Protein-2 (VAMP2) Mediates cAMP-Stimulated Renin Release in Mouse Juxtaglomerular Cells. *J. Biol. Chem.* **2011**, *286*, 28608–28618. [CrossRef] [PubMed]
56. Mendez, M.; Gaisano, H.Y. Role of the SNARE Protein SNAP23 on cAMP-Stimulated Renin Release in Mouse Juxtaglomerular Cells. *Am. J. Physiol. Ren. Physiol.* **2013**, *304*, F498–F504. [CrossRef] [PubMed]
57. Goda, Y. SNAREs and Regulated Vesicle Exocytosis. *Proc. Natl. Acad. Sci. USA* **1997**, *94*, 769–772. [CrossRef] [PubMed]
58. Kurtz, A. Control of Renin Synthesis and Secretion. *Am. J. Hypertens.* **2012**, *25*, 839–847. [CrossRef] [PubMed]
59. Mendez, M. Renin Release: Role of SNAREs. *Am. J. Physiol. Regul. Integr. Comp. Physiol.* **2014**, *307*, R484–R486. [CrossRef] [PubMed]
60. Latham, C.F.; Osborne, S.L.; Cryle, M.J.; Meunier, F.A. Arachidonic Acid Potentiates Exocytosis and Allows Neuronal SNARE Complex to Interact with Munc18a. *J. Neurochem.* **2006**, *100*, 1543–1554. [CrossRef]
61. Chen, Y.A.; Scheller, R.H. SNARE-Mediated Membrane Fusion. *Nat. Rev. Mol. Cell Biol.* **2001**, *2*, 98–106. [CrossRef]
62. Lang, T.; Bruns, D.; Wenzel, D.; Riedel, D.; Holroyd, P.; Thiele, C.; Jahn, R. SNAREs are Concentrated in Cholesterol-Dependent Clusters that Define Docking and Fusion Sites for Exocytosis. *EMBO J.* **2001**, *20*, 2202–2213. [CrossRef] [PubMed]
63. Owicki, J.C.; McConnell, H.M. Theory of Protein-Lipid and Protein-Protein Interactions in Bilayer Membranes. *Proc. Natl. Acad. Sci. USA* **1979**, *76*, 4750–4754. [CrossRef]
64. Weber, T.; Zemelman, B.; McNew, J.; Westermann, B.; Gmachl, M.; Parlati, F.; Söllner, T.H.; Rothman, J.E. SNAREpins: Minimal Machinery for Membrane Fusion. *Cell* **1998**, *92*, 759–772. [CrossRef]
65. Wang, Y.; Tang, B.L. SNARES in Neurons–Beyond Synaptic Vesicle Exocytosis. *Mol Membr Biol.* **2006**, *23*, 377–384. [CrossRef]
66. Söllner, T.; Whiteheart, S.; Brunner, M.; Erdjument-Bromage, H.; Geromanos, S.; Tempst, P.; Rothman, J.E. SNAP Receptors Implicated in Vesicle Targeting and Fusion. *Nature* **1993**, *362*, 318–324. [CrossRef] [PubMed]
67. Nichols, B.J.; Ungermann, C.; Pelham, H.R.B.; Wickner, W.T.; Haas, A. Homotypic Vacuolar Fusion Mediated by t- and v-SNAREs. *Nature* **1997**, *387*, 199–202. [CrossRef] [PubMed]
68. Jahn, R.; Scheller, R.H. SNARES–Engines for Membrane Fusion. *Nat. Rev. Mol. Cell Biol.* **2006**, *7*, 631–643. [CrossRef] [PubMed]
69. Teng, F.Y.H.; Wang, Y.; Tang, B.L. The Syntaxins. *Genome Biol.* **2001**, *2*, 1–7. [CrossRef] [PubMed]
70. Connell, E.; Darios, F.; Broersen, K.; Gatsby, N.; Peak-Chew, S.; Rickman, C.; Davletov, B. Mechanism of Arachidonic Acid Action on Syntaxin–Munc18. *EMBO Rep.* **2007**, *8*, 414–419. [CrossRef] [PubMed]
71. Darios, F.; Davletov, B. Omega-3 and Omega-6 Fatty Acids Stimulate Cell Membrane Expansion by Acting on Syntaxin 3. *Nature* **2006**, *440*, 813–817. [CrossRef] [PubMed]

72. Darios, F.; Connell, E.; Davletov, B. Phospholipases and Fatty Acid Signalling in Exocytosis. *J. Physiol.* **2007**, *585*, 699–704. [CrossRef]
73. Chattopadhyay, S.; Sun, P.; Wang, P.; Abonyo, B.; Cross, N.L.; Liu, L. Fusion of Lamellar Body with Plasma Membrane Is Driven by the Dual Action of Annexin II Tetramer and Arachidonic Acid. *J. Biol. Chem.* **2003**, *278*, 39675–39683. [CrossRef]
74. Darios, F.; Ruipérez, V.; López, I.; Villanueva, J.; Gutierrez, L.M.; Davletov, B. Alpha-Synuclein Sequesters Arachidonic Acid to Modulate SNARE-Mediated Exocytosis. *EMBO Rep.* **2010**, *11*, 528–533. [CrossRef] [PubMed]
75. Rickman, C.; Davletov, B. Arachidonic Acid Allows SNARE Complex Formation in the Presence of Munc18. *Chem. Biol.* **2005**, *12*, 545–553. [CrossRef]
76. Gerber, S.H.; Suüdhof, T.C. Molecular Determinants of Regulated Exocytosis. *Diabetes* **2002**, *51*, S3–S11. [CrossRef] [PubMed]
77. Rizo, J.; Südhof, T.C. Snares and munc18 in Synaptic Vesicle Fusion. *Nat. Rev. Neurosci.* **2002**, *3*, 641–653. [CrossRef]
78. Bossi, G.; Trambas, C.; Booth, S.; Clark, R.; Stinchcombe, J.; Griffiths, G.M. The Secretory Synapse: The Secrets of a Serial Killer. *Immunol. Rev.* **2002**, *189*, 152–160. [CrossRef] [PubMed]
79. Sheetz, M.P. Cell Control by Membrane–Cytoskeleton Adhesion. *Nat. Rev. Mol. Cell Biol.* **2001**, *2*, 392–396. [CrossRef] [PubMed]
80. Finger, F.P.; White, J.G. Fusion and Fission: Membrane Trafficking in Animal Cytokinesis. *Cell* **2002**, *108*, 727–730. [CrossRef]
81. Bednarek, S.Y.; Falbel, T.G. Membrane Trafficking during Plant Cytokinesis. *Traffic* **2002**, *3*, 621–629. [CrossRef] [PubMed]
82. Spiliotis, E.T.; Nelson, W.J. Spatial Control of Exocytosis. *Curr. Opin. Cell Biol.* **2003**, *15*, 430–437. [CrossRef]
83. Drubin, D.G.; Nelson, W. Origins of Cell Polarity. *Cell* **1996**, *84*, 335–344. [CrossRef]
84. Juhl, K.; Høy, M.; Olsen, H.L.; Bokvist, K.; Efanov, A.M.; Hoffmann, E.K.; Gromada, J. cPLA2alpha-Evoked Formation of Arachidonic Acid and Lysophospholipids Is Required for Exocytosis in Mouse Pancreatic Beta-Cells. *Am. J. Physiol. Endocrinol. Metab.* **2003**, *285*, E73–E81. [CrossRef] [PubMed]
85. Jones, P.M.; Persaud, S.J. Arachidonic Acid as a Second Messenger in Glucose-Induced Insulin Secretion from Pancreatic Beta-Cells. *J. Endocrinol.* **1993**, *137*, 7–14. [CrossRef] [PubMed]
86. Persaud, S.J.; Muller, D.; Belin, V.D.; Kitsou-Mylona, I.; Asare-Anane, H.; Papadimitriou, A.; Burns, C.J.; Huang, G.C.; Amiel, S.A.; Jones, P.M. The Role of Arachidonic Acid and Its Metabolites in Insulin Secretion from Human Islets of Langerhans. *Diabetes* **2007**, *56*, 197–203. [CrossRef]
87. Ahrén, B.; Magrum, L.J.; Havel, P.J.; Greene, S.F.; Phinney, S.D.; Johnson, P.R.; Stern, J.S. Augmented Insulinotropic Action of Arachidonic Acid through the Lipoxygenase Pathway in the Obese Zucker Rat. *Obes. Res.* **2000**, *8*, 475–480. [CrossRef]
88. Jayaraja, S.; Dakhama, A.; Yun, B.; Ghosh, M.; Lee, H.; Redente, E.F.; Uhlson, C.L.; Murphy, R.C.; Leslie, C.C. Cytosolic Phospholipase A2 Contributes to Innate Immune Defense against Candida Albicans Lung Infection. *BMC Immunol.* **2016**, *17*, 27. [CrossRef] [PubMed]
89. Suram, S.; Silveira, L.J.; Mahaffey, S.; Brown, G.D.; Bonventre, J.V.; Williams, D.L.; Gow, N.A.R.; Bratton, D.L.; Murphy, R.C.; Leslie, C.C. Cytosolic Phospholipase A2α and Eicosanoids Regulate Expression of Genes in Macrophages Involved in Host Defense and Inflammation. *PLoS ONE* **2013**, *8*, e69002. [CrossRef]
90. Clancy, R.M.; Dahinden, C.A.; Hugli, T.E. Complement-Mediated Arachidonate Metabolism. *Prog. Biochem. Pharmacol.* **1985**, *20*, 120–131.
91. Duvall, M.G.; Fuhlbrigge, M.E.; Reilly, R.B.; Walker, K.H.; Kılıç, A.; Levy, B.D. Human NK Cell Cytoskeletal Dynamics and Cytotoxicity Are Regulated by LIM Kinase. *J. Immunol.* **2020**, *205*, 801–810. [CrossRef]
92. Putta, P.; Smith, A.H.; Chaudhuri, P.; Guardia-Wolff, R.; Rosenbaum, M.A.; Graham, L.M. Activation of the Cytosolic Calcium Independent Phospholipase A2 β Isoform Contributes to TRPC6 Externalization via Release of Arachidonic Acid. *J. Biol. Chem.* **2021**, *297*, 101180. [CrossRef] [PubMed]
93. Sheppe, A.E.F.; Edelmann, M.J. Roles of Eicosanoids in Regulating Inflammation and Neutrophil Migration as an Innate Host Response to Bacterial Infections. *Infect. Immun.* **2021**, *89*, e00095-21. [CrossRef] [PubMed]
94. Cheng, Q.; Wang, Z.; Ma, R.; Chen, Y.; Yan, Y.; Miao, S.; Jiao, J.; Cheng, X.; Kong, L.; Ye, D. Lipoxin A4 Protects against Lipopolysaccharide-Induced Sepsis by Promoting Innate Response Activator B Cells Generation. *Int. Immunopharmacol.* **2016**, *39*, 229–235. [CrossRef] [PubMed]
95. Das, U.N. Inflammatory bowel disease as a disorder of an imbalance between pro- and anti-inflammatory molecules and deficiency of resolution bioactive lipids. *Lipids Health Dis.* **2016**, *15*, 1–7. [CrossRef] [PubMed]
96. Herrera, B.S.; Kantarci, A.; Zarrough, A.; Hasturk, H.; Leung, K.P.; Van Dyke, T.E. LXA4 Actions Direct Fibroblast Function and Wound Closure. *Biochem. Biophys. Res. Commun.* **2015**, *464*, 1072–1077. [CrossRef]
97. Babbin, B.A.; Laukoetter, M.G.; Nava, P.; Koch, S.; Lee, W.Y.; Capaldo, C.T.; Peatman, E.; Severson, E.A.; Flower, R.J.; Perretti, M.; et al. Annexin A1 Regulates Intestinal Mucosal Injury, Inflammation, and Repair. *J. Immunol.* **2008**, *181*, 5035–5044. [CrossRef] [PubMed]
98. Das, U.N. Bioactive Lipids as Modulators of Immune Check Point Inhibitors. *Med. Hypotheses* **2020**, *135*, 109473. [CrossRef]
99. Das, U.N. Essential Fatty Acids in Health and Disease. *J. Assoc. Physicians India* **1999**, *47*, 906–911.
100. Poorani, R.; Bhatt, A.N.; Dwarakanath, B.; Das, U.N. COX-2, Aspirin and Metabolism of Arachidonic, Eicosapentaenoic and Docosahexaenoic Acids and their Physiological and Clinical Significance. *Eur. J. Pharmacol.* **2016**, *785*, 116–132. [CrossRef]
101. Morise, T.; Miyamori, I.; Ikeda, M.; Takeda, Y.; Koshida, H.; Yasuhara, S.; Takeda, R. The Role of Prostacyclin (PGI2) in the Regulation of Blood Pressure and Aldosterone Response to Angiotensin II. *Folia Endocrinol. Jpn.* **1983**, *59*, 148–156. [CrossRef]

102. Schölkens, B.A. Antihypertensive Effect of Prostacyclin (PGI2) in Experimental Hypertension and its Influence on Plasma Renin Activity in Rats. *Prostaglandins Med.* **1978**, *1*, 359–372. [CrossRef]
103. Frölich, J.C. Prostacyclin in Hypertension. *Int. Soc. Hypertens.* **1990**, *8*, S73–S78.
104. Das, U.N. Essential Fatty Acids and their Metabolites Could Function as Endogenous HMG-CoA Reductase and ACE Enzyme Inhibitors, Anti-Arrhythmic, Anti-Hypertensive, Anti-Atherosclerotic, Anti-Inflammatory, Cytoprotective, and Cardioprotective Molecules. *Lipids Health Dis.* **2008**, *7*, 37. [CrossRef] [PubMed]
105. Graciano, M.F.R.; Leonelli, M.; Curi, R.; Carpinelli, A. Omega-3 Fatty Acids Control Productions of Superoxide and Nitrogen Oxide and Insulin Content in INS-1E Cells. *J. Physiol. Biochem.* **2016**, *72*, 699–710. [CrossRef]
106. Tanaka, S.; Ito, S.; Shimamoto, C.; Matsumura, H.; Inui, T.; Marunaka, Y.; Nakahari, T. Nitric Oxide Synthesis Stimulated by Arachidonic Acid Accumulation via PPARα in Acetylcholine-Stimulated Gastric Mucous Cells. *Exp. Physiol.* **2021**, *106*, 1939–1949. [CrossRef]
107. Berra-Romani, R.; Faris, P.; Negri, S.; Botta, L.; Genova, T.; Moccia, F. Arachidonic Acid Evokes an Increase in Intracellular Ca^{2+} Concentration and Nitric Oxide Production in Endothelial Cells from Human Brain Microcirculation. *Cells* **2019**, *8*, 689. [CrossRef]
108. Head, B.P.; Patel, H.H.; Insel, P.A. Interaction of Membrane/Lipid Rafts with the Cytoskeleton: Impact on Signaling and Function. *Biochim. Biophys. Acta Biomembr.* **2013**, *1838*, 532–545. [CrossRef]
109. Head, B.P.; Patel, H.; Roth, D.M.; Murray, F.; Swaney, J.S.; Niesman, I.R.; Farquhar, M.G.; Insel, P.A. Microtubules and Actin Microfilaments Regulate Lipid Raft/Caveolae Localization of Adenylyl Cyclase Signaling Components. *J. Biol. Chem.* **2006**, *281*, 26391–26399. [CrossRef] [PubMed]
110. Liu, Z.; Persson, S.; Zhang, Y. The Connection of Cytoskeletal Network with Plasma Membrane and the Cell Wall. *J. Integr. Plant Biol.* **2015**, *57*, 330–340. [CrossRef]
111. Luna, E.J.; Hitt, A.L. Cytoskeleton–Plasma Membrane Interactions. *Science* **1992**, *258*, 955–964. [CrossRef] [PubMed]
112. Di Giaimo, R.; Penna, E.; Pizzella, A.; Cirillo, R.; Perrone-Capano, C.; Crispino, M. Cross Talk at the Cytoskeleton-Plasma Membrane Interface: Impact on Neuronal Morphology and Functions. *Int. J. Mol. Sci.* **2020**, *21*, 9133. [CrossRef]
113. Saarikangas, J.; Zhao, H.; Lappalainen, P. Regulation of the Actin Cytoskeleton-Plasma Membrane Interplay by Phosphoinositides. *Physiol. Rev.* **2010**, *90*, 259–289. [CrossRef] [PubMed]
114. Raucher, D.; Stauffer, T.; Chen, W.; Shen, K.; Guo, S.; York, J.D.; Sheetz, M.P.; Meyer, T. Phosphatidylinositol 4,5-Bisphosphate Functions as a Second Messenger that Regulates Cytoskeleton-Plasma Membrane Adhesion. *Cell* **2000**, *100*, 221–228. [CrossRef]
115. Ammar, M.R.; Kassas, N.; Chasserot-Golaz, S.; Bader, M.-F.; Vitale, N. Lipids in Regulated Exocytosis: What are They Doing? *Front. Endocrinol.* **2013**, *4*, 125. [CrossRef] [PubMed]
116. Das, U.N.; Ramesh, G.; Kumar, G.S.; Madhavi, N.; Kumar, K.V.; Sagar, P.S.; Koratkar, R.; Padma, M. Free Radicals, Lipid Peroxidation and Essential Fatty Acids in Patients with Pneumonia, Septicemia and Collagen Vascular Diseases. *J. Nutr. Med.* **1992**, *3*, 117–127. [CrossRef]
117. Das, U. Essential Fatty Acid Metabolism in Patients with Essential Hypertension, Diabetes Mellitus and Coronary Heart Disease. *Prostaglandins Leukot. Essent. Fat. Acids* **1995**, *52*, 387–391. [CrossRef]
118. Das, U.N. Bioactive Lipids and Vascular Disease. *Eur. J. Clin. Nutr.* **2021**, *75*, 1528–1531. [CrossRef]
119. Das, U. Cross Talk Among Leukocytes, Platelets, and Endothelial Cells and its Relevance to Atherosclerosis and Coronary Heart Disease. *Curr. Nutr. Food Sci.* **2009**, *5*, 75–93. [CrossRef]
120. Cornwell, D.G.; Panganamala, R.V. Atherosclerosis: An Intracellular Deficiency in Essential Fatty Acids. *Prog. Lipid Res.* **1981**, *20*, 365–376. [CrossRef]
121. Das, U.N. Essential Fatty Acids -A Review. *Curr. Pharm. Biotechnol.* **2006**, *7*, 467–482. [CrossRef]
122. Das, U.N. Essential fatty Acids Enhance Free Radical Generation and Lipid Peroxidation to Induce Apoptosis of Tumor Cells. *Clin. Lipidol.* **2011**, *6*, 463–489. [CrossRef]
123. Tateishi, N.; Kakutani, S.; Kawashima, H.; Shibata, H.; Morita, I. Dietary Supplementation of Arachidonic Acid Increases Arachidonic Acid and Lipoxin A4 Contents in Colon, but Does not Affect Severity or Prostaglandin E2 Content in Murine Colitis Model. *Lipids Health Dis.* **2014**, *13*, 30. [CrossRef]
124. Tateishi, N.; Kaneda, Y.; Kakutani, S.; Kawashima, H.; Shibata, H.; Morita, I. Dietary Supplementation with Arachidonic Acid Increases Arachidonic Acid Content in Paw, but Does not Affect Arthritis Severity or Prostaglandin E2 Content in Rat Adjuvant-Induced Arthritis Model. *Lipids Health Dis.* **2015**, *14*, 3. [CrossRef]
125. Kakutani, S.; Ishikura, Y.; Tateishi, N.; Horikawa, C.; Tokuda, H.; Kontani, M.; Kawashima, H.; Sakakibara, Y.; Kiso, Y.; Shibata, H.; et al. Supplementation of Arachidonic Acid-Enriched Oil Increases Arachidonic Acid Contents in Plasma Phospholipids, but Does not Increase their Metabolites and Clinical Parameters in Japanese Healthy Elderly Individuals: A Randomized Controlled Study. *Lipids Health Dis.* **2011**, *10*, 241. [CrossRef] [PubMed]
126. Bathina, S.; Das, U. Resolvin D1 Decreases Severity of Streptozotocin-Induced Type 1 Diabetes Mellitus by Enhancing BDNF Levels, Reducing Oxidative Stress, and Suppressing Inflammation. *Int. J. Mol. Sci.* **2021**, *22*, 1516. [CrossRef] [PubMed]
127. Haworth, O.; Cernadas, M.; Yang, R.; Serhan, C.N.; Levy, B.D. Resolvin E1 Regulates Interleukin 23, Interferon-Gamma and Lipoxin A4 to Promote the Resolution of Allergic Airway Inflammation. *Nat. Immunol.* **2008**, *9*, 873–879. [CrossRef] [PubMed]
128. Bathina, S.; Gundala, N.K.; Rhenghachar, P.; Polavarapu, S.; Hari, A.D.; Sadananda, M.; Das, U.N. Resolvin D1 Ameliorates Nicotinamide-Streptozotocin-Induced Type 2 Diabetes Mellitus by its Anti-inflammatory Action and Modulating PI3K/Akt/mTOR Pathway in the Brain. *Arch. Med. Res.* **2020**, *51*, 492–503. [CrossRef]

129. Bathina, S.; Das, U.N. PUFAs, BDNF and Lipoxin A4 Inhibit Chemical-Induced Cytotoxicity of RIN5F Cells in Vitro and Streptozotocin-Induced Type 2 Diabetes Mellitus in Vivo. *Lipids Health Dis.* **2019**, *18*, 214. [CrossRef]
130. Gundala, N.K.V.; Naidu, V.G.M.; Das, U.N. Arachidonic Acid and Lipoxin A4 Attenuate Alloxan-Induced Cytotoxicity to RIN5F Cells in Vitro and Type 1 Diabetes Mellitus in Vivo. *Biofactors* **2017**, *43*, 251–271. [CrossRef] [PubMed]
131. Harris, W.S.; Mozaffarian, D.; Rimm, E.; Kris-Etherton, P.; Rudel, L.L.; Appel, L.J.; Engler, M.M.; Engler, M.B.; Sacks, F. Omega-6 Fatty Acids and Risk for Cardiovascular Disease: A Science Advisory from the American Heart Association Nutrition SubCommittee of the Council on Nutrition, Physical Activity, and Metabolism; Council on Cardiovascular Nursing; and Council on Epidemiology and Prevention. *Circulation* **2009**, *119*, 902–907.
132. Nelson, G.J.; Schmidt, P.C.; Bartolini, G.; Kelley, D.S.; Kyle, D. The Effect of Dietary Arachidonic Acid on Platelet Function, Platelet Fatty Acid Composition, and Blood Coagulation in Humans. *Lipids* **1997**, *32*, 421–425. [CrossRef] [PubMed]
133. Wilborn, C.M.; Roberts, C.; Kerksick, M.I.; Taylor, L.; Campbell, B.; Kreider, R. Changes in Whole Blood and Clinical Safety Markers over 50 Days of Concomitant Arachidonic Acid Supplementation and Resistance Training. In Proceedings of the International Society of Sports Nutrition Archived, Clearwater Beach, FL, USA, 9–11 June 2006.
134. De Souza, E.O.; Lowery, R.P.; Wilson, J.M.; Sharp, M.H.; Mobley, C.B.; Fox, C.D.; Lopez, H.L.; Shields, K.A.; Rauch, J.T.; Healy, J.C.; et al. Effects of Arachidonic Acid Supplementation on Acute Anabolic Signaling and Chronic Functional Performance and Body Composition Adaptations. *PLoS ONE* **2016**, *11*, e0155153. [CrossRef] [PubMed]

nutrients

MDPI

Review

The Role of Ketogenic Diet in the Treatment of Neurological Diseases

Damian Dyńka [1], Katarzyna Kowalcze [1] and Agnieszka Paziewska [1,2,*]

1 Institute of Health Sciences, Faculty of Medical and Health Sciences, Siedlce University of Natural Sciences and Humanities, 08-110 Siedlce, Poland
2 Department of Neuroendocrinology, Centre of Postgraduate Medical Education, 01-813 Warsaw, Poland
* Correspondence: agnieszka.paziewska@uph.edu.pl

Abstract: Over a hundred years of study on the favourable effect of ketogenic diets in the treatment of epilepsy have contributed to a long-lasting discussion on its potential influence on other neurological diseases. A significant increase in the number of scientific studies in that field has been currently observed. The aim of this paper is a widespread, thorough analysis of the available scientific evidence in respect of the role of the ketogenic diet in the therapy of neurological diseases such as: epilepsy, Alzheimer's disease (AD), Parkinson's disease (PD), multiple sclerosis (MS) and migraine. A wide range of the mechanisms of action of the ketogenic diet has been demonstrated in neurological diseases, including, among other effects, its influence on the reduction in inflammatory conditions and the amount of reactive oxygen species (ROS), the restoration of the myelin sheath of the neurons, the formation and regeneration of mitochondria, neuronal metabolism, the provision of an alternative source of energy for neurons (ketone bodies), the reduction in glucose and insulin concentrations, the reduction in amyloid plaques, the induction of autophagy, the alleviation of microglia activation, the reduction in excessive neuronal activation, the modulation of intestinal microbiota, the expression of genes, dopamine production and the increase in glutamine conversion into GABA. The studies discussed (including randomised controlled studies), conducted in neurological patients, have stressed the effectiveness of the ketogenic diet in the treatment of epilepsy and have demonstrated its promising therapeutic potential in Alzheimer's disease (AD), Parkinson's disease (PD), multiple sclerosis (MS) and migraine. A frequent advantage of the diet was demonstrated over non-ketogenic diets (in the control groups) in the therapy of neurological diseases, with simultaneous safety and feasibility when conducting the nutritional model.

Keywords: ketogenic diet; neurological diseases; epilepsy; Alzheimer's disease (AD); Parkinson's disease (PD); multiple sclerosis (MS); migraine; brain; neurone; neuroinflammation; ketone bodies; ketogenic; neurotransmitters; neuroplasticity; treatment; prevention; inflammatory; anti-inflammatory; low carb; high fat; nutrition

Citation: Dyńka, D.; Kowalcze, K.; Paziewska, A. The Role of Ketogenic Diet in the Treatment of Neurological Diseases. *Nutrients* **2022**, *14*, 5003. https://doi.org/10.3390/nu14235003

Academic Editor: Sareen Gropper

Received: 28 October 2022
Accepted: 21 November 2022
Published: 24 November 2022

1. Introduction

The extreme and dynamic increase in incidences of neurological diseases (including neurodegenerative ones) constitutes a real problem for many millions of people worldwide. This phenomenon is all the more so alarming when taking into account the results of the Global Burden of Disease (GBD) study. It has been shown, indeed, that widely understood neurological diseases are the second most frequent cause of death worldwide [1]. Moreover, they are the most frequent cause of disability and, therefore, an increase in the DALY (disability-adjusted life years) index, i.e., years of life lost due to health damage or preterm death. For comparison, attention is called to the fact that the problem is far greater than the declared COVID-19 pandemic, and for that reason, it is worthwhile to take a more detailed look at this issue [2]. It seems justified to pose the question of whether this can be associated with an even greater epidemic of neurological diseases in the coming years. Although possible post-COVID-19 complications in the form of neurological disorders (including neurodegenerative ones) have been reported [3–5], it has been, however, demonstrated that a positive COVID-19 test result causes no greater risk of neurological disease occurrence than other respiratory tract infections. One exception is ischaemic stroke [6], but the possibility of a twice higher risk of epilepsy development has also been suggested [7]. Taking into account the increasing incidence of neurological diseases, a need is seen for a complex approach, looking for new solutions or an analysis of the already existing ones in order to refine or draw new conclusions based on the most recent data. One of the ways of such a complex therapeutic approach to neurological diseases is the nutritional aspect. Taking into account over a hundred years of study and clinical practice, the ketogenic diet is worth particular attention in this respect.

The ketogenic diet is a method of nutrition leading to the increased production of ketone bodies (β-hydroxybutyrate, acetoacetate and acetone) in the organism and, thus, to a condition of ketosis. This effect comes about by obtaining the greatest energy share from fats and minimising the consumption of carbohydrates. Importantly, a ketosis condition can also be obtained, among other methods, through fasting or the observation of a very low-calorie diet (not necessarily with a predominance of fats); however, that condition will not be nutritional ketosis [8,9]. Keeping in mind this minor reservation, it is difficult to refine a universal definition of the ketogenic diet, although the most accurate definition seems to be the one mentioned above. Transforming the theoretical considerations into practice, the diet can be expressed as a share of the macronutrients within the following range: fats 60–90% (usually 70–75%) of the whole energy content of the diet, carbohydrates below 50 g daily (which usually accounts for 5–10% of the whole energy content of the diet) and protein 1.0–1.2 to 1.7 g per kg body weight (which usually accounts for about 20% of the whole energy value of the diet) [10–15]. The ketogenic diet, as a rule, should imitate a fasting condition in the body, leading, however, to no negative effects of starvation. Both fasting and the observation of the ketogenic diet lead to increased oxidation of fatty acids and the use of the ketone bodies produced in that process (β-hydroxybutyrate, acetoacetate and acetone) as the main energy substrate. Therefore, the diet distinguishes itself from other diets since it is not based on energy obtained from glucose but on that from ketone bodies produced from fats [8,16–20].

The property of mimicking a fasting condition is of particular importance in the context of the subject of this paper. This is associated with the fact that even before the "creation" of the ketogenic diet, a fast was proposed in order to help neurology patients with epilepsy, after which an evident improvement was observed. That practice has a very long tradition in history since it can be found that many sources, even those from the 5th century before Christ, have described the favourable effects of fasting in cases of epilepsy. They suggested that fasting is an effective method of treatment for epileptic seizures [21]. Other historical data mention that Hippocrates suggested a need for the limitation of calories in the treatment of epilepsy [22]. Some mentions of the topic can also be found in biblical texts. The Gospel according to Saint Mark describes how Jesus cured a boy of epilepsy, saying that it could be treated with fast and prayer. Interestingly, the whole scene of the cure is shown in the painting "The Transfiguration" by Raphael [23–25]. In the less distant past, i.e., in 1911, doctors from Paris reported the benefits resulting from the observation of fasting during the treatment of epilepsy based on their own medical practice [26]. The 20th century was the time at which fasting and epilepsy were increasingly and frequently the subjects of scientific research and not only clinical practice [22]. The real breakthrough was the moment at which it was discovered that a ketogenic diet could show fast-like therapeutic properties in epilepsy but without the negative consequences of starvation (i.e., malnutrition). The paper by Wilder in 1921, which was the beginning of ketogenic diet use in the treatment of epilepsy, can serve as a point of reference [22,27,28]. The proportions of macronutrients in a clinical ketogenic diet were then proposed as 10–15 g of carbohydrates daily and 1 g of protein per kg body weight; the rest of the energy share should be provided by fats [29]. The extremely favourable results observed meant that from that time on, the use of the ketogenic diet in epilepsy became increasingly widespread, to such an extent that it was recommended in textbooks as the standard therapy of that disease [25]. Over the years, the diet has been increasingly extensively studied in various diseases; however, its main range of use for over a hundred years has included neurological diseases (initially used in epilepsy). Taking into account its primary use and the many years of practice, it seems justified to study its potential application in other neurological diseases (including neurodegenerative ones). The perspective of over a hundred years since its breakthrough application means that a meticulous analysis of current knowledge concerning not only epilepsy but also other neurological diseases seems to be of great importance for the development of science in this respect. The intensification of studies on the effect of the ketogenic diet on neurological diseases is well reflected in Figure 1 showing the number of publications in the PubMed search engine after entering the words "ketogenic diet neurological disease". A significant increase in the number of publications can be seen at the beginning of the 21st century, with a significantly growing tendency in the subsequent years.

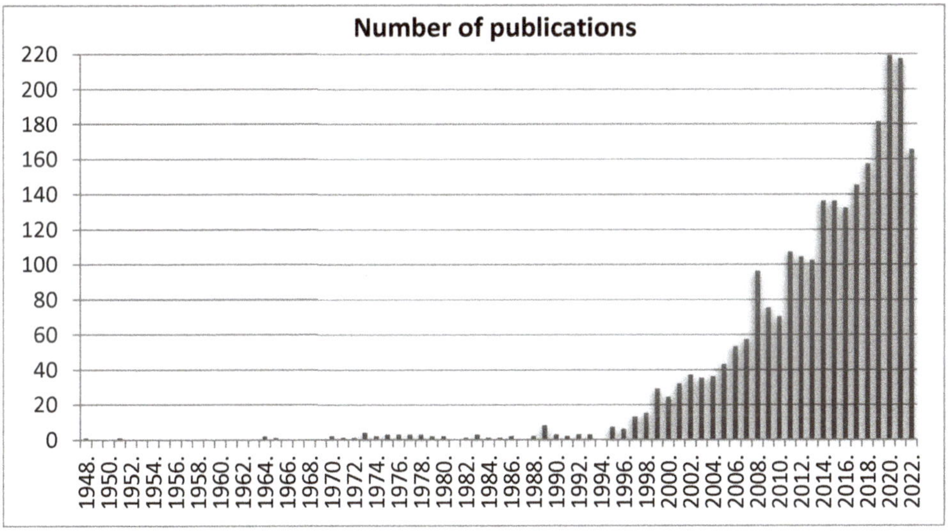

Figure 1. The number of publications for the entry "ketogenic diet neurological disease" in the PubMed base in the period from 1 January 1948 to 17 November 2022. Date of search: 17 November 2022.

2. The Range and Mechanisms of the Ketogenic Diet Effect in Neurological Diseases

The potential mechanisms of the ketogenic diet are multifaceted. The extremely wide range of its effect on neurological diseases is described in a publication of 2021, in which the authors thoroughly analysed 170 studies and described the range of its effect in 14 various aspects, e.g., on neuroprotection, neuroplasticity, neuroinflammation, function of neurotransmitters, epigenetics, nociception, changes in cell energetics and metabolism, and other aspects [30]. An important piece of information is the fact that the brain can function excellently, taking energy from ketone bodies as the energy substrate [31]. Frequently, that is the only option for energy provision when glucose is not readily available [32]. The cerebral cells, in fact, contain monocarboxylate transporters (MCTs), through which ketone bodies are transported in order to provide energy, similar to many other body cells [33]. The wide and, in the first place, favourable range of the ketogenic diet effect was presented in a widespread meta-analysis in 2021, based on 49 animal studies from the years 1979–2020. Strong neuronal protection was demonstrated against acute central nervous system damage as a result of a reduction in the death rate, damage and dysfunction of neurons [34]. Moreover, the frequently used calorie deficit, together with a ketogenic diet, also shows an additional neuroprotective potential. It increases, in fact, the number of neuroprotective factors, i.e., the brain-derived neurotrophic factor (BDNF) and the glial cell line-derived neurotrophic factor (GDNF), neutrophin-3 (NT-3), and molecular chaperones. The deficit also improves mitochondrial function (and thus increases the efficiency of energy production and reduces reactive oxygen species (ROS) production). It also shows an anti-inflammatory potential in the result of, among other effects, the inhibition of the activities of cyclooxygenase-2 (COX-2) and inducible nitric oxide synthase (iNOS) and in the result of a blockade of the synthesis of proinflammatory interleukins (IL-1β, IL-2, IL-4, IL-6) and tumour necrosis factor alpha (TNFα). It also reduces the level of the central component of the inflammatory process, i.e., transcriptional nuclear factor kappaB (NFκB) [35,36].

Many neurological diseases (including neurodegenerative ones) are characterised by glucose metabolism disturbances in the neurons. Although not the only one, the most characteristic example is traumatic cerebral damage, which includes brain energetic collapse due to major mechanical trauma. The group at particular risk are sportsmen practising contact sports (i.e., martial arts, American football). They are thus at risk of repeated brain concussion episodes, which soon lead to chronic post-traumatic encephalopathy. Extremely important is the fact that in the situation of brain trauma and energetic collapse, the number of MCT channels (which transport ketone bodies to cells) in the brain cells increases by 85%; the number of β-hydroxybutyrate-metabolising enzymes also increases [37–42]. Based on that, we can say that the brain demands another source of energy, which can be provided by ketone bodies. Most recent studies also show that in patients with traumatic cerebral damage, in whom a ketogenic diet was applied, no clinical adverse effects were noted, and its safety and applicability were found [43]. The application of the ketogenic diet, through the mediation of β-hydroxybutyrate, can reduce demyelination, the death of oligodendrocytes (producing myelin) and the degeneration of axons caused by glucose deficiency [44]. The widely understood damage can also concern the mitochondrial structures at the level of respiratory chain complexes. It has been demonstrated that β-hydroxybutyrate, on one hand, provides components participating in the reconstruction of the respiratory chain; on the other, energy acquisition from ketone bodies is possible even in the case of damage to the first complex of the respiratory chain [45,46]. For brain functioning, even in healthy individuals, an increase in β-hydroxybutyrate concentration can be far more favourable in comparison to glucose. This has been confirmed in a study in which hydroxybutyrate infusion in healthy individuals (until 5.5 mmol/L concentrations are obtained), compared with the lack of such infusions, reduced cerebral glucose utilisation by 14% and caused an increase of cerebral blood flow by as much as 30%, with unchanged oxygen consumption. The authors directly suggest a possibility of the neuroprotective effect of ketone bodies [47].

In many neurodegenerative diseases, the problems include the development of amyloid plaques, which, in turn, are strongly correlated with high glucose concentration, diabetes mellitus and insulin resistance [48,49]. Such problems most frequently accompany high-carbohydrate diets [50]. Taking into account the strongly hypoglycaemic effect of a ketogenic diet and the evident insulin-concentration-reducing effect [51,52], it seems reasonable that a ketogenic diet would have a favourable effect on the prevention of the deposition of amyloid plaques or a reduction in their number. It has been found that studies have actually demonstrated such an effect of the diet, which, however, goes beyond the effect on glycaemia and may be associated, i.a., with autophagy as well. The condition of ketosis can promote macroautophagy in the brain through the activation of sirtuin 1 (SIRT1) and hypoxia-induced factor 1α (HIF-1α) and the inhibition of the mTORC1 complex. This can prevent neurodegenerative disorders, among other effects, through the elimination of protein aggregates or damaged mitochondria [53,54]. It thus prevents the accumulation of autophagosomes and improves the survival of cortical neurons, leading to a reduction in cerebral damage [55–57]. Another study has also revealed that ketogenic diets can effectively reduce the amount of beta-amyloid and also other redundant by-products of metabolism in cerebral tissue [58].

Among other potential mechanisms of influence of the ketogenic diet on the nervous system and neurological diseases, its anti-inflammatory activity can be mentioned, which is multifaceted [59]. A neuronal inflammatory condition can result, i.a., from trauma, ischaemia, degeneration or infection [60]. The ketogenic diet can exert effects on the regulation of both central and peripheral inflammatory mechanisms [61]. Its effect has been shown in a reduction in microglia activation and the decreased expression of proinflammatory cytokines, i.a., IL-1β, IL-6 and TNF-α in the hippocampus. Moreover, it can inhibit neuritis through the suppression of the cyclooxygenase-2 (COX-2)-dependent pathway by peroxisome proliferator-activated receptor γ (PPARγ) activation [62–64]. Moreover, astrocytes (glial cells) present in the brain have the properties of ketone body production, and for that reason, it is suggested that this fact can exert a neuroprotective effect [65]. The possible (only anti-inflammatory) mechanisms of ketogenic diet action are much more numerous [66,67]; therefore, their anti-inflammatory potential possibly results from their synergy.

We also discuss the possible effect of a ketogenic diet on the blood–brain barrier (BBB). Although no unequivocal evidence is currently available, based on current knowledge, it is justified to look at that topic from a future-oriented perspective. Similar to the intestinal barrier, here, a break of membrane integrity can also occur. This happens, among other conditions, in epilepsy, Alzheimer's disease (AD) and other neurological diseases [68–70]. In the period of an increased concentration of ketones in the blood, the permeability of the blood–brain barrier for β-hydroxybutyrate also increases [71]. The condition of ketosis can cause the restoration of the integrity of the blood–brain barrier as a result, i.a., of an increased content of connexin-43 (Cx43) in building the barrier and of monocarboxylate transporters (MCTs) and (glucose transporter) GLUT transporters [8,72,73]. Additionally, the ketogenic diet promotes the outflow of the above-mentioned amyloid plaques across the blood–brain barrier because it increases the concentration of proteins participating in the elimination of amyloid plaques, i.e., LDL receptor-related protein 1 (LRP1), glycoprotein P (P-gp) and phosphatidylinositol binding clathrin assembly protein (PICALM) [74].

Another possible mechanism of ketogenic diet activity in the nervous system includes, among other effects, an ability to change the cerebral metabolism of glutamine. As Yudkoff et al. believe, the condition of ketosis can intensify the metabolism of astrocytes, which results in the increased conversion of glutamate into glutamine. This can contribute to a reduction in the main stimulatory neurotransmitter (glutamate) concentration, with a simultaneous increase of the main inhibitory neurotransmitter (GABA) level, resulting in the intensified conversion of glutamine into GABA [75,76]. Thus, a decrease in excitotoxicity and greater mood improvement occur as the result of physiological nervous quietening. On the other hand, excitotoxicity inhibition, increased resistance to stress and an effect on synaptic plasticity can also result from the β-hydroxybutyrate effect increasing mitochondrial respiration, leading to a change in brain-derived neurotrophic factor (BDNF) expression [77]. That is one of the reasons why in individuals on ketogenic diets, frequently, a greater quietening, better concentration, mood improvement and an increase in cognitive abilities are observed [78,79].

The remaining possible favourable mechanisms of the ketogenic diet's influence on the nervous system include, i.a., the effect on mitochondria (particularly important in neurological diseases), which has been increasingly widely discussed in the latest publications [80]. Apart from the earlier described indirect effect on mitochondria, through action at the gene level, the diet shows an ability to promote the biogenesis of mitochondria in the hippocampus [81]. Another mechanism, increasingly frequently mentioned in publications, is the effect of the ketogenic diet on the prevention and treatment of neurological diseases (including neurodegenerative ones) through a change of intestinal microbiota [82]. This results from a possible influence of intestinal dysbiosis on the development of such diseases, while on the other hand, it is known that a ketogenic diet significantly affects the remodelling of the intestinal microbiota [83]. Thus, increasingly, the subject of the potential indirect favourable effect of the ketogenic diet on neurological diseases through the modulation of intestinal microbiota is discussed. This is all the more justified when taking into account the direct connection of the intestine with the brain by the gut–brain axis, through which microbiota can affect brain processes (that concern, i.a., neurotransmission) and vice versa [84]. The ketogenic diet can also affect the processes of neurogenesis, that is, brain regeneration, the development of new nervous cells and their linking in neuronal networks [85,86]. Potential mechanism of ketogenic diet effect in neurological diseases are illustrated in Figure 2. Keeping in mind all the mentioned processes and the remaining mechanisms of the possible effect of the ketogenic diet on the prophylaxis and treatment of neurological diseases (including neurodegenerative ones), the arguments suggesting an unquestionable need for further science development in this field and an extension of the knowledge in this respect seem to be well-grounded. This would allow us to discover the existing (including those not yet known) neuroprotective mechanisms of the ketogenic diet.

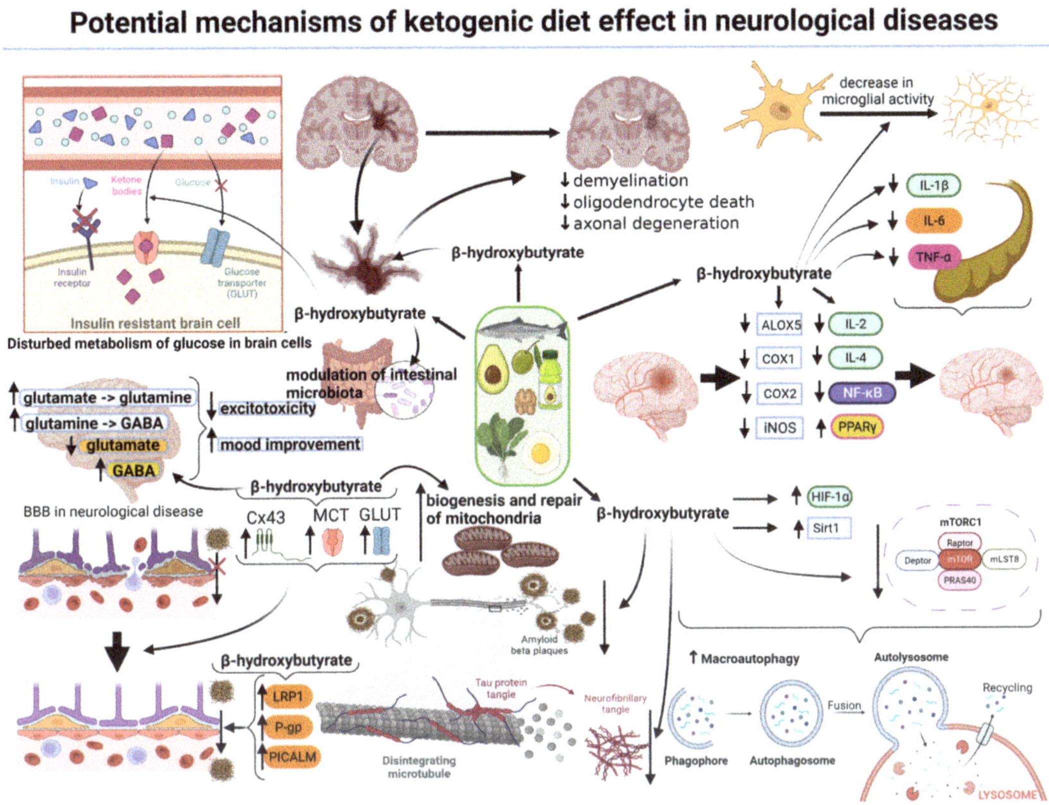

Figure 2. Potential mechanisms of the ketogenic diet effect in neurological diseases. GABA: Gamma-Aminobutyric Acid; BBB: blood–brain barrier; Cx43: connexin-43; MCT: monocarboxylate transporters; GLUT: glucose transporters; LRP1: LDL receptor-related protein 1; P-gp: glycoprotein P; PICALM: phosphatidylinositol binding clathrin assembly protein; IL-1β: interleukin-1β; IL-6: interleukin-6; TNF-α: tumor necrosis factor α; ALOX5: arachidonate 5-lipoxygenase gene; COX1: cyclooxygenase 1; COX2: cyclooxygenase 2; iNOS: inducible nitric oxide synthase; IL-2: interleukin-2; IL-4: interleukin 4; NF-κB: nuclear factor kappa-light-chain-enhancer of activated B cells; PPARγ: peroxisome proliferator-activated receptor γ; HIF-1α: hypoxia-induced factor 1α; Sirt1: sirtuin 1; mTORC1: mammalian target of rapamycin complex 1; mLST8: mammalian lethal with SEC13 protein 8; PRAS40: proline-rich Akt substrate of 40 kDa; mTOR: mammalian target of rapamycin. The above figure was created with BioRender.com, accessed on 23 November 2022. Agreement number: TF24OJVLEL.

3. The Role of the Ketogenic Diet in the Treatment of Epilepsy

Epilepsy is a neurological disorder associated with constant recurrent seizure attacks. The disease itself is not fraught with a great risk of death, and in most patients, the prognosis (measured based on the absence of seizures) is favourable. It is, however, associated with a significant impairment in the quality of life. Although it concerns all age groups (and either sex), it is one of the most frequent neurological disorders of childhood (it occurs slightly more frequently in males). About 50 million people struggle with the disease worldwide. The mean incidence of active epilepsy is 6.38 cases per 1000 population. It affects, to a greater extent, the populations of countries with low and medium incomes. Hence, millions of people worldwide are affected. It is worth stressing, however, that not all individuals with seizures have epilepsy since seizures can also occur, e.g., due to acute injury to the central nervous system [87–90]. Currently, as a rule, epileptic patients are prescribed drugs, which, however, give no benefit to some patients diagnosed with so-called drug-resistant epilepsy. This concerns about one-third of all epileptic patients, including 7–20% of children and 30–40% of adults [91,92]. In such cases, a ketogenic diet may be the only rescue, which has been confirmed, among other studies, by an extensive meta-analysis in 2020 [93].

Although a hundred years ago, the ketogenic diet was the main therapeutic option for epilepsy, after 1940, pharmacological treatments became the main therapeutic method. An antiepileptic drug called Dilantin superseded the use of the ketogenic diet to a significant extent. Its effectiveness in reducing epileptic seizures caused doctors to be more willing to reach for that method of therapy, less and less frequently ordering ketogenic diets [25]. That resulted, of course, from obtaining similar results with a lower amount of work (since the institution of the diet requires a significantly wider medical approach than prescribing a drug). As mentioned above, pharmacotherapy has, however, failed and still fails in about 1/3 of cases. For that reason, the ketogenic diet, after over a hundred years, still is one of the standard therapeutic options in the case of drug-resistant epilepsy when pharmacotherapy fails. A renaissance of the ketogenic diet was observed in 1997 after the movie "First do no harm", presenting the experience of Charlie Abrahams (son of Jim Abrahams, a Hollywood film producer) with the ketogenic diet, which proved to be the last resort for epileptic seizures in the boy [14,22]. The available data show that the ketogenic diet has the properties of reducing (to a smaller or greater extent) epileptic seizures, frequently by several score percentage points or 50–90% of patients; a 90% reduction in seizures may be experienced by about 27% of patients. In some patients, complete remission was observed [94–99]. In 2022, an extensive study was conducted on 160 pediatric patients (mean age five years and nine months) with epilepsy treated with the ketogenic diet, and the effects were monitored for 3, 6, 12 and 24 months after its institution. An absence of seizures was observed (depending on the duration) in 13.7% of children after three months, 12.5% of children after six months, 14.4% after 12 months and 10.6% after 24 months. On the other hand, a reduction of ≥50% was observed in 41.9% of children after three months, 37.5% after six months, 28.7% after 12 months and 16.2% after 24 months [100]. A meta-analysis of 2022, based on the studies conducted on epileptic children, demonstrated a reduction in the number of seizures by at least 50%, together with a complete absence of seizures in as many as 48.31% of children. It was demonstrated that children on a ketogenic diet had a 5.6 times greater chance for seizure frequency reduction by at least 50% compared with a control group [101]. Another extensive study revealed that in patients observing a ketogenic diet for epilepsy, a reduction in seizure frequency by ≥50% occurred in 35–56.1% of the participants compared with 6–18% in the control group [102]. In 2020, a meta-analysis focused on the effect of the ketogenic diet on babies (<2 years) with epilepsy, based on a total number of 534 subjects. It was demonstrated that in as many as 33% of cases, a complete regression of seizures occurred, while in 59% of the babies, a significant reduction of seizure episodes by ≥50% was achieved. The authors have concluded that based on the results obtained, it can be said that the ketogenic diet is effective and safe in babies with drug-resistant epilepsy [103]. Currently, instead of the classic ketogenic diet, increasingly frequently, the modified Atkins diet (MAD) is used, which also is a model of the ketogenic diet. In one meta-analysis, the two nutritional models were compared in respect of better effectiveness. In the group of patients on the classic ketogenic diet, the reductions in seizure frequency by ≥50% after 1, 3, 6, 12 and 24 months were (as a percentage of the patients) 62%, 60%, 52%, 42% and 46%, respectively, while after 1, 3, 6 and 12 months in the patients using the modified Atkins diet, the percentages were 55%, 47%, 42%, and 29%, respectively. In view of that, the authors concluded that the effectiveness of both models was similar [104]. The authors of another meta-analysis drew a similar conclusion. They mentioned, however, a slight advantage of the modified Atkins diet in respect of the lower number of potential adverse events. Furthermore, the whole effect can also be indirectly influenced by the fact that MAD is, as a rule, tastier, thus encouraging better compliance with is requirements [93].

Two other meta-analyses also demonstrated similar results [102,105]. The ketogenic diet is also frequently prescribed in a version enriched with medium-chain triglycerides (MCTs) in view of their high ketogenicity. This was checked in one randomised controlled study, and it was revealed that the therapeutic effects were similarly favourable in both versions of the diet. The mean percentage of initial seizures was similar in the third, sixth and twelfth months. In the classic version of the diet, in the third month, it reached 66.5%; in the 6th month—48.5% and in the 12th month—40.8%. In the MCT-enriched version, it was 68.9% in the third month, 67.6% in the sixth month and 53.2% in the twelfth month [106]. The possible mechanisms of action of the ketogenic diet and the reduction in the number of epileptic seizures, in spite of over a hundred years of studies in this respect, are still not fully elucidated. Some potential mechanisms are, however, known and have an influence on the reduction of seizures; it is possible that they contribute to the observed therapeutic effect through synergistic actions [14,107]. The first of them is the possible anticonvulsant effect of the ketone bodies themselves, which was demonstrated in some studies [108–112] but not confirmed in other studies [113,114]. A study of 2019 has shed new light on that issue, presenting new molecular foundations of the anticonvulsant action of ketone bodies. It turns out that the anticonvulsant effectiveness of the ketogenic diet is positively correlated with the serum concentration of β-hydroxybutyrate, which has an ability to directly activate the potassium voltage-gated channel subfamily Q (KCNQ1/3) channels [115]. Another potential mechanism of ketone bodies' action concerns the effect on neuronal metabolism (including mitochondrial function) and synaptic function. That mechanism postulates that it is glucose, which is readily available for neurons, the diffusion of which through the blood–brain barrier is quite easy (since the transport takes place in the endothelium of cerebral capillaries), that is indispensable for the initiation of neuronal seizure activity [116]. The ketogenic diet is supposed to reduce seizure frequency by decreasing the availability of glucose and, hence, energy and, in particular, the rate of its consumption by neurons. That has been confirmed in studies demonstrating a direct effect of glucose on the seizure threshold [117,118]. Some effects can be exerted by medium-chain triglycerides (MCTs), which are particularly frequently and willingly used in ketogenic diets. Apart from the properties of increasing the concentration of ketones in the body, they have the ability to increase decanoic acid concentration in plasma. The acid directly inhibits the α-amino-3-hydroxy-5-methyl-4-isoxazolepropionic acid receptor (AMPA) and, owing to that, shows an even better anticonvulsant effect than ketones [81,119,120]. In a publication in 2020, the extraordinary effectiveness of medium-chain fatty acids was confirmed since the addition of MCT oil to diets twice daily for three months was sufficient to reduce the number of seizures by 42% in adult patients with incurable epilepsy [121]. In the state of ketosis, an increase in mitochondrial metabolism is also observed, leading to an increase in ATP production, which activates ATP-sensitive potassium channels (KATPs), reducing, in turn, neuronal excitability [81,119,120]. It has also been demonstrated that one of the ketone bodies, that is, acetoacetate (Acac), inhibits the voltage-dependent Ca^{2+} channels (VDCCs) and decreases excitatory postsynaptic currents (EPSCs) at sites showing epileptic activity and, thus, inhibits convulsions in vivo [122]. Another potential mechanism is supposed to be the effect of ketones increasing the synthesis of the neurotransmitter GABA (through a reduction in aspartate concentration), the higher concentration of which can exert an inhibitory effect on convulsion initiation, and the synthesis of the neurotransmitter A1 adenosine, which can also show some anticonvulsant activity, reducing as well the level of excitatory glutamate [123–126]. Other possible mechanisms include the modulation of intestinal microbiota [127], the reduction of proinflammatory cytokine levels [128], an epigenetic effect of β-hydroxybutyrate through the inhibition of class I histone deacetylases and an effect on the transcription of genes (particularly those associated with the widely understood antioxidant factors) [129,130]. The most recent publications also discuss the above-mentioned mechanisms and suggest that, most likely, a synergy of all of them enables the antiepileptic effect of the ketogenic diet [131].

4. The Role of the Ketogenic Diet in the Therapy of Alzheimer's Disease (AD)

Alzheimer's disease (AD) is the most frequent cause of dementia worldwide. It is estimated that 50–70% of dementia cases are caused by this disease. The problem of dementia is very serious since it is estimated that by 2050, its incidence will triple. Currently, about 10 million new cases are noted annually. Taking into account the approximate current number of over 55 million people affected by the disease worldwide, its tripled incidence in a not-too-distant future seems to foreshadow a real calamity [132,133]. Alzheimer's disease (AD) develops for many years, frequently not showing earlier any specific symptoms. Therefore, the use of adequate early diagnostic procedures and the institution of therapeutic management are frequently difficult. Early symptoms in the form of minor memory disturbances are frequently not regarded as an onset of Alzheimer's disease (AD). With time, as a result of progressing brain neurodegenerative processes, the disturbances exacerbate, starting a new stage of disease advancement. Among other symptoms, problems occur with performing everyday activities, memorizing the meaning of words, disorientation and disorders of mood and sleep. As a result, the patient affected by the disease is frequently unable to function unassisted due to the occurrence of neurological and psychiatric symptoms [134–136]. The mortality rate due to Alzheimer's disease (AD) has increased between 2000 and 2018 by as much as 146.2%. The disease has become the fifth most frequent cause of death among elderly people in America, which illustrates the scale of the problem [137]. In the brain of patients with Alzheimer's disease (AD), increased amounts are found of β-amyloid (βA) and hyperphosphorylated tau protein (tau-p), which aggregate to form intracellular neurofibrillary tangles [138]. The factors contributing to Alzheimer's disease (AD) development are multifaceted. They include, among other factors: depression and long-term stress, diabetes mellitus, hypertension, dyslipidaemia, obesity, cardiovascular diseases, traumatic cerebral injury, hyperhomocysteinaemia, oral cavity diseases, loss of hearing, sleep disorders, low physical activity, tobacco smoking, alcohol consumption, vitamin D deficiencies, inadequate diet, air pollution, poor education level and avoidance of social contacts [135].

The ketogenic diet can exert a favourable effect on Alzheimer's disease (AD) in many mechanisms [139]. First, it is worth mentioning that the disease is frequently called type 3 diabetes mellitus, which involves the brain. This results from the fact that it shows molecular and biochemical features that are observed in type 1 and type 2 diabetes. Moreover, diabetic patients are at a significantly higher risk of AD development [140,141]. An important argument to support this is the fact of brain insulin resistance occurrence in patients with the disease. Insulin resistance developing in the brain of AD patients and disturbed signalling processes are well-grounded in the literature [141–143]. Socio-economic progress has contributed to people increasingly choosing processed products that belong to the western diet pattern. This, undoubtedly, is the cause of the development of not only diabetes mellitus but also, as demonstrated in a study in 2022, of AD, leading to brain insulin resistance and the progression of neurodegenerative processes [144]. In the course of AD, a reduction is observed in the amounts of GLUT1 and GLUT3, the two main glucose transporters in the brain. This is correlated with tau protein hyperphosphorylation and the density of neurofibrillary tangles in the brain, i.e., the typical signs of AD [145]. Insulin resistance and a reduced amount of glucose transporters in the brain cause neurons to have significantly hindered access to energy sources, so brain functioning disturbances are not surprising. A then-instituted, adequately composed and customised ketogenic diet can, in such cases, exert its effect on at least two main domains. On one hand, an induction of the ketosis state and an increase in ketone bodies concentration would provide an alternative source of energy for the brain (ketone bodies). In view of that, energy generation from glucose (problematic in the brain of AD patients) will not be indispensable anymore for the normal work of the brain since it can function excellently using ketone bodies, the transport of which (contrary to glucose) into the brain is not impaired in AD patients [146,147]. On the other hand, the ketogenic diet (particularly in combination with calorie deficits) would act as a causal treatment. It shows the ability to reduce insulin and glucose concentrations, and thus, insulin resistance, which is the cause of brain function disorders in AD, will be consistently reduced. The ketogenic diet can reduce insulin resistance through a number of mechanisms [52,148,149]. The most common method of determination of insulin resistance is the calculation of the Homeostatic Model Assessment–Insulin Resistance (HOMA-IR) index. Many studies have demonstrated a significant reduction of the index value due to the application of the ketogenic diet, which is a direct confirmation of its effect on insulin resistance. It was demonstrated, among other findings, that after 12 weeks, the HOMA-IR index value was reduced by 62.5% (from the initial 3.73 value to 1.4) [150]. In another study, just four weeks were enough to reduce the index by almost half (45.9%) [151]. The effect of the ketogenic diet on disturbed brain bioenergetics in AD and possible mechanisms of reduction of amyloid plaques has been increasingly suggested [152]. Animal studies have already demonstrated that, compared with a standard diet, the ketogenic diet results in the reduced deposition of amyloid plaques in the hippocampus, the decreased activation of the microglia and the improvement of cognitive functions, including learning and spatial memory [67]. In another study, a 25% reduction in amyloid plaques was found in mice with AD after the application of the diet [58]. The ketogenic diet can also act by exerting an effect on the expression of genes associated with neurodegenerative diseases such as AD, including genes associated with the metabolism of the hippocampus, and it prevents disorders of oxidative phosphorylation [153–155]. In AD, the function of mitochondria is also abnormal, leading to a gradual loss of their ability to produce energy. That phenomenon is related to inflammatory processes and the accumulation of amyloid plaques. It was demonstrated, however, that a ketogenic diet is able to induce the formation of new mitochondria through the activation of mitogenesis-regulating pathways. It also reduces the inflammatory condition in the brain, resulting from, among other factors, the excessive production of reactive oxygen species (ROS) by dysfunctional mitochondria [156–159]. Evidence shows that diet supplementation with medium-chain fatty acids from MCTs, similar to this case, seems to be extremely helpful, and this has been mentioned in an increasing number of publications [36]. The fatty acids show high ketogenicity, and the body is able to transform them in a simple process into ketones. A study in 2018 in patients with mild and moderate AD demonstrated that supplementation with a 30 g daily dose of MCTs contributed to the doubling of the uptake of ketones in the brain and increasing the brain's total energy metabolism. Ketones produced from MCTs compensated for glucose deficiency in the brains of AD patients, proportional to the concentration of ketones in plasma [160]. MCT oil can maintain or improve cognitive functions in AD patients in a significant majority of cases, at about 80% [161]. It has also been demonstrated that supplementation with MCTs contributed to, among other results, an improvement of the working memory and cognitive functions in patients with AD as well as in individuals without dementia [162,163]. In an animal model, the effect of MCTs was also demonstrated on mitochondrial function improvement and the alleviation of the unfavourable action of β-amyloid on cortical neurons and the reduction of its total amount [58,164–167]. Moreover, the use of MCT oil seems helpful since it facilitates the maintenance of the high ketogenicity of the diet, even with a slightly increased consumption of carbohydrates (in relation to the ketogenic diet without MCT oil) [168]. A double-blind, placebo-controlled study also demonstrated that increased serum β-hydroxybutyrate concentrations contributed to an improvement in cognitive functions and memory [169]. In the case of AD, supplementation with exogenous β-hydroxybutyrate for 20 months also produced an improvement in cognitive functions, mood and everyday functioning [170]. The first randomised controlled study in patients with an unequivocal diagnosis of AD, assessing the effect of the ketogenic diet on the disease, was published in 2021. It compared the effect of the diet with that of a standard diet based on low-fat content. Compared with the low-fat diet, in the group of patients on the ketogenic diet, an improvement in cognitive function by 2.12 ± 8.70 points on the Addenbrooke's Cognitive Examination III (ACE-III) scale, an improvement in everyday functioning by 3.12 ± 5.01 points on the Alzheimer's Disease Cooperative Study Activities of Daily Living Inventory (ADCS-ADL) scale and an improvement in the quality of life by

3.37 ± 6.86 points on the Quality of Life in AD (QOL-AD) scale were demonstrated. It was also noted that the adverse effects were mild, while changes in cardiovascular risk parameters were mostly favourable [171]. Importantly, in spite of frequent accusations of problems with compliance with ketogenic diet requirements, half of the patients decided to continue it after the 12 weeks of study duration. The significant change in the quality of life of AD patients may be even more pronounced than the effect of drugs, including cholinesterase inhibitors, which exert an inconsistent influence on the quality of life [172,173]. Taking these results into account, the low-fat recommendations in AD should definitely be verified and challenged with the current results of scientific research.

5. The Role of the Ketogenic Diet in the Therapy of Parkinson's Disease (PD)

Parkinson's disease (PD) is a frequently observed neurodegenerative disease of the brain, the incidence of which has doubled since 1990. It develops particularly in elderly people, namely, in 1% of individuals aged over 60 years, although it is increasingly diagnosed in persons at a young age. It is a significant cause of disability worldwide since in 2019, it was responsible for 5.8 million disability-adjusted life years (increased by 81% in relation to 2000). At least 53 million people worldwide struggle with that disease, and in 2019, it was the cause of 329 thousand deaths, which was an increase of over 100% in relation to the year 2000. The disease is manifested with motor sluggishness, tremor, equilibrium disturbances and dysaesthesia or neuropsychiatric signs. The cause of these signs is damage to the neurons in the substantia nigra responsible, inter alia for dopamine production [1,174–179]. In this case, the ketogenic diet also may prove effective, and that effect is more and more frequently the subject of studies.

The ketogenic diet can affect Parkinson's disease (PD) through several mechanisms resulting from the nature of the disease. Although specific unequivocal causes have still not been established, a persistent inflammatory condition of the nervous system, mitochondrial dysfunction, reactive oxygen species (ROS) excess, a reduced ability to produce dopamine, abnormal cerebral glucose metabolism and the accumulation of damaged proteins, so-called Lewy bodies composed of misfolded α-synuclein, have been observed [180,181]. It has been found that the ketogenic diet can affect each of the aspects mentioned. The anti-inflammatory effect in all neurodegenerative diseases has been described earlier, and in this respect, the diet has a multifaceted activity [182]. However, studies are available on the anti-inflammatory effect of the diet strictly in PD. Among other studies, a publication in 2022 demonstrated that the anti-inflammatory effect of the ketogenic diet in the disease is related to a modulation of the Akt/GSK-3β/CREB signalling pathway, mediated by the acetylation of metabotropic glutamate receptor 5 (mGluR5) promoter region histones in a rat Parkinson's disease model [183]. This study confirmed mainly the neuroprotective effect of preventive ketosis compared to receiving KD as a therapeutic diet in the lipopolysaccharide (LPS)-induced rat PD model. After the induction of PD (with LPS), the model showed an increased regulation of proinflammatory mediators (TNF-α, IL-1 and IL-6), the loss of dopaminergic neurons, a reduction in mGluR5+ microglial cells, an increase in TSPO+ microglial cells, a reduction in H3K9 acetylation in the mGluR5 promoter region, and mGluR5 mRNA reduction with a decrease in the phosphorylation levels of the Akt/GSK-3/CREB pathway. These disturbances were improved by the dietary intervention of preventive KD in particular. PET imaging enabled the noninvasive detection and monitoring of the anti-inflammatory effect on PD (via the KD diet) related to histone acetylation or the DNA methylation of the mGluR5 gene.

KD suppressed the inflammatory response (neuroinflammation) relevant to microglial activation and had a neuroprotective influence. The anti-inflammatory effect of KD on PD was related to the modulation of the mGluR5/Akt/GSK-3β/CREB signalling pathway by increasing the level of histone acetylation of the mGluR5 promoter region.

In addition, the pathological processes of neuroinflammation connected with PD are supposed to ameliorate by the multiple neuroprotective mechanisms of KD-induced ketosis involving the inhibition of proinflammatory mediator gene expression, the inhibition of the

NLRP3 inflammasome assembly, epigenetic adaptations associated with calorie restriction, polyunsaturated fatty acids, ROS reduction, and the gut microbiome. Ketone bodies serve not only as an energy substrate but also as a signalling molecule. Finally, microglial cells can be modulated by various epigenetic mechanisms (DNA methylation and histone acetylation) and, thus, regulate neuroinflammation, resulting in neuroprotection [183].

Damaged mitochondria, ROS excess and abnormal glucose metabolism are closely interrelated, and the ketogenic diet can have an influence on those as well. Ketone bodies show an ability to rebuild new mitochondria to increase mitochondrial respiration and the production of ATP molecules. The reduction of free radicals thus also results from the improved efficiency of the respiratory chain in mitochondria [45,81,184]. The main ketone body, β-hydroxybutyrate, can also reduce the dopaminergic neurodegeneration and mitochondrial deficit observed in PD [38,185]. The neuroprotection of dopaminergic neurons seems to be of great importance, taking into account the scale of the problem of dopamine deficiency. Levodopa (L-DOPA), i.e., a dopamine precursor, has been, for years, one of the main drugs prescribed for PD. That drug can, however, contribute to the increased aggregation of α-synuclein, which, abnormally tangled, causes the formation of the Lewy bodies present in PD [186]. It has been found, however, that when using levodopa together with a ketogenic diet, the results can be far more favourable than those of the treatment with the drug alone (through an improvement of its bioavailability) [187–189]. Studies on PD animal models have clearly demonstrated that ketone bodies reduce dopaminergic neuronal death (BHB administered in vitro to cortical neurons and subcutaneously infused in mice) [38,190], decrease the number of proinflammatory cells in the brain and improve motor functions [62]. Another potentially possible mechanism of the ketogenic diet's influence on PD is the indirect effect mediated by changes in the intestinal microbiome. It is known that diet significantly influences intestinal microbiome remodelling; on the other hand, it is known that the microbiome plays a great role in the pathogenesis and course of PD [191]. In 2018, a randomised controlled study was conducted on 47 patients with PD (38 of whom completed the study), who were divided into two groups: a group in which the ketogenic diet was applied and a group on a low-fat and high-carbohydrate diet. Although, after eight weeks, an improvement was observed in both groups, it was more pronounced in the group of patients on the ketogenic diet. In Part I of the Unified Parkinson's Disease Rating Scale (UPDRS), the results in the ketogenic diet group improved by 41% in relation to the initial values (-4.58 ± 2.17 points), while those in the low-fat group improved by 11% (-0.99 ± 3.63 points). That was true of non-motor signs, while, in respect of motor signs, both groups presented a similar significant improvement. The adverse effects observed in both groups included increased sensations of hunger in the low-fat group and periodical exacerbations of tremor and/or stiffness in the ketogenic diet group [192]. A study in 2022 conducted on 16 patients with PD observing a ketogenic diet for 12 weeks also demonstrated its favourable effects. A significant improvement was noted on the Parkinson Anxiety Scale (PAS). At the same time, an improvement was observed in body mass, BMI, waist circumference, the concentrations of C-reactive protein (CRP), glycated haemoglobin, fasting insulin, HDL cholesterol and triglycerides. On the Center for Epidemiologic Studies Depression Scale (CESD-R-20), no major changes in depression symptoms were found [193]. In another study, among other findings, it was demonstrated that in PD patients, ketogenic diet application resulted in a mean reduction in Unified Parkinson Disease Rating Scale (UPDRS) result values by 10.72 points (a reduction by almost half compared with the initial results). An improvement was observed in respect of posture equilibrium, gait, resting tremor, mood and energy level, and it was achieved in just 28 days [188]. Another study comparing the ketogenic diet with a high-carbohydrate diet in PD patients demonstrated that in spite of the absence of motor function differences in the UPDRS between the groups, in the group of patients on the ketogenic diet, better results were observed in respect of short-term memory and verbal fluency [194]. In 2022, a case report was published on a 69-year-old woman with PD and mild depression and anxiety symptoms, to whom the ketogenic diet had been applied. A reduction in depression symptoms by 8 points on the Center for Epidemiologic Studies Depression Scale (CESDR) (from 42 to 34) and an improvement in the Parkinson Anxiety Scale (PAS) by 6 points (from 23 to 17) were noted. A significant improvement occurred in all health biomarkers, including a reduction in cardiovascular disease risk. On the other hand, in the UPDRS, an increase was observed from 24 to 33 points [195]. An improvement in UPDRS also occurred in all five study participants with PD, strictly complying with ketogenic diet requirements (fats 90%, proteins 8%, carbohydrates 2%) for 28 days [194]. Włodarek also referred to these results in his publication in 2019 [36]. Another study assessed the effect of the ketogenic diet on the quality of voice, which is decreased in PD patients. A Voice Handicap Index (VHI) test was used, and an improvement was observed in voice quality parameters in patients with PD, suggesting a possible alternative therapy for the improvement of that parameter in PD patients [196]. Taking into account the wide range of the effect on many aspects and potential therapeutic possibilities of the ketogenic diet in PD, further studies in this respect seem extremely necessary.

6. The Role of the Ketogenic Diet in the Therapy of Multiple Sclerosis (MS)

Multiple sclerosis (MS) is a neurodegenerative, inflammatory disease of the central nervous system (affecting the brain and spinal cord) of autoimmune origin. It concerns about 2.8 million people of either sex worldwide, including young individuals. It is also the main cause of disability among young people. Its incidence is unfortunately increasing; in 2013, it affected 2.3 million patients. It consists of damage to the myelin sheaths protecting neurons, thus causing disorders in the transmission of nerve impulses. Its manifestations take various forms, not infrequently different in different individuals (it can take a progressive form as well as a relapsing–remitting form). The frequently present symptoms include tingling sensations, limb weakening, problems with equilibrium, fatigue, dizziness, vision disorders and dysaesthesia [197–200].

Currently, a body of objective evidence that suggests a potentially favourable effect of the ketogenic diet in the treatment of MS is available. It can affect the course and prophylaxis of the disease while simultaneously offering safety of use and feasibility [201]. Taking into account the demyelination processes observed in MS, an effect of the ketogenic diet is suggested for the possible reconstruction and repair of the myelin sheaths. A possible even greater influence on these processes seems to be exerted by the ketogenic diet in the Mediterranean model, which has been suggested by the authors of a publication in 2022 [202]. The ketogenic diet affects the concentration of the brain-derived neurotrophic factor (BDNF), which is the main neurotrophic growth factor produced by neurons participating in myelin repair. The diet acts through a ketone body, i.e., β-hydroxybutyrate, which penetrates the blood–brain barrier, and through its effect on mitochondrial respiration and NF-KB, indirectly increasing BDNF synthesis through the activation of p300/EP300 histone acetyltransferase. Moreover, an inverse relationship has been demonstrated between serum glucose concentration and the amount of BDNF [203–206]. The Mediterranean model of the diet could exert an even greater effect in view of the increased amount of polyphenols in such a diet. It was shown that they activate the nuclear CREB factor and, thus, can additionally increase the amount of BDNF [207]. Another study also concerned the Mediterranean model of the ketogenic diet in 26 patients with MS. After four months on the diet, a significant intensification of the sensation of satiety (with similar values of ghrelin) and an increase in lean body mass and paraoxonase 1 (PON1) levels were observed. The authors explicitly suggest the favourable effect of the Mediterranean (isocaloric) ketogenic diet on the metabolism of their patients and associate the increase in the sensation of satiety with the reduction in inflammatory conditions and oxidation processes based on the observed changes of the studied parameters [208]. Another mechanism of action of the diet in MS therapy is its effect on the serum neurofilament light chain (sNfL), which is associated with multiple sclerosis (MS) and can serve as a marker of that disease. This was observed in a study on patients with relapsing–remitting MS. It was noted that the diet, six months after its institution, decreased sNfL levels, thus showing a neuroprotective effect in

MS [209]. An extensive anti-inflammatory effect was also observed, specifically in patients with the disease. This was demonstrated in a six-month-long randomised controlled study of 60 patients. Compared with the control group, in the ketogenic diet group, a significantly reduced expression was noted of the arachidonate 5-lipoxygenase (ALOX5) gene, which encodes the enzymes for the biosynthesis of proinflammatory eicosanoids. Compared, however, to the results of the same individuals before and after the institution of dietary therapy, a significantly impaired expression was noted of other proinflammatory enzymes, i.e., cyclooxygenase 1 (COX-1) and cyclooxygenase 2 (COX-2), and an inverse correlation was observed between the expression of proinflammatory genes and patients' quality of life in the Multiple Sclerosis Quality of Life-54 (MSQOL-54) scale [210]. In a study in 2022, among 65 participants with recurrent MS, in whom the ketogenic diet was applied for six months, some promising results were observed. Benefits were noted, among other findings, in the form of improvement in neurological disability, quality of life, depression or inflammatory conditions. A reduction occurred, by almost half, in the fatigue and depression results reported by the study participants. An improvement occurred not only in mental health but also in physical health. In the participants, an improvement was also seen in the mean values of the disability status in the Expanded Disability Status Scale (EDSS), from 2.3 ± 0.9 to 1.9 ± 1.1 points [211]. That has also been confirmed by an earlier study in 19 patients on the ketogenic diet for three months and 16 patients using it for six months. An improvement occurred in the results of depression and fatigue, body mass decreased, and the level of serologic proinflammatory adipokines was reduced, with good tolerance and safety of the diet [212]. No significant clinical improvement was, however, observed in another study on MS patients using ketogenic diets enriched with MCTs, but an evident reduction of fasting glucose and insulin levels was noted [213]. Taking into account the nature of the ketogenic diet, mimicking fasting, the possible additional mechanisms of its action are known and have been demonstrated in another randomised controlled study. It has been shown that the fast-mimicking diet increases regeneration in the oligodendrocyte precursors and remyelination in axons and reduces the symptoms of autoimmunisation and, thus, the symptoms of MS. Moreover, it is able to reduce the concentrations of proinflammatory cytokines, T helper type 1 cells (TH1), T helper type 17 cells (TH17) and antigen-presenting cells (APCs). The study has also provided evidence of potential benefits resulting from the ketogenic diet in the treatment of patients with relapsing–remitting MS (RRMS) [214]. Taking into account the multifaceted favourable effect of the ketogenic diet in MS, further studies could effectively lead to a change in the therapeutic approach to the disease.

7. Ketogenic Diet in the Therapy of Migraine

Migraine is the most frequently occurring neurological disease. It occurs in 12% of the world population, and chronic migraine (CM) is observed in 1–2% of people worldwide. Among the individuals struggling with episodic migraine, 2.5% of patients develop its chronic form. It is manifested with frequent paroxysmal headaches, which are frequently very intense. It is, therefore, a disease that significantly impairs patients' quality of life and willingness to function on a daily basis. Individuals with migraine are at a higher risk of the development of other symptoms, including psychic disorders, sleep disorders or cardiovascular manifestations [215].

The ketogenic diet in the case of migraine shows significant potential benefits, which have been observed in an increasing number of recent clinical studies. Although the causes of migraine have not been unequivocally established, disorders in mitochondrial functioning and related problems with ATP production are supposed to belong to the possible mechanisms [216]. In such cases, the cause may be a disturbed metabolism of the cerebral cells, for which the possibility of taking energy from glucose is reduced; thus, ketone bodies seem to be a promising option here since they provide an alternative form of energy to the brain. However, looking not at the theory but at the actual results of studies in patients suffering from migraine, interesting conclusions can be drawn. A randomised controlled study comparing the effect of a very-low-calorie ketogenic diet with that of a very-low-calorie non-ketogenic diet demonstrated an advantage of the ketogenic diet. The study was conducted on 35 patients with migraine and overweight. The participants on the ketogenic diet, compared with the non-ketogenic diet, demonstrated a mean reduction of −3.73 migraine days monthly and −3.02 migraine attacks monthly. The percentage of patients with at least a 50% reduction in the mean number of days with migraine in the ketogenic group was 74.28%, compared with only 8.57% in the non-ketogenic group. It was suggested that diet can be a useful therapeutic strategy for migraine [217]. Another study in 2022, conducted on 23 patients with migraine and overweight, revealed a reduction in the number of days with headaches from 12.5 ± 9.5 days monthly on average, before the institution of the diet, to 6.7 ± 8.6 days on average after its institution. Furthermore, a reduction in the number of days taking drugs on an ad hoc basis was observed—from 11.06 ± 9.37 days monthly on average, before the institution of the ketogenic diet, to 4.93 ± 7.99 days after the institution. Moreover, body mass, fatty tissue mass and, thus, BMI values were also reduced in these patients. The favourable influence goes beyond the effect of body weight loss, and the authors concluded that other mechanisms must underlie the therapeutic activity [218]. The aim of another study in 2022 was an assessment of the effect of the ketogenic diet on treatment-refractory migraine in 22 patients (in the first study) and 31 subjects (in the second study). Very importantly, the ketogenic diet was compared with a low-carbohydrate diet. This means that the effects observed resulted strictly from the status of ketosis and not from the fact of the minimisation of carbohydrate contents in the diet. In the first group of 22 patients on the ketogenic diet, a significant reduction was observed in the frequency of migraine attacks, headache intensity and the amount of drugs taken. For comparison, in the low-carbohydrate diet group, no major changes were noted. In the study of the second group of 31 patients, similar effects were observed, which was evidence of the effectiveness of the ketogenic diet in the case of treatment-resistant migraine. A relationship was also observed between the production of ketones and their effect on headaches [219]. Another study also analysed the effect of the nutritional model on drug-resistant migraine. Patients with drug-resistant migraine were subjected to dietary intervention in the form of a ketogenic diet lasting three months. Very promising results were obtained since the mean number of days with pain decreased from 30 to 7.5 and the mean duration of pain episodes decreased to 5.5 h. Initially, 83% of the study participants reported maximal pain levels (on a pain rating scale). After the institution of the ketogenic diet, 55% of the study participants observed a reduction in pain intensity [220]. In 2022, for the first time, a study was conducted in order to systematise the data concerning the effect of ketogenic diets on migraine. The authors concluded that in most studies analysed, ketogenic diets reduced the number of attacks and their intensity in patients suffering from migraine [221].

8. Limitations and Potential Adverse Health Impacts of the Ketogenic Diet

The main limitations include a low number of available studies and, particularly, a low number of qualitative studies. Although the highest-quality body of evidence suggests the effectiveness of the ketogenic diet in the treatment of epilepsy, the situation is not similar in the case of the remaining neurological diseases. This results from the fact that the development of research in the field of neurological diseases other than epilepsy is a relatively new phenomenon. All meta-analyses and systematic reviews concerning Alzheimer's disease, Parkinson's disease, multiple sclerosis and migraine begin in 2020 (according to the PubMed search engine). This fact justifies the need for the exploration of this new research domain.

Potential adverse effects of the use of the ketogenic diet are possible, but they rarely occur and usually result from inadequately adjusted diets. It should be kept in mind that the therapeutic approach to patients should be maximally individualised. Such effects were described in a meta-analysis in 2020, involving a total number of 932 participants [93]. It demonstrated that in epileptic children on ketogenic diets, gastrointestinal signs occurred most frequently. Among other effects, diarrhoea, constipation and vomiting were observed.

The signs also occurred in children treated by other methods; therefore, they could not be ascribed strictly to the ketogenic diet. The remaining adverse effects, less frequently described in the meta-analysis, included body mass loss, infections, nausea, dysphagia, and lethargy. In adults, on the other hand, among other symptoms and signs, the following were rarely observed: headache, abdominal pain, irregular menstruation, drowsiness, and nephrolithiasis. In the study by Arora et al., no side effects of the ketogenic diet were found in adults with traumatic brain injury. The diet proved safe [43]. In a randomised controlled study on patients with Alzheimer's disease, no major adverse effects were observed [171]. In a study conducted on patients with Parkinson's disease, some of them presented periodical exacerbations of tremor and/or stiffness, increased irritability, and excessive hunger or thirst [192]. It should be kept in mind, however, that the period of the so-called "keto flu" is frequently regarded as an adverse effect. That period is, in fact, a transient episode frequently occurring at the beginning of treatment with the diet, and it poses no threat [222].

9. Summary

Taking into account all the data discussed, the ketogenic diet is undoubtedly a very promising model in the therapy of the above-mentioned neurological diseases. More than a hundred years of studies on the ketogenic diet's effect on neurological diseases (starting with epilepsy) means that they belong to the main fields of research related to the therapeutic potential of the diet. This results from its very wide, pleiotropic effect on the body as well as from a number of (including those not yet known) mechanisms of action on the nervous system. Its favourable activity in neurological diseases, demonstrated in clinical studies, is related to the following: reducing the production of reactive oxygen species (ROS); reducing neuronal inflammatory conditions; the reconstruction of neuronal myelin sheaths; the repair of damaged mitochondria and the formation of new mitochondria and, thus, the effect on the disturbed neuronal metabolism in a number of neurological diseases; the provision of an alternative energy source for neurons in the form of ketone bodies; a reduction in glucose and insulin concentrations; the induction of autophagy; the reduction of microglia stimulation; the reduction of the excitatory postsynaptic current (EPSC) through action on voltage-dependent Ca^{2+} channels (VDCC); intestinal microbiota modulation and gene expression (epigenetic origin); assistance in the production of indispensable dopamine; and an increase in glutamine conversion into the neurotransmitter GABA. Together, with all the mentioned mechanisms, it is not surprising that the ketogenic diet in clinical studies shows a favourable effect on a number of neurological diseases, including epilepsy, Alzheimer's disease (AD), Parkinson's disease (PD), multiple sclerosis (MS) and migraine, which has been demonstrated in this paper. This is shown in Table 1. At present, a large body of indirect evidence suggests the effectiveness of the diet, while in the last few years, studies have increasingly aimed at demonstrating its actual direct effect on neurological patients. The currently available scientific data suggest a promising influence of the nutritional model in the therapy of neurological diseases. Despite that, there is a definite need to continue the further development of research in this field. Keeping in mind the exponentially growing incidence of neurological diseases, this approach could contribute in the future to a potential application of the diet in the clinical therapy of neurological patients, thus increasing the quality of life and lifespan of millions of people worldwide.

Table 1. Review of selected randomised controlled studies concerning the effect of the ketogenic diet on the discussed neurological diseases.

Disease	No. of Patients	Intervention in the Study Group	Intervention in the Control Group	Result	Reference
Alzheimer's disease	26	12-week ketogenic diet; 58% fat (including 26% saturated and 32% unsaturated), 29% protein, 7% roughage and 6% carbohydrates net. + a multivitamin preparation *	12-week diet according to New Zealand's principles of healthy nutrition; 11% fat (3% saturated, 8% unsaturated), 19% protein, 8% roughage, and 62% carbohydrates net. + a multivitamin preparation *	Compared with the control group, the ketogenic diet improved the following: cognitive functions (by 2.12 pts on the ACE-III scale), everyday functioning (by 3.13 pts on the ADCS-ADL scale), quality of life (by 3.37 pts on the QOL-AD scale)	[171]
Parkinson's disease	47	8-week ketogenic diet, 1750 kcal, 152 g fat (67 g saturated), 75 g protein, 16 g carbohydrates net and 11 g roughage + a possibility of additional "loading" of 500 kcal (50 g fat (22 g saturated), 6 g protein, 5 g carbohydrates net and 4 g roughage)	8-week low-fat diet 1750 kcal, 42 g fat (10 g saturated), 75 g protein, 246 g carbohydrates net and 33 g roughage + a possibility of additional "loading" of 500 kcal (4 g fat (1 g saturated), 6 g protein, 102 g carbohydrates net and 11 g roughage)	Improvement in Part I of the UPDRS by 48% compared with 11% improvement in the control group, a greater improvement of non-motor signs in the ketogenic group, improvement of motor signs in both groups	[192]
Multiple sclerosis	60	6-month ketogenic diet, >160 g fat, ≤100 g protein, <50 g carbohydrates net, high (unspecified) roughage consumption	6-month diet according to the principles of a healthy diet in the German population	Reduced expression of proinflammatory ALOX5 compared with the control group, impaired expression also of other proinflammatory enzymes (COX1, COX2), significant inverse correlation between expression of proinflammatory ALOX5 and COX2 and the MSQoL-54 marker	[211]
Migraine	35	4-week very-low-calorie ketogenic diet (VLCKD), 20 g fat, ≥75 g protein, 30–50 g carbohydrates + a preparation with microelements	4-week very-low-calorie non-ketogenic diet (VLCnKD), 20 g fat, ≅50 g protein, ≥70 g carbohydrates + a preparation with microelements	Reduction in the mean number of days with migraine monthly by 3.73 days, compared with the control group, in the number of migraine attacks over that time by 3.02; a greater percentage of patients in the ketogenic group, in whom a reduction by at least 50% of the days with migraine, was obtained (74.28% vs. 8.57% of the participants)	[218]
Epilepsy	145	12-month ketogenic diet enriched with MCT, 70–75% energy from fat (30% long-chain fatty acids, 40–45% MCT), 10% energy from protein, 15% energy from carbohydrates	12-month classic ketogenic diet; in most cases, the fat to carbohydrates and protein ratio was 4:1; protein was maintained in amounts recommended by the WHO	Effectiveness of both types of ketogenic diets: a reduction of the mean number of initial attacks in the 3rd, 6th and 12th months in the classic version was 66.5%,48.5%, and 40,5%, respectively, while in the MCT version, it was 68.9%, 67.6%, and 53.2%, respectively	[106]

* Multivitamin and Mineral Boost, Clinicians Ltd., Auckland, New Zealand. ACE-III: Addenbrooke's Cognitive Examination III; ADCS-ADL: Alzheimer's Disease Cooperative Study Activities of Daily Living Inventory; QOL-AD: Quality of Life in AD; UPDRS: Unified Parkinson's Disease Rating Scale; ALOX5: arachidonate 5-lipoxygenase; COX1: cyclooxygenase 1; COX2: cyclooxygenase 2; MSQOL-54: Multiple Sclerosis Quality of Life-54; MCT: medium-chain triglycerides; WHO: World Health Organization.

Author Contributions: Writing—original draft preparation, D.D.; writing—review and editing, A.P., K.K. and D.D.; supervision, A.P. and K.K.; funding acquisition A.P. and K.K. All authors have read and agreed to the published version of the manuscript.

Funding: The publication was financed by the Polish Ministry of Education and Science with funds from the Institute of Health Sciences Faculty of Medical and Health Sciences, Siedlce University of Natural Sciences and Humanities.

Institutional Review Board Statement: Not applicable.

Informed Consent Statement: Not applicable.

Data Availability Statement: Not applicable.

Acknowledgments: The authors would like to thank Piotr Słomski for translating and proof-reading the publication and for his scientific commitment.

Conflicts of Interest: The authors declare no conflict of interest.

References

1. GBD 2016 Neurology Collaborators. Global, regional, and national burden of neurological disorders, 1990–2016: A systematic analysis for the Global Burden of Disease Study 2016. *Lancet Neurol.* **2019**, *18*, 459–480. [CrossRef] [PubMed]
2. Feigin, V.L. The Evolution of Neuroepidemiology: Marking the 40-Year Anniversary of Publishing Studies on Epidemiology of Neurological Disorders. *Neuroepidemiology* **2022**, *56*, 2–3. [CrossRef] [PubMed]
3. Shehata, G.; Lord, K.; Grudzinski, M.; Elsayed, M.; Abdelnaby, R.; Elshabrawy, H. Neurological Complications of COVID-19: Underlying Mechanisms and Management. *Int. J. Mol. Sci.* **2021**, *22*, 4081. [CrossRef] [PubMed]
4. Patone, M.; Handunnetthi, L.; Saatci, D.; Pan, J.; Katikireddi, S.V.; Razvi, S.; Hunt, D.; Mei, X.W.; Dixon, S.; Zaccardi, F.; et al. Neurological complications after first dose of COVID-19 vaccines and SARS-CoV-2 infection. *Nat. Med.* **2021**, *27*, 2144–2153. [CrossRef] [PubMed]
5. Taquet, M.; Sillett, R.; Zhu, L.; Mendel, J.; Camplisson, I.; Dercon, Q.; Harrison, P.J. Neurological and psychiatric risk trajectories after SARS-CoV-2 infection: An analysis of 2-year retrospective cohort studies including 1284437 patients. *Lancet Psychiatry* **2022**, *9*, 815–827. [CrossRef]
6. Zarifkar, P.; Peinkhofer, C.; Benros, M.E.; Kondziella, D. Frequency of Neurological Diseases After COVID-19, Influenza A/B and Bacterial Pneumonia. *Front. Neurol.* **2022**, *13*, 904796. [CrossRef]
7. Wise, J. COVID-19: Increased risk of some neurological and psychiatric disorders remains two years after infection, study finds. *BMJ* **2022**, *378*, o2048. [CrossRef]
8. Puchalska, P.; Crawford, P.A. Multi-dimensional Roles of Ketone Bodies in Fuel Metabolism, Signaling, and Therapeutics. *Cell Metab.* **2017**, *25*, 262–284. [CrossRef]
9. Gupta, L.; Khandelwal, D.; Kalra, S.; Gupta, P.; Dutta, D.; Aggarwal, S. Ketogenic diet in endocrine disorders: Current perspectives. *J. Postgrad. Med.* **2017**, *63*, 242–251.
10. Phinney, S.D.; Bistrian, B.R.; Wolfe, R.R.; Blackburn, G.L. The human metabolic response to chronic ketosis without caloric restriction: Physical and biochemical adaptation. *Metabolism* **1983**, *32*, 757–768. [CrossRef]
11. Phinney, S.D. Ketogenic diets and physical performance. *Nutr. Metab.* **2004**, *1*, 2. [CrossRef] [PubMed]
12. Paoli, A.; Rubini, A.; Volek, J.S.; Grimaldi, K.A. Beyond weight loss: A review of the therapeutic uses of very-low-carbohydrate (ketogenic) diets. *Eur. J. Clin. Nutr.* **2013**, *67*, 789–796. [CrossRef] [PubMed]
13. Gregory, R.M.; Hamdan, H.; Torisky, D.M.; Akers, J.D. A Low-Carbohydrate Ketogenic Diet Combined with 6-Weeks of Crossfit Training Improves Body Composition and Performance. *Int. J. Sport. Exerc. Med.* **2017**, *3*, 54. [CrossRef]
14. Wilson, J.; Lowery, R. *The Ketogenic Bible*; Victory Belt Publishing Inc.: Las Vegas, NV, USA, 2017; ISBN-13 978-1-628601-04-6.
15. Valenzuela, P.L.; Castillo García, A.; Lucia, A.; Naclerio, F. Ketogenic diets in strength-trained individuals: A narrative review. *Nutrients* **2021**, *13*, 3083. [CrossRef]
16. Reichard, G.A., Jr.; Owen, O.E.; Haff, A.C.; Paul, P.; Bortz, W.M. Ketone-body production and oxidation in fasting obese humans. *J. Clin. Investig.* **1974**, *53*, 508–515. [CrossRef]
17. Steinhauser, M.L.; Olenchock, B.A.; O'Keefe, J.; Lun, M.; Pierce, K.A.; Lee, H.; Pantano, L.; Klibanski, A.; Shulman, G.I.; Clish, C.B.; et al. The circulating metabolome of human starvation. *J. Clin. Investig.* **2018**, *3*, e121434. [CrossRef]
18. Alharbi, A.; Al-Sowayan, N. The Effect of Ketogenic-Diet on Health. *Food Nutr. Sci.* **2020**, *11*, 301–313. [CrossRef]
19. Dong, T.A.; Sandesara, P.B.; Dhindsa, D.S.; Mehta, A.; Arneson, L.C.; Dollar, A.L.; Taub, P.R.; Sperling, L.S. Intermittent Fasting: A Heart Healthy Dietary Pattern? *Am. J. Med.* **2020**, *133*, 901–907. [CrossRef]
20. Williams, M.; Turos, E. The Chemistry of the Ketogenic Diet: Updates and Opportunities in Organic Synthesis. *Int. J. Mol. Sci.* **2021**, *22*, 5230. [CrossRef]
21. Temkin, O. *The Falling Sickness: A History of Epilepsy from the Greeks to the Beginnings of Modern Neurology*; Johns Hopkins University Press: Baltimore, MD, USA, 1994.
22. Wheless, J.W. History of the ketogenic diet. *Epilepsia* **2008**, *49*, 3–5. [CrossRef]

23. Livingston, S. *Comprehensive Management of Epilepsy in Infancy, Childhood, and Adolescence*; Charles C Thomas: Springfield, IL, USA, 1972; pp. 378–405.
24. Huisjen, D. *Today's Parallel Bible*; Zondervan Corp: Grand Rapids, MI, USA, 2000; Volume 14–29, pp. 2306–2308.
25. Wheless, J.W. *History and Origin of the Ketogenic Diet, en Epilepsy and the Ketogenic Diet*; Stafstrom, C.E., Rho, J.M., Eds.; Humana Press: Totowa, NJ, USA, 2004; pp. 31–50.
26. Guelpa, G.; Marie, A. La lutte contre l'e'pilepsie par la de' sintoxication et par la re'e'ducation alimentaire. *Rev. De Ther. Med. Chir.* **1911**, *78*, 8–13.
27. Wilder, R.M. The effects of ketonemia on the course of epilepsy. *Mayo Clin. Bull.* **1921**, *2*, 307.
28. Woodyatt, R.T. Objects and method of diet adjustment in diabetes. *Arch. Intern. Med.* **1921**, *28*, 125. [CrossRef]
29. Peterman, M.G. The ketogenic diet in epilepsy. *JAMA* **1925**, *84*, 1979–1983. [CrossRef]
30. Field, R.; Field, T.; Pourkazemi, F.; Rooney, K. Ketogenic diets and the nervous system: A scoping review of neurological outcomes from nutritional ketosis in animal studies. *Nutr. Res. Rev.* **2021**, *35*, 268–281. [CrossRef]
31. Jensen, N.J.; Wodschow, H.Z.; Nilsson, M.; Rungby, J. Effects of Ketone Bodies on Brain Metabolism and Function in Neurodegenerative Diseases. *Int. J. Mol. Sci.* **2020**, *21*, 8767. [CrossRef]
32. Shaito, A.; Hasan, H.; Habashy, K.J.; Fakih, W.; Abdelhady, S.; Ahmad, F.; Zibara, K.; El-Yazbi, A.F.; Kobeissy, F.H.; Eid, A.H. Western diet aggravates neuronal insult inpost-traumatic brain injury: Proposed pathways for interplay. *EBioMedicine* **2020**, *57*, 102829. [CrossRef]
33. Felmlee, M.A.; Jones, R.S.; Rodriguez-Cruz, V.; Follman, K.E.; Morris, M.E. Monocarboxylate Transporters (SLC16): Function, Regulation, and Role in Health and Disease. *Pharmacol. Rev.* **2020**, *72*, 466–485. [CrossRef]
34. Gambardella, I.; Ascione, R.; D'Agostino, D.P.; Ari, C.; Worku, B.; Tranbaugh, R.F.; Ivascu, N.; Villena-Vargas, J.; Girardi, L.N. Systematic Review—Neuroprotection of ketosis in acute injury of the mammalian central nervous system: A meta-analysis. *J. Neurochem.* **2021**, *158*, 105–118. [CrossRef]
35. Maalouf, M.; Rho, J.M.; Mattson, M.P. The neuroprotective properties of calorie restriction, the ketogenic diet, and ketone bodies. *Brain Res. Rev.* **2009**, *59*, 293–315. [CrossRef]
36. Włodarek, D. Role of Ketogenic Diets in Neurodegenerative Diseases (Alzheimer's Disease and Parkinson's Disease). *Nutrients* **2019**, *11*, 169. [CrossRef] [PubMed]
37. Suzuki, M.; Suzuki, M.; Kitamura, Y.; Mori, S.; Sato, K.; Dohi, S.; Sato, T.; Matsuura, A.; Hiraide, A. β-Hydroxybutyrate, a Cerebral Function Improving Agent, Protects Rat Brain Against Ischemic Damage Caused by Permanent and Transient Focal Cerebral Ischemia. *Jpn. J. Pharmacol.* **2002**, *89*, 36–43. [CrossRef] [PubMed]
38. Tieu, K.; Perier, C.; Caspersen, C.; Teismann, P.; Wu, D.C.; Yan, S.D.; Naini, A.; Vila, M.; Jackson-Lewis, V.; Ramasamy, R.; et al. D-beta-hydroxybutyrate rescues mitochondrial respiration and mitigates features of Parkinson disease. *J. Clin. Investig.* **2003**, *112*, 892–901. [CrossRef] [PubMed]
39. Prins, M.; Giza, C. Induction of Monocarboxylate Transporter 2 Expression and Ketone Transport following Traumatic Brain Injury in Juvenile and Adult Rats. *Dev. Neurosci.* **2006**, *28*, 447–456. [CrossRef]
40. Barkhoudarian, G.; Hovda, D.A.; Giza, C.C. The Molecular Pathophysiology of Concussive Brain Injury. *Clin. Sports Med.* **2011**, *30*, 33–48. [CrossRef] [PubMed]
41. Wang, Y.; Liu, N.; Zhu, W.; Zhang, K.; Si, J.; Bi, M.; Lv, X.; Wang, J. Protective effect of β-hydroxybutyrate on glutamate induced cell death in HT22 cells. *Int. J. Clin. Exp. Med.* **2016**, *9*, 23433–23439.
42. Dilimulati, D.; Zhang, F.; Shao, S.; Lv, T.; Lu, Q.; Cao, M.; Jin, Y.; Jia, F.; Zhang, X. Ketogenic Diet Modulates Neuroinflammation via Metabolites from Lactobacillus reuteri After Repetitive Mild Traumatic Brain Injury in Adolescent Mice. *Cell. Mol. Neurobiol.* 2022; 1–17, *online ahead of print*. [CrossRef]
43. Arora, N.; Litofsky, N.S.; Golzy, M.; Aneja, R.; Staudenmyer, D.; Qualls, K.; Patil, S. Phase I single center trial of ketogenic diet for adults with traumatic brain injury. *Clin. Nutr. ESPEN* **2022**, *47*, 339–345. [CrossRef] [PubMed]
44. Mu, J.; Wang, T.; Li, M.; Guan, T.; Guo, Y.; Zhang, X.; Zhang, G.; Kong, J. Ketogenic diet protects myelin and axons in diffuse axonal injury. *Nutr. Neurosci.* **2022**, *25*, 1534–1547. [CrossRef]
45. Kim, D.Y.; Vallejo, J.; Rho, J.M. Ketones prevent synaptic dysfunction induced by mitochondrial respiratory complex inhibitors. *J. Neurochem.* **2010**, *114*, 130–141. [CrossRef]
46. Kim, D.Y.; Abdelwahab, M.G.; Lee, S.H.; O'Neill, D.; Thompson, R.J.; Duff, H.J.; Sullivan, P.G.; Rho, J.M. Ketones Prevent Oxidative Impairment of Hippocampal Synaptic Integrity through KATP Channels. *PLoS ONE* **2015**, *10*, e0119316. [CrossRef]
47. Svart, M.; Gormsen, L.C.; Hansen, J.; Zeidler, D.; Gejl, M.; Vang, K.; Aanerud, J.; Moeller, N. Regional cerebral effects of ketone body infusion with 3-hydroxybutyrate in humans: Reduced glucose uptake, unchanged oxygen consumption and increased blood flow by positron emission tomography. A randomized, controlled trial. *PLoS ONE* **2018**, *13*, e0190556. [CrossRef]
48. Prapong, T.; Buss, J.; Hsu, W.H.; Heine, P.; West Greenlee, H.; Uemura, E. Amyloid beta-peptide decreases neuronal glucose uptake despite causing increase in GLUT3 mRNA transcription and GLUT3 translocation to the plasma membrane. *Exp. Neurol.* **2002**, *174*, 253–258. [CrossRef] [PubMed]
49. Akhtar, M.W.; Sanz-Blasco, S.; Dolatabadi, N.; Parker, J.; Chon, K.; Lee, M.S.; Soussou, W.; McKercher, S.R.; Ambasudhan, R.; Nakamura, T.; et al. Elevated glucose and oligomeric β-amyloid disrupt synapses via a common pathway of aberrant protein S-nitrosylation. *Nat. Commun.* **2016**, *7*, 10242. [CrossRef] [PubMed]
50. Henderson, S.T. High carbohydrate diets and Alzheimer's disease. *Med. Hypotheses* **2004**, *62*, 689–700. [CrossRef]

51. Choi, Y.J.; Jeon, S.-M.; Shin, S. Impact of a Ketogenic Diet on Metabolic Parameters in Patients with Obesity or Overweight and with or without Type 2 Diabetes: A Meta-Analysis of Randomized Controlled Trials. *Nutrients* **2020**, *12*, 2005. [CrossRef] [PubMed]
52. Yuan, X.; Wang, J.; Yang, S.; Gao, M.; Cao, L.; Li, X.; Hong, D.; Tian, S.; Sun, C. Effect of the ketogenic diet on glycemic control, insulin resistance, and lipid metabolism in patients with T2DM: A systematic review and meta-analysis. *Nutr. Diabetes* **2020**, *10*, 38. [CrossRef] [PubMed]
53. McCarty, M.F.; DiNicolantonio, J.J.; O'Keefe, J.H. Ketosis may promote brain macroautophagy by activating Sirt1 and hypoxia-inducible factor-1. *Med. Hypotheses* **2015**, *85*, 631–639. [CrossRef]
54. Loos, B.; Klionsky, D.J.; Wong, E. Augmenting brain metabolism to increase macro- and chaperone-mediated autophagy for decreasing neuronal proteotoxicity and aging. *Prog. Neurobiol.* **2017**, *156*, 90–106. [CrossRef]
55. Camberos-Luna, L.; Gerónimo-Olvera, C.; Montiel, T.; Rincon-Heredia, R.; Massieu, L. The Ketone Body, β-Hydroxybutyrate Stimulates the Autophagic Flux and Prevents Neuronal Death Induced by Glucose Deprivation in Cortical Cultured Neurons. *Neurochem. Res.* **2016**, *41*, 600–609. [CrossRef]
56. Montiel, T.; Montes-Ortega, L.A.; Flores-Yáñez, S.; Massieu, L. Treatment with the Ketone Body D-β-hydroxybutyrate Attenuates Autophagy Activated by NMDA and Reduces Excitotoxic Neuronal Damage in the Rat Striatum In Vivo. *Curr. Pharm. Des.* **2020**, *26*, 1377–1387. [CrossRef]
57. Liśkiewicz, D.; Liśkiewicz, A.; Nowacka-Chmielewska, M.M.; Grabowski, M.; Pondel, N.; Grabowska, K.; Student, S.; Barski, J.J.; Małecki, A. Differential Response of Hippocampal and Cerebrocortical Autophagy and Ketone Body Metabolism to the Ketogenic Diet. *Front. Cell. Neurosci.* **2021**, *15*, 733607. [CrossRef] [PubMed]
58. Van der Auwera, I.; Wera, S.; Van Leuven, F.; Henderson, S.T. A ketogenic diet reduces amyloid beta 40 and 42 in a mouse model of Alzheimer's disease. *Nutr. Metab.* **2005**, *2*, 28. [CrossRef]
59. Koh, S.; Dupuis, N.; Auvin, S. Ketogenic diet and Neuroinflammation. *Epilepsy Res.* **2020**, *167*, 106454. [CrossRef]
60. Disabato, D.J.; Quan, N.; Godbout, J.P. Neuroinflammation: The devil is in the details. *J. Neurochem.* **2016**, *139* (Suppl. S2), 136–153. [CrossRef] [PubMed]
61. McDonald, T.J.W.; Cervenka, M.C. Ketogenic Diets for Adult Neurological Disorders. *Neurotherapeutics* **2018**, *15*, 1018–1031. [CrossRef] [PubMed]
62. Yang, X.; Cheng, B. Neuroprotective and Anti-inflammatory Activities of Ketogenic Diet on MPTP-induced Neurotoxicity. *J. Mol. Neurosci.* **2010**, *42*, 145–153. [CrossRef]
63. Jeong, E.A.; Jeon, B.T.; Shin, H.J.; Kim, N.; Lee, D.H.; Kim, H.J.; Kang, S.S.; Cho, G.J.; Choi, W.S.; Roh, G.S. Ketogenic diet-induced peroxisome proliferator-activated receptor-γ activation decreases neuroinflammation in the mouse hippocampus after kainic acid-induced seizures. *Exp. Neurol.* **2011**, *232*, 195–202. [CrossRef]
64. Fu, S.; Wang, J.; Xue, W.; Liu, H.; Liu, B.; Zeng, Y.; Li, S.; Huang, B.; Lv, Q.; Wang, W.; et al. Anti-inflammatory effects of BHBA inboth in vivo and in vitro Parkinson's disease models are mediated by GPR109A-dependent mechanisms. *J. Neuroinflamm.* **2015**, *12*, 9. [CrossRef]
65. Guzmán, M.; Blázquez, C. Ketone body synthesis in the brain: Possible neuroprotective effects. *Prostaglandins Leukot. Essent. Fat. Acids* **2004**, *70*, 287–292. [CrossRef]
66. Jiang, Z.; Yin, X.; Wang, M.; Chen, T.; Wang, Y.; Gao, Z.; Wang, Z. Effects of Ketogenic Diet on Neuroinflammation in Neurodegenerative Diseases. *Aging Dis.* **2022**, *13*, 1146. [CrossRef]
67. Xu, Y.; Jiang, C.; Wu, J.; Liu, P.; Deng, X.; Zhang, Y.; Peng, B.; Zhu, Y. Ketogenic diet ameliorates cognitive impairment and neuroinflammation in a mouse model of Alzheimer's disease. *CNS Neurosci. Ther.* **2022**, *28*, 580–592. [CrossRef] [PubMed]
68. Friedman, A.; Kaufer, D. Blood-Brain Barrier Breakdown and Blood-Brain Communication in Neurological and Psychiatric Diseases. *Cardiovasc. Psychiatry Neurol.* **2011**, *2011*, 431470. [CrossRef]
69. Xiao, M.; Xiao, Z.J.; Yang, B.; Lan, Z.; Fang, F. Blood-Brain Barrier: More Contributor to Disruption of Central Nervous System Homeostasis Than Victim in Neurological Disorders. *Front. Neurosci.* **2020**, *14*, 764. [CrossRef] [PubMed]
70. Kurz, C.; Walker, L.; Rauchmann, B.; Perneczky, R. Dysfunction of the blood–brain barrier in Alzheimer's disease: Evidence from human studies. *Neuropathol. Appl. Neurobiol.* **2022**, *48*, e12782. [CrossRef]
71. Hasselbalch, S.G.; Knudsen, G.M.; Jakobsen, J.; Hageman, L.P.; Holm, S.; Paulson, O.B. Blood-brain barrier permeability of glucose and ketone bodies during short-term starvation in humans. *Am. J. Physiol. Metab.* **1995**, *268*, E1161–E1166. [CrossRef] [PubMed]
72. White, H.; Venkatesh, B. Clinical review: Ketones and brain injury. *Crit. Care* **2011**, *15*, 219. [CrossRef] [PubMed]
73. Ho, C.-F.; Chan, K.-W.; Yeh, H.-I.; Kuo, J.; Liu, H.-J.; Wang, C.-Y. Ketone bodies upregulate endothelial connexin 43 (Cx43) gap junctions. *Vet. J.* **2013**, *198*, 696–701. [CrossRef]
74. Versele, R.; Corsi, M.; Fuso, A.; Sevin, E.; Businaro, R.; Gosselet, F.; Fenart, L.; Candela, P. Ketone Bodies Promote Amyloid-β1–40 Clearance in a Human in Vitro Blood–Brain Barrier Model. *Int. J. Mol. Sci.* **2020**, *21*, 934. [CrossRef]
75. Yudkoff, M.; Daikhin, Y.; Horyn, O.; Nissim, I.; Nissim, I. Ketosis and brain handling of glutamate, glutamine, and GABA. *Epilepsia* **2008**, *49*, 73–75. [CrossRef]
76. Lund, T.M.; Risa, O.; Sonnewald, U.; Schousboe, A.; Waagepetersen, H.S. Availability of neurotransmitter glutamate is diminished when β-hydroxybutyrate replaces glucose in cultured neurons. *J. Neurochem.* **2009**, *110*, 80–91. [CrossRef]

77. Marosi, K.; Kim, S.W.; Moehl, K.; Scheibye-Knudsen, M.; Cheng, A.; Cutler, R.; Camandola, S.; Mattson, M.P. 3-Hydroxybutyrate regulates energy metabolism and induces BDNF expression in cerebral cortical neurons. *J. Neurochem.* **2016**, *139*, 769–781. [CrossRef] [PubMed]
78. Garcia-Penas, J.J. Epilepsia, cognicion y dieta cetogenica [Epilepsy, cognition and ketogenic diet]. *Rev. Neurol.* **2018**, *66*, S71–S75. (In Spanish) [PubMed]
79. Włodarczyk, A.; Cubała, W.J.; Stawicki, M. Ketogenic diet for depression: A potential dietary regimen to maintain euthymia? *Prog. Neuro-Psychopharmacol. Biol. Psychiatry* **2021**, *109*, 110257. [CrossRef] [PubMed]
80. Tao, Y.; Leng, S.X.; Zhang, H. Ketogenic diet: An effective treatment approach for neurodegenerative diseases. *Curr. Neuropharmacol.* 2022; *online ahead of print*. [CrossRef]
81. Bough, K.J.; Wetherington, J.; Hassel, B.; Pare, J.F.; Gawryluk, J.W.; Greene, J.G.; Shaw, R.; Smith, Y.; Geiger, J.D.; Dingledine, R.J. Mitochondrial biogenesis in the anticonvulsant mechanism of the ketogenic diet. *Ann. Neurol.* **2006**, *60*, 223–235. [CrossRef] [PubMed]
82. Lim, J.-M.; Letchumanan, V.; Tan, L.T.-H.; Hong, K.-W.; Wong, S.-H.; Ab Mutalib, N.-S.; Lee, L.-H.; Law, J.W.-F. Ketogenic Diet: A Dietary Intervention via Gut Microbiome Modulation for the Treatment of Neurological and Nutritional Disorders (a Narrative Review). *Nutrients* **2022**, *14*, 3566. [CrossRef]
83. Attaye, I.; van Oppenraaij, S.; Warmbrunn, M.V.; Nieuwdorp, M. The Role of the Gut Microbiota on the Beneficial Effects of Ketogenic Diets. *Nutrients* **2021**, *14*, 191. [CrossRef]
84. O'Mahony, S.M.; Clarke, G.; Borre, Y.E.; Dinan, T.G.; Cryan, J.F. Serotonin, tryptophan metabolism and the brain-gut-microbiomeaxis. *Behav. Brain Res.* **2015**, *277*, 32–48. [CrossRef]
85. Silva, M.C.; Rocha, J.; Pires, C.S.; Ribeiro, L.C.; Brolese, G.; Leite, M.C.; Almeida, L.M.V.; Tramontina, F.; Ziegler, D.R.; Gonçalves, C.A. Transitory gliosis in the CA3 hippocampal region in rats fed on a ketogenic diet. *Nutr. Neurosci.* **2005**, *8*, 259–264. [CrossRef]
86. Maalouf, M.; Sullivan, P.G.; Davis, L.; Kim, D.Y.; Rho, J.M. Ketones inhibit mitochondrial production of reactive oxygen species production following glutamate excitotoxicity by increasing NADH oxidation. *Neuroscience* **2007**, *145*, 256–264. [CrossRef]
87. Beghi, E.; Hesdorffer, D. Prevalence of epilepsy—An unknown quantity. *Epilepsia* **2014**, *55*, 963–967. [CrossRef]
88. Fiest, K.M.; Sauro, K.M.; Wiebe, S.; Patten, S.B.; Kwon, C.S.; Dykeman, J.; Pringsheim, T.; Lorenzetti, D.L.; Jetté, N. Prevalence and incidence of epilepsy: A systematic review and meta-analysis of international studies. *Neurology* **2017**, *89*, 296–303. [CrossRef] [PubMed]
89. Beghi, E. The Epidemiology of Epilepsy. *Neuroepidemiology* **2020**, *54*, 185–191. [CrossRef] [PubMed]
90. WHO. Epilepsy. Available online: https://www.who.int/news-room/fact-sheets/detail/epilepsy (accessed on 22 September 2022).
91. García-Ramos, R.; García Pastor, A.; Masjuan, J.; Sánchez, C.; Gil, A. FEEN: Informe sociosantario FEEN sobre la epilepsia en España [Feen report on epilepsy in Spain]. *Neurologia* **2011**, *26*, 548–555. (In Spanish) [CrossRef] [PubMed]
92. Xue-Ping, W.; Hai-Jiao, W.; Li-Na, Z.; Xu, D.; Ling, L. Risk factors for drug-resistant epilepsy. *Medicine* **2019**, *98*, e16402. [CrossRef] [PubMed]
93. Martin-McGill, K.J.; Bresnahan, R.; Levy, R.G.; Cooper, P.N. Ketogenic diets for drug-resistant epilepsy. *Cochrane Database Syst. Rev.* **2020**, *2020*, CD001903. [CrossRef]
94. Barborka, C.J. Epilepsy in adults: Results of treatment by ketogenic diet in one hundred cases. *Arch. Neurol. Psychiatry* **1930**, *23*, 904–914. [CrossRef]
95. Bastible, C. The ketogenic treatment of epilepsy. *Ir. J. Med. Sci.* **1931**, *6*, 506–520. [CrossRef]
96. Klein, P.; Janousek, J.; Barber, A.; Weissberger, R. Ketogenic diet treatment in adults with refractory epilepsy. *Epilepsy Behav.* **2010**, *19*, 575–579. [CrossRef]
97. Schoeler, N.E.; Wood, S.; Aldridge, V.; Sander, J.W.; Cross, J.H.; Sisodiya, S.M. Ketogenic dietary therapies for adults with epilepsy: Feasibility and classification of response. *Epilepsy Behav.* **2014**, *37*, 77–81. [CrossRef]
98. Klein, P.; Tyrlikova, I.; Mathews, G.C.; Wu, H.Z.; Ahmad, K.E.; Tan, K.; Blair, N.F. Dietary treatment in adults with refractory epilepsy: A review. *Neurology* **2014**, *83*, 1978–1985. [CrossRef]
99. Ułamek-Kozioł, M.; Czuczwar, S.J.; Januszewski, S.; Pluta, R. Ketogenic Diet and Epilepsy. *Nutrients* **2019**, *11*, 2510. [CrossRef]
100. Herrero, J.R.; Villarroya, E.C.; Peñas, J.J.G.; Alcolea, B.G.; Fernández, B.G.; Macfarland, L.A.P.; Pedrón-Giner, C. Ketogenic dietary therapies for epilepsy: Experience in 160 patients over 18 years. 2022, 96, 511–522. *An. De Pediatría* **2022**, *96*, 511–522. [CrossRef]
101. Pizzo, F.; Collotta, A.D.; Di Nora, A.; Costanza, G.; Ruggieri, M.; Falsaperla, R. Ketogenic diet in pediatric seizures: A randomized controlled trial review and meta-analysis. *Expert Rev. Neurother.* **2022**, *22*, 169–177. [CrossRef] [PubMed]
102. Sourbron, J.; Klinkenberg, S.; van Kuijk, S.M.J.; Lagae, L.; Lambrechts, D.; Braakman, H.M.H.; Majoie, M. Ketogenic diet for the treatment of pediatric epilepsy: Review and meta-analysis. *Child's Nerv. Syst.* **2020**, *36*, 1099–1109. [CrossRef]
103. Lyons, L.; Schoeler, N.E.; Langan, D.; Cross, H. Use of ketogenic diet therapy in infants with epilepsy: A systematic review and meta-analysis. *Epilepsia* **2020**, *61*, 1261–1281. [CrossRef]
104. Rezaei, S.; Abdurahman, A.A.; Saghazadeh, A.; Badv, R.S.; Mahmoudi, M. Short-term and long-term efficacy of classical ketogenic diet and modified Atkins diet in children and adolescents with epilepsy: A systematic review and meta-analysis. *Nutr. Neurosci.* **2019**, *22*, 317–334. [CrossRef] [PubMed]
105. Henderson, C.B.; Filloux, F.M.; Alder, S.C.; Lyon, J.L.; Caplin, D.A. Efficacy of the Ketogenic Diet as a Treatment Option for Epilepsy: Meta-analysis. *J. Child Neurol.* **2006**, *21*, 193–198. [CrossRef]

106. Neal, E.G.; Chaffe, H.; Schwartz, R.H.; Lawson, M.S.; Edwards, N.; Fitzsimmons, G.; Whitney, A.; Cross, J.H. A randomized trial of classical and medium-chain triglyceride ketogenic diets in the treatment of childhood epilepsy. *Epilepsia* **2009**, *50*, 1109–1117. [CrossRef]
107. Meira, I.D.; Romão, T.T.; Prado, H.J.P.D.; Krüger, L.T.; Pires, M.E.P.; Da Conceição, P.O. Ketogenic Diet and Epilepsy: What We Know So Far. *Front. Neurosci.* **2019**, *13*, 5. [CrossRef]
108. Thio, L.L.; Wong, M.; Yamada, K.A. Ketone bodies do not directly alter excitatory or inhibitory hippocampal synaptic transmission. *Neurology* **2000**, *54*, 325. [CrossRef]
109. Likhodii, S.S.; Burnham, W.M. Ketogenic diet: Does acetone stop seizures? *Med. Sci. Monit.* **2002**, *8*, HY19-24. [PubMed]
110. Rho, J.M.; Anderson, G.D.; Donevan, S.D.; White, H.S. Acetoacetate, Acetone, and Dibenzylamine (a Contaminant in l-(+)-β-Hydroxybutyrate) Exhibit Direct Anticonvulsant Actions in Vivo. *Epilepsia* **2002**, *43*, 358–361. [CrossRef] [PubMed]
111. Gasior, M.; Rogawski, M.A.; Hartman, A. Neuroprotective and disease-modifying effects of the ketogenic diet. *Behav. Pharmacol.* **2006**, *17*, 431–439. [CrossRef]
112. D'Agostino, D.P.; Pilla, R.; Held, H.E.; Landon, C.S.; Puchowicz, M.; Brunengraber, H.; Ari, C.; Arnold, P.; Dean, J.B. Therapeutic ketosis with ketone ester delays central nervous system oxygen toxicity seizures in rats. *Am. J. Physiol. Integr. Comp. Physiol.* **2013**, *304*, R829–R836. [CrossRef]
113. Likhodii, S.S.; Serbanescu, I.; Cortez, M.A.; Murphy, P.; Snead, O.C.; Burnham, W.M. Anticonvulsant properties of acetone, a brain ketone elevated by the ketogenic diet. *Ann. Neurol.* **2003**, *54*, 219–226. [CrossRef] [PubMed]
114. Thio, L.L.; Rensing, N.; Maloney, S.; Wozniak, D.F.; Xiong, C.; Yamada, K.A. A ketogenic diet does not impair rat behavior or long-term potentiation. *Epilepsia* **2010**, *51*, 1619–1623. [CrossRef] [PubMed]
115. Manville, R.; Papanikolaou, M.; Abbott, G.W. M-Channel Activation Contributes to the Anticonvulsant Action of the Ketone Body β-Hydroxybutyrate. *J. Pharmacol. Exp. Ther.* **2020**, *372*, 148–156. [CrossRef] [PubMed]
116. Greene, A.E.; Todorova, M.T.; Seyfried, T.N. Perspectives on the metabolic management of epilepsy through dietary reduction of glucose and elevation of ketone bodies. *J. Neurochem.* **2003**, *86*, 529–537. [CrossRef] [PubMed]
117. Huttenlocher, P.R. Ketonemia and Seizures: Metabolic and Anticonvulsant Effects of Two Ketogenic Diets in Childhood Epilepsy. *Pediatr. Res.* **1976**, *10*, 536–540. [CrossRef]
118. Garriga-Canut, M.; Schoenike, B.; Qazi, R.; Bergendahl, K.; Daley, T.J.; Pfender, R.M.; Morrison, J.F.; Ockuly, J.C.; Stafstrom, C.E.; Sutula, T.P.; et al. 2-Deoxy-D-glucose reduces epilepsy progression by NRSF-CtBP–dependent metabolic regulation of chromatin structure. *Nat. Neurosci.* **2006**, *9*, 1382–1387. [CrossRef]
119. Mironov, S.; Richter, D. Intracellular signalling pathways modulate KATP channels in inspiratory brainstem neurones and their hypoxic activation: Involvement of metabotropic receptors, G-proteins and cytoskeleton. *Brain Res.* **2000**, *853*, 60–67. [CrossRef] [PubMed]
120. Li, D.-P.; Chen, S.-R.; Pan, H.-L. Adenosine inhibits paraventricular pre-sympathetic neurons through ATP-dependent potassium channels. *J. Neurochem.* **2010**, *113*, 530–542. [CrossRef] [PubMed]
121. Rasmussen, E.; Patel, V.; Tideman, S.; Frech, R.; Frigerio, R.; Narayanan, J. Efficacy of supplemental MCT oil on seizure reduction of adult drug-resistant epilepsy—A single-center open-label pilot study. *Nutr. Neurosci.* 2022; 1–5, *online ahead of print*. [CrossRef]
122. Kadowaki, A.; Sada, N.; Juge, N.; Wakasa, A.; Moriyama, Y.; Inoue, T. Neuronal inhibition and seizure suppression by acetoacetate and its analog, 2-phenylbutyrate. *Epilepsia* **2017**, *58*, 845–857. [CrossRef] [PubMed]
123. Szot, P.; Weinshenker, D.; Rho, J.M.; Storey, T.W.; Schwartzkroin, P.A. Norepinephrine is required for the anticonvulsant effect of the ketogenic diet. *Dev. Brain Res.* **2001**, *129*, 211–214. [CrossRef] [PubMed]
124. Wang, Z.J.; Bergqvist, C.; Hunter, J.V.; Jin, D.; Wang, D.-J.; Wehrli, S.; Zimmerman, R.A. In vivo measurement of brain metabolites using two-dimensional double-quantum MR spectroscopy? Exploration of GABA levels in a ketogenic diet. *Magn. Reson. Med.* **2003**, *49*, 615–619. [CrossRef] [PubMed]
125. Yudkoff, M.; Daikhin, Y.; Melø, T.M.; Nissim, I.; Sonnewald, U.; Nissim, I. The Ketogenic Diet and Brain Metabolism of Amino Acids: Relationship to the Anticonvulsant Effect. *Annu. Rev. Nutr.* **2007**, *27*, 415–430. [CrossRef]
126. Gavrilovici, C.; Rho, J.M. Metabolic epilepsies amenable to ketogenic therapies: Indications, contraindications, and underlying mechanisms. *J. Inherit. Metab. Dis.* **2021**, *44*, 42–53. [CrossRef]
127. Olson, C.A.; Vuong, H.E.; Yano, J.M.; Liang, Q.Y.; Nusbaum, D.J.; Hsiao, E.Y. The Gut Microbiota Mediates the Anti-Seizure Effects of the Ketogenic Diet. *Cell* **2018**, *173*, 1728–1741.e13. [CrossRef]
128. Dupuis, N.; Curatolo, N.; Benoist, J.-F.; Auvin, S. Ketogenic diet exhibits anti-inflammatory properties. *Epilepsia* **2015**, *56*, e95–e98. [CrossRef]
129. Shimazu, T.; Hirschey, M.D.; Newman, J.; He, W.; Shirakawa, K.; Le Moan, N.; Grueter, C.A.; Lim, H.; Saunders, L.R.; Stevens, R.D.; et al. Suppression of Oxidative Stress by β-Hydroxybutyrate, an Endogenous Histone Deacetylase Inhibitor. *Science* **2013**, *339*, 211–214. [CrossRef]
130. Wang, X.; Wu, X.; Liu, Q.; Kong, G.; Zhou, J.; Jiang, J.; Zhu, Q. Ketogenic metabolism inhibits histone deacetylase (HDAC) and reduces oxidative stress after spinal cord injury in rats. *Neuroscience* **2017**, *366*, 36–43. [CrossRef] [PubMed]
131. Ko, A.; Kwon, H.E.; Kim, H.D. Updates on the ketogenic diet therapy for pediatric epilepsy. *Biomed. J.* **2022**, *45*, 19–26. [CrossRef] [PubMed]
132. Patterson, C. *World Alzheimer Report 2018*; Alzheimer's Disease International: London, UK, 2018.
133. WHO. Dementia. Available online: https://www.who.int/news-room/fact-sheets/detail/dementia (accessed on 20 September 2022).

134. Scheltens, P.; Blennow, K.; Breteler, M.M. Alzheimer's disease. *Lancet* **2016**, *388*, 505–517. [CrossRef]
135. Zhang, X.-X.; Tian, Y.; Wang, Z.-T.; Ma, Y.-H.; Tan, L.; Yu, J.-T. The Epidemiology of Alzheimer's Disease Modifiable Risk Factors and Prevention. *J. Prev. Alzheimer's Dis.* **2021**, *8*, 313–321. [CrossRef]
136. Barczak, A. Wczesne rozpoznanie choroby Alzheimera—Wskazówki dla lekarza POZ. *Lek. POZ* **2022**, *8*, 57–62.
137. Alzheimer's Association. Alzheimer's Disease Facts and Figures. *Alzheimer Dement.* **2020**, *16*, 391–460. [CrossRef]
138. Winblad, B.; Amouyel, P.; Andrieu, S.; Ballard, C.; Brayne, C.; Brodaty, H.; Cedazo-Minguez, A.; Dubois, B.; Edvardsson, D.; Feldman, H.; et al. Defeating Alzheimer's disease and other dementias: A priority for European science and society. *Lancet Neurol.* **2016**, *15*, 455–532. [CrossRef]
139. Lilamand, M.; Mouton-Liger, F.; Paquet, C. Ketogenic diet therapy in Alzheimer's disease: An updated review. *Curr. Opin. Clin. Nutr. Metab. Care* **2021**, *24*, 372–378. [CrossRef]
140. de la Monte, S.M.; Wands, J.R. Alzheimer's Disease is Type 3 Diabetes—Evidence Reviewed. *J. Diabetes Sci. Technol.* **2008**, *2*, 1101–1113. [CrossRef]
141. Talbot, K.; Wang, H.-Y.; Kazi, H.; Han, L.-Y.; Bakshi, K.P.; Stucky, A.; Fuino, R.L.; Kawaguchi, K.R.; Samoyedny, A.J.; Wilson, R.S.; et al. Demonstrated brain insulin resistance in Alzheimer's disease patients is associated with IGF-1 resistance, IRS-1 dysregulation, and cognitive decline. *J. Clin. Investig.* **2012**, *122*, 1316–1338. [CrossRef]
142. Ferreira, L.S.S.; Fernandes, C.S.; Vieira, M.N.N.; De Felice, F.G. Insulin Resistance in Alzheimer's Disease. *Front. Neurosci.* **2018**, *12*, 830. [CrossRef] [PubMed]
143. Sędzikowska, A.; Szablewski, L. Insulin and Insulin Resistance in Alzheimer's Disease. *Int. J. Mol. Sci.* **2021**, *22*, 9987. [CrossRef] [PubMed]
144. Mietelska-Porowska, A.; Domańska, J.; Want, A.; Więckowska-Gacek, A.; Chutorański, D.; Koperski, M.; Wojda, U. Induction of Brain Insulin Resistance and Alzheimer's Molecular Changes by Western Diet. *Int. J. Mol. Sci.* **2022**, *23*, 4744. [CrossRef] [PubMed]
145. Liu, Y.; Liu, F.; Iqbal, K.; Grundke-Iqbal, I.; Gong, C.-X. Decreased glucose transporters correlate to abnormal hyperphosphorylation of tau in Alzheimer disease. *FEBS Lett.* **2008**, *582*, 359–364. [CrossRef]
146. Lying-Tunell, U.; Lindblad, B.S.; Malmlund, H.O.; Persson, B. Cerebral blood flow and metabolic rate of oxygen, glucose, lactate, pyruvate, ketone bodies and amino acids. *Acta Neurol. Scand.* **1981**, *63*, 337–350. [CrossRef]
147. Castellano, C.-A.; Nugent, S.; Paquet, N.; Tremblay, S.; Bocti, C.; Lacombe, G.; Imbeault, H.; Turcotte, E.; Fulop, T.; Cunnane, S.C. Lower Brain 18F-Fluorodeoxyglucose Uptake But Normal 11C-Acetoacetate Metabolism in Mild Alzheimer's Disease Dementia. *J. Alzheimer's Dis.* **2015**, *43*, 1343–1353. [CrossRef]
148. Muscogiuri, G.; El Ghoch, M.; Colao, A.; Hassapidou, M.; Yumuk, V.; Busetto, L. Obesity Management Task Force (OMTF) of the European Association for the Study of Obesity (EASO). European Guidelines for Obesity Management in Adults with a Very Low-Calorie Ketogenic Diet: A Systematic Review and Meta-Analysis. *Obes. Facts* **2021**, *14*, 222–245, PMCID: PMC8138199. [CrossRef] [PubMed]
149. Sakr, H.F.; Sirasanagandla, S.R.; Das, S.; Bima, A.I.; Elsamanoudy, A.Z. Low-Carbohydrate Ketogenic Diet for Improvement of Glycemic Control: Mechanism of Action of Ketosis and Beneficial Effects. *Curr. Diabetes Rev.* 2022; *ahead of print*. [CrossRef]
150. Michalczyk, M.M.; Klonek, G.; Maszczyk, A.; Zajac, A. The Effects of a Low Calorie Ketogenic Diet on Glycaemic Control Variables in Hyperinsulinemic Overweight/Obese Females. *Nutrients* **2020**, *12*, 1854. [CrossRef]
151. Johnstone, A.M.; Horgan, G.W.; Murison, S.D.; Bremner, D.M.; Lobley, G.E. Effects of a high-protein ketogenic diet on hunger, appetite, and weight loss in obese men feeding ad libitum. *Am. J. Clin. Nutr.* **2008**, *87*, 44–55. [CrossRef]
152. Taylor, M.K.; Sullivan, D.K.; Keller, J.E.; Burns, J.M.; Swerdlow, R.H. Potential for Ketotherapies as Amyloid-Regulating Treatment in Individuals at Risk for Alzheimer's Disease. *Front. Neurosci.* **2022**, *16*, 899612. [CrossRef]
153. Hernandez, A.R.; Hernandez, C.M.; Truckenbrod, L.M.; Campos, K.T.; McQuail, J.A.; Bizon, J.L.; Burke, S.N. Age and Ketogenic Diet Have Dissociable Effects on Synapse-Related Gene Expression Between Hippocampal Subregions. *Front Aging Neurosci.* **2019**, *11*, 239. [CrossRef] [PubMed]
154. Ling, Y.; Wang, D.-D.; Sun, Y.-X.; Zhao, D.-J.; Ni, H. Neuro-Behavioral Status and the Hippocampal Expression of Metabolic Associated Genes in Wild-Type Rat Following a Ketogenic Diet. *Front. Neurol.* **2019**, *10*, 65. [CrossRef] [PubMed]
155. Koppel, S.J.; Pei, D.; Wilkins, H.M.; Weidling, I.W.; Wang, X.; Menta, B.W.; Perez-Ortiz, J.; Kalani, A.; Manley, S.; Novikova, L.; et al. A ketogenic diet differentially affects neuron and astrocyte transcription. *J. Neurochem.* **2021**, *157*, 1930–1945. [CrossRef]
156. Hughes, S.D.; Kanabus, M.; Anderson, G.; Hargreaves, I.P.; Rutherford, T.; Donnell, M.O.; Cross, J.H.; Rahman, S.; Eaton, S.; Heales, S.J.R. The ketogenic diet component decanoic acid increases mitochondrial citrate synthase and complex I activity in neuronal cells. *J. Neurochem.* **2014**, *129*, 426–433. [CrossRef] [PubMed]
157. Wilkins, H.M.; Swerdlow, R.H. Amyloid precursor protein processing and bioenergetics. *Brain Res. Bull.* **2017**, *133*, 71–79. [CrossRef] [PubMed]
158. Hasan-Olive, M.; Lauritzen, K.H.; Ali, M.; Rasmussen, L.J.; Storm-Mathisen, J.; Bergersen, L.H. A Ketogenic Diet Improves Mitochondrial Biogenesis and Bioenergetics via the PGC1α-SIRT3-UCP2 Axis. *Neurochem. Res.* **2019**, *44*, 22–37. [CrossRef] [PubMed]
159. Seira, O.; Kolehmainen, K.; Liu, J.; Streijger, F.; Haegert, A.; Lebihan, S.; Boushel, R.; Tetzlaff, W. Ketogenesis controls mitochondrial gene expression and rescues mitochondrial bioenergetics after cervical spinal cord injury in rats. *Sci. Rep.* **2021**, *11*, 16359. [CrossRef]

160. Croteau, E.; Castellano, C.-A.; Richard, M.A.; Fortier, M.; Nugent, S.; Lepage, M.; Duchesne, S.; Whittingstall, K.; Turcotte, E.; Bocti, C.; et al. Ketogenic Medium Chain Triglycerides Increase Brain Energy Metabolism in Alzheimer's Disease. *J. Alzheimer's Dis.* **2018**, *64*, 551–561. [CrossRef]
161. Juby, A.G.; Blackburn, T.E.; Mager, D.R. Use of medium chain triglyceride (MCT) oil in subjects with Alzheimer's disease: A randomized, double-blind, placebo-controlled, crossover study, with an open-label extension. *Alzheimer's Dementia Transl. Res. Clin. Interv.* **2022**, *8*, e12259. [CrossRef]
162. Reger, M.A.; Henderson, S.T.; Hale, C.; Cholerton, B.; Baker, L.D.; Watson, G.; Hyde, K.; Chapman, D.; Craft, S. Effects of β-hydroxybutyrate on cognition in memory-impaired adults. *Neurobiol. Aging* **2004**, *25*, 311–314. [CrossRef]
163. Ota, M.; Matsuo, J.; Ishida, I.; Hattori, K.; Teraishi, T.; Tonouchi, H.; Ashida, K.; Takahashi, T.; Kunugi, H. Effect of a ketogenic meal on cognitive function in elderly adults: Potential for cognitive enhancement. *Psychopharmacology* **2016**, *233*, 3797–3802. [CrossRef] [PubMed]
164. Kashiwaya, Y.; Takeshima, T.; Mori, N.; Nakashima, K.; Clarke, K.; Veech, R.L. d-β-Hydroxybutyrate protects neurons in models of Alzheimer's and Parkinson's disease. *Proc. Natl. Acad. Sci. USA* **2000**, *97*, 5440–5444. [CrossRef] [PubMed]
165. Studzinski, C.M.; MacKay, W.A.; Beckett, T.L.; Henderson, S.T.; Murphy, M.P.; Sullivan, P.G.; Burnham, W.M. Induction of ketosis may improve mitochondrial function and decrease steady-state amyloid-β precursor protein (APP) levels in the aged dog. *Brain Res.* **2008**, *1226*, 209–217. [CrossRef] [PubMed]
166. Kashiwaya, Y.; Bergman, C.; Lee, J.-H.; Wan, R.; King, M.T.; Mughal, M.R.; Okun, E.; Clarke, K.; Mattson, M.P.; Veech, R.L. A ketone ester diet exhibits anxiolytic and cognition-sparing properties, and lessens amyloid and tau pathologies in a mouse model of Alzheimer's disease. *Neurobiol. Aging* **2013**, *34*, 1530–1539. [CrossRef]
167. Nafar, F.; Clarke, J.; Mearow, K. Coconut oil protects cortical neurons from amyloid beta toxicity by enhancing signaling of cell survival pathways. *Neurochem. Int.* **2017**, *105*, 64–79. [CrossRef]
168. Neal, E.G.; Cross, J.H. Efficacy of dietary treatments for epilepsy. *J. Hum. Nutr. Diet.* **2010**, *23*, 113–119. [CrossRef]
169. Henderson, S.T.; Vogel, J.L.; Barr, L.J.; Garvin, F.; Jones, J.J.; Costantini, L.C. Study of the ketogenic agentAC-1202 in mild to moderate Alzheimer's disease: A randomized, double-blind, placebo-controlled, multicenter trial. *Nutr. Metab. (Lond)* **2009**, *6*, 31. [CrossRef]
170. Newport, M.T.; VanItallie, T.B.; Kashiwaya, Y.; King, M.T.; Veech, R.L. A new way to produce hyperketonemia: Use of ketone ester in a case of Alzheimer's disease. *Alzheimer's Dement.* **2015**, *11*, 99–103. [CrossRef]
171. Phillips, M.C.L.; Deprez, L.M.; Mortimer, G.M.N.; Murtagh, D.K.J.; McCoy, S.; Mylchreest, R.; Gilbertson, L.J.; Clark, K.M.; Simpson, P.V.; McManus, E.J.; et al. Randomized crossover trial of a modified ketogenic diet in Alzheimer's disease. *Alzheimer's Res. Ther.* **2021**, *13*, 51. [CrossRef]
172. Raina, P.; Santaguida, P.; Ismaila, A.; Patterson, C.; Cowan, D.; Levine, M.; Booker, L.; Oremus, M. Effectiveness of Cholinesterase Inhibitors and Memantine for Treating Dementia: Evidence Review for a Clinical Practice Guideline. *Ann. Intern. Med.* **2008**, *148*, 379–397. [CrossRef]
173. Siemers, E.; Holdridge, K.C.; Sundell, K.L.; Liu-Seifert, H. Function and clinical meaningfulness of treatments for mild Alzheimer's disease. *Alzheimer's Dementia Diagn. Assess. Dis. Monit.* **2016**, *2*, 105–112. [CrossRef] [PubMed]
174. Shulman, L.M. Understanding disability in Parkinson's disease. *Mov. Disord.* **2010**, *25* (Suppl. S1), S131–S135. [CrossRef] [PubMed]
175. Simon, D.K.; Tanner, C.M.; Brundin, P. Parkinson Disease Epidemiology, Pathology, Genetics, and Pathophysiology. *Clin Geriatr Med.* **2020**, *36*, 1–12. [CrossRef]
176. Sveinbjornsdottir, S. The clinical symptoms of Parkinson's disease. *J. Neurochem.* **2016**, *139*, 318–324. [CrossRef] [PubMed]
177. Lotankar, S.; Prabhavalkar, K.S.; Bhatt, L.K. Biomarkers for Parkinson's Disease: Recent Advancement. *Neurosci. Bull.* **2017**, *33*, 585–597. [CrossRef] [PubMed]
178. World Health Organization. *Global Health Estimates 2020: Disease Burden by Cause, Age, Sex by Country and by Region, 2000–2019*; World Health Organization: Geneva, Switzerland, 2020.
179. Bloem, B.R.; Okun, M.S.; Klein, C. Parkinson's disease. *Lancet* **2021**, *397*, 2284–2303. [CrossRef]
180. Evans, J.R.; Mason, S.L.; Williams-Gray, C.H.; Foltynie, T.; Brayne, C.; Robbins, T.W.; Barker, R.A. The natural history of treated Parkinson's disease in an incident, community based cohort. *J. Neurol. Neurosurg. Psychiatry* **2011**, *82*, 1112–1118. [CrossRef]
181. Shaafi, S.; Najmi, S.; Aliasgharpour, H.; Mahmoudi, J.; Sadigh-Etemad, S.; Farhoudi, M.; Baniasadi, N. The efficacy of the ketogenic diet on motor functions in Parkinson's disease: A rat model. *Iran. J. Neurol.* **2016**, *15*, 63–69.
182. Pietrzak, D.; Kasperek, K.; Rękawek, P.; Piątkowska-Chmiel, I. The Therapeutic Role of Ketogenic Diet in Neurological Disorders. *Nutrients.* **2022**, *14*, 1952. [CrossRef]
183. Zhu, Y.; Tang, X.; Cheng, Z.; Dong, Q.; Ruan, G. The Anti-Inflammatory Effect of Preventive Intervention with Ketogenic Diet Mediated by the Histone Acetylation of mGluR5 Promotor Region in Rat Parkinson's Disease Model: A Dual-Tracer PET Study. *Park. Dis.* **2022**, *2022*, 3506213. [CrossRef]
184. Masino, S.A.; Rho, J.M. Mechanisms of Ketogenic Diet Action. In *Jasper's Basic Mechanisms of the Epilepsies*; Noebels, J.L., Avoli, M., Rogawski, M.A., Olsen, R.W., Delgado-Escueta, A.V., Eds.; National Center for Biotechnology Information: Bethesda, MD, USA, 2012; pp. 1157–1182.
185. Norwitz, N.G.; Hu, M.T.; Clarke, K. The Mechanisms by Which the Ketone Body D-β-Hydroxybutyrate May Improve the Multiple Cellular Pathologies of Parkinson's Disease. *Front Nutr.* **2019**, *6*, 63. [CrossRef] [PubMed]

186. Aureli, C.; Cassano, T.; Masci, A.; Francioso, A.; Martire, S.; Cocciolo, A.; Chichiarelli, S.; Romano, A.; Gaetani, S.; Mancini, P.; et al. 5-S-cysteinyldopamine neurotoxicity: Influence on the expression of α-synuclein and ERp57 in cellular and animal models of Parkinson's disease. *J. Neurosci. Res.* **2014**, *92*, 347–358. [CrossRef] [PubMed]
187. Cooke, M.S.; Evans, M.D.; Dizdaroglu, M.; Lunec, J. Oxidative DNA damage: Mechanisms, mutation, and disease. *FASEB J.* **2003**, *17*, 1195–1214. [CrossRef] [PubMed]
188. VanItallie, T.B.; Nonas, C.; di Rocco, A.; Boyar, K.; Hyams, K.; Heymsfield, S.B. Treatment of Parkinson Disease with Diet-Induced Hyperketonemia: A Feasibility Study. *Neurology* **2005**, *64*, 728–730. [CrossRef] [PubMed]
189. Caputi, V.; Giron, M.C. Microbiome-Gut-Brain Axis and Toll-Like Receptors in Parkinson's Disease. *Int. J. Mol. Sci.* **2018**, *19*, 1689. [CrossRef] [PubMed]
190. Irfannuddin, I.; Sarahdeaz, S.F.P.; Murti, K.; Santoso, B.; Koibuchi, N. The effect of ketogenic diets on neurogenesis and apoptosis in the dentate gyrus of the male rat hippocampus. *J. Physiol. Sci.* **2021**, *71*, 3. [CrossRef]
191. Sampson, T.R.; Debelius, J.W.; Thron, T.; Janssen, S.; Shastri, G.G.; Ilhan, Z.E.; Challis, C.; Schretter, C.E.; Rocha, S.; Gradinaru, V.; et al. Gut Microbiota Regulate Motor Deficits and Neuroinflammation in a Model of Parkinson's Disease. *Cell* **2016**, *167*, 1469–1480.e12. [CrossRef]
192. Phillips, M.C.; Murtagh, D.K.; Gilbertson, L.J.; Asztely, F.J.; Lynch, C.D. Low-fat versus ketogenic diet in Parkinson's disease: A pilot randomized controlled trial. *Mov. Disord.* **2018**, *33*, 1306–1314. [CrossRef]
193. Tidman, M.M.; White, D.; White, T. Effects of an low carbohydrate/healthy fat/ketogenic diet on biomarkers of health and symptoms, anxiety and depression in Parkinson's disease: A pilot study. *Neurodegener. Dis. Manag.* **2022**, *12*, 57–66. [CrossRef]
194. Krikorian, R.; Shidler, M.D.; Summer, S.S.; Sullivan, P.G.; Duker, A.P.; Isaacson, R.S.; Espay, A.J. Nutritional ketosis for mild cognitive impairment in Parkinson's disease: A controlled pilot trial. *Clin. Park. Relat. Disord.* **2019**, *1*, 41–47. [CrossRef]
195. Tidman, M. Effects of a Ketogenic Diet on Symptoms, Biomarkers, Depression, and Anxiety in Parkinson's Disease: A Case Study. *Cureus* **2022**, *14*, e23684. [CrossRef] [PubMed]
196. Koyuncu, H.; Fidan, V.; Toktas, H.; Binay, O.; Celik, H. Effect of ketogenic diet versus regular diet on voice quality of patients with Parkinson's disease. *Acta Neurol. Belg.* **2021**, *121*, 1729–1732. [CrossRef] [PubMed]
197. Cameron, M.H.; Lord, S. Postural Control in Multiple Sclerosis: Implications for Fall Prevention. *Curr. Neurol. Neurosci. Rep.* **2010**, *10*, 407–412. [CrossRef]
198. Hebert, J.R.; Corboy, J.R.; Manago, M.M.; Schenkman, M. Effects of Vestibular Rehabilitation on Multiple Sclerosis–Related Fatigue and Upright Postural Control: A Randomized Controlled Trial. *Phys. Ther.* **2011**, *91*, 1166–1183. [CrossRef] [PubMed]
199. The Multiple Sclerosis International Federation (MSIF). Atlas of MS. In *Mapping Multiple Sclerosis around the World Key Epidemiology Findings*, 3rd ed.; MSIF: London, UK, 2020.
200. Updated Atlas of MS Shows Over 2.8 Million People Worldwide Have Multiple Sclerosis—With Nearly 1 Million in the US. Available online: https://www.nationalmssociety.org/About-the-Society/News/Updated-Atlas-of-MS-Shows-Over-2-8-million-People (accessed on 20 September 2022).
201. Lin, W.-S.; Lin, S.-J.; Liao, P.-Y.; Suresh, D.; Hsu, T.-R.; Wang, P.-Y. Role of Ketogenic Diets in Multiple Sclerosis and Related Animal Models: An Updated Review. *Adv. Nutr. Int. Rev. J.* **2022**, *13*, 2002–2014. [CrossRef]
202. Di Majo, D.; Cacciabaudo, F.; Accardi, G.; Gambino, G.; Giglia, G.; Ferraro, G.; Candore, G.; Sardo, P. Ketogenic and Modified Mediterranean Diet as a Tool to Counteract Neuroinflammation in Multiple Sclerosis: Nutritional Suggestions. *Nutrients* **2022**, *14*, 2384. [CrossRef]
203. Cellerino, A.; Carroll, P.; Thoenen, H.; Barde, Y.A. Reduced Size of Retinal Ganglion Cell Axons and Hypomyelination in Mice Lacking Brain-Derived Neurotrophic Factor. *Mol. Cell. Neurosci.* **1997**, *9*, 397–408. [CrossRef]
204. Veer, A.V.; Du, Y.; Fischer, T.Z.; Boetig, D.R.; Wood, M.R.; Dreyfus, C.F. Brain-derived neurotrophic factor effects on oligodendrocyte progenitors of the basal forebrain are mediated through trkB and the MAP kinase pathway. *J. Neurosci. Res.* **2009**, *87*, 69–78. [CrossRef]
205. Lee, D.-H.; Geyer, E.; Flach, A.-C.; Jung, K.; Gold, R.; Flügel, A.; Linker, R.A.; Lühder, F. Central nervous system rather than immune cell-derived BDNF mediates axonal protective effects early in autoimmune demyelination. *Acta Neuropathol.* **2012**, *123*, 247–258. [CrossRef]
206. Paoli, A.; Cenci, L.; Pompei, P.; Sahin, N.; Bianco, A.; Neri, M.; Caprio, M.; Moro, T. Effects of Two Months of Very Low Carbohydrate Ketogenic Diet on Body Composition, Muscle Strength, Muscle Area, and Blood Parameters in Competitive Natural Body Builders. *Nutrients* **2021**, *13*, 374. [CrossRef]
207. Vauzour, D. Polyphenols and brain health. *OCL* **2017**, *24*, A202. [CrossRef]
208. Benlloch, M.; López-Rodríguez, M.M.; Cuerda-Ballester, M.; Drehmer, E.; Carrera, S.; Ceron, J.J.; Tvarijonaviciute, A.; Chirivella, J.; Fernández-García, D.; de la Rubia Ortí, J.E. Satiating effect of a ketogenic diet and its impact on muscle improvement and oxidation state in multiple sclerosis patients. *Nutrients* **2019**, *11*, 1156. [CrossRef] [PubMed]
209. Bock, M.; Steffen, F.; Zipp, F.; Bittner, S. Impact of Dietary Intervention on Serum Neurofilament Light Chain in Multiple Sclerosis. *Neurol. Neuroimmunol. Neuroinflamm.* **2021**, *9*, e1102. [CrossRef] [PubMed]
210. Bock, M.; Karber, M.; Kuhn, H. Ketogenic diets attenuate cyclooxygenase and lipoxygenase gene expression in multiple sclerosis. *eBioMedicine* **2018**, *36*, 293–303. [CrossRef] [PubMed]

211. Brenton, J.N.; Lehner-Gulotta, D.; Woolbright, E.; Banwell, B.; Bergqvist, A.G.C.; Chen, S.; Coleman, R.; Conaway, M.; Goldman, M.D. Phase II study of ketogenic diets in relapsing multiple sclerosis: Safety, tolerability and potential clinical benefits. *J. Neurol. Neurosurg. Psychiatry* **2022**, *93*, 637–644. [CrossRef]
212. Brenton, J.N.; Banwell, B.; Bergqvist, A.C.; Lehner-Gulotta, D.; Gampper, L.; Leytham, E.; Coleman, R.; Goldman, M.D. Pilot study of a ketogenic diet in relapsing-remitting MS. *Neurol. Neuroimmunol. Neuroinflamm.* **2019**, *6*, e565. [CrossRef]
213. Lee, J.E.; Titcomb, T.J.; Bisht, B.; Rubenstein, L.M.; Louison, R.; Wahls, T.L. A Modified MCT-Based Ketogenic Diet Increases Plasma β-Hydroxybutyrate but Has Less Effect on Fatigue and Quality of Life in People with Multiple Sclerosis Compared to a Modified Paleolithic Diet: A Waitlist-Controlled, Randomized Pilot Study. *J. Am. Coll. Nutr.* **2021**, *40*, 13–25. [CrossRef]
214. Choi, I.Y.; Piccio, L.; Childress, P.; Bollman, B.; Ghosh, A.; Brandhorst, S.; Suarez, J.; Michalsen, A.; Cross, A.H.; Morgan, T.E.; et al. A Diet Mimicking Fasting Promotes Regeneration and Reduces Autoimmunity and Multiple Sclerosis Symptoms. *Cell Rep.* **2016**, *15*, 2136–2146. [CrossRef]
215. Burch, R.C.; Buse, D.C.; Lipton, R.B. Migraine. *Neurol. Clin.* **2019**, *37*, 631–649. [CrossRef]
216. Roos-Araujo, D.; Stuart, S.; Lea, R.A.; Haupt, L.M.; Griffiths, L.R. Epigenetics and migraine; complex mitochondrial interactions contributing to disease susceptibility. *Gene* **2014**, *543*, 1–7. [CrossRef]
217. Di Lorenzo, C.; Pinto, A.; Ienca, R.; Coppola, G.; Sirianni, G.; Di Lorenzo, G.; Parisi, V.; Serrao, M.; Spagnoli, A.; Vestri, A.; et al. A Randomized Double-Blind, Cross-Over Trial of very Low-Calorie Diet in Overweight Migraine Patients: A Possible Role for Ketones? *Nutrients* **2019**, *11*, 1742. [CrossRef]
218. Valente, M.; Garbo, R.; Filippi, F.; Antonutti, A.; Ceccarini, V.; Tereshko, Y.; Di Lorenzo, C.; Gigli, G.L. Migraine Prevention through Ketogenic Diet: More than Body Mass Composition Changes. *J. Clin. Med.* **2022**, *11*, 4946. [CrossRef] [PubMed]
219. Lovati, C.; D'Alessandro, C.M.; Della Ventura, S.; Muzio, F.; Pantoni, L. Ketogenic diet in refractory migraine: Possible efficacy and role of ketone bodies—A pilot experience. *Neurol. Sci.* **2022**, *43*, 6479–6485. [CrossRef] [PubMed]
220. Bongiovanni, D.; Benedetto, C.; Corvisieri, S.; Del Favero, C.; Orlandi, F.; Allais, G.; Sinigaglia, S.; Fadda, M. Effectiveness of ketogenic diet in treatment of patients with refractory chronic migraine. *Neurol. Sci.* **2021**, *42*, 3865–3870. [CrossRef] [PubMed]
221. Caminha, M.C.; Moreira, A.B.; Matheus, F.C.; Rieger, D.K.; Moreira, J.D.; Dalmarco, E.M.; Demarchi, I.G.; Lin, K. Efficacy and tolerability of the ketogenic diet and its variations for preventing migraine in adolescents and adults: A systematic review. *Nutr. Rev.* **2022**, *80*, 1634–1647. [CrossRef]
222. Bostock, E.C.S.; Kirkby, K.C.; Taylor, B.V.; Hawrelak, J.A. Consumer Reports of "Keto Flu" Associated With the Ketogenic Diet. *Front. Nutr.* **2020**, *7*, 20. [CrossRef] [PubMed]

nutrients MDPI

Review

Effect of the Ketogenic Diet on the Prophylaxis and Treatment of Diabetes Mellitus: A Review of the Meta-Analyses and Clinical Trials

Damian Dyńka [1], Katarzyna Kowalcze [1], Filip Ambrozkiewicz [2] and Agnieszka Paziewska [1,3,*]

[1] Institute of Health Sciences, Faculty of Medical and Health Sciences, Siedlce University of Natural Sciences and Humanities, 08-110 Siedlce, Poland
[2] Laboratory of Translational Cancer Genomics, Biomedical Center, Faculty of Medicine in Pilsen, Charles University, Alej Svobody 1665/76, 32300 Pilsen, Czech Republic
[3] Department of Neuroendocrinology, Centre of Postgraduate Medical Education, 01-813 Warsaw, Poland
* Correspondence: agnieszka.paziewska@uph.edu.pl

Abstract: The exponentially growing frequency of diagnosing diabetes mellitus means that a verification of the previous dietetic approach to treating the disease seems justified. The simultaneous growth of interest in the ketogenic diet and the development of knowledge in this field have contributed to the increasingly frequent application of the ketogenic diet in diabetes treatment. This paper also deals with that issue; its aim includes an extensive analysis of the influence of the ketogenic diet on the prophylaxis and treatment of diabetes. The paper has been prepared based on a wide, meticulous analysis of the available literature on the subject. Among other findings, a favorable effect of that nutrition model has been demonstrated on the values of glycated hemoglobin, glucose, insulin, or other metabolic parameters in diabetes patients. The effect of the ketogenic diet on the pharmacotherapy of type 1 and type 2 diabetes has been presented and compared with the standard nutritional management plan recommended for that disease. Further research is needed in this field, especially studies with a long follow-up period. The discussed articles report interesting therapeutic advantages to the ketogenic diet in comparison with standard diets.

Keywords: ketogenic diet; ketosis; diabetes mellitus; type 1 diabetes mellitus; type 2 diabetes mellitus; glycaemia; glycated hemoglobin; HbA1c; HOMA-IR; metabolic syndrome; chronic diseases; obesity; lipid profile; lipidaemia; prophylaxis; treatment; nutrition recommendation; nutrition intervention; ketogenic; dietary patterns; carbohydrates; PRISMA; Cochrane

Citation: Dyńka, D.; Kowalcze, K.; Ambrozkiewicz, F.; Paziewska, A. Effect of the Ketogenic Diet on the Prophylaxis and Treatment of Diabetes Mellitus: A Review of the Meta-Analyses and Clinical Trials. *Nutrients* **2023**, *15*, 500. https://doi.org/10.3390/nu15030500

Academic Editor: Sareen Gropper

Received: 16 December 2022
Revised: 14 January 2023
Accepted: 16 January 2023
Published: 18 January 2023

1. Introduction

Ongoing progress in medical science has contributed to the increasingly frequent research activities conducted into the ketogenic diet. Starting from 1921, when the diet was used for the first time, the scope of studies on the ketogenic nutrition method has been continuously increasing. Its primary application concerned the treatment of epilepsy and was initially developed over a hundred years ago [1]. The significant effects of that diet, observed in the treatment of epilepsy, even in drug-resistant varieties, have received much attention from both the scientific establishment and the public. Currently, it is still widely used in the treatment of epilepsy; not infrequently, its efficacy exceeds that of pharmacotherapy [2–4]. The extraordinary therapeutic effects seen in drug-resistant epilepsy gave a motive for the development of research into the effect of the ketogenic diet in many other domains [5–7]. One very interesting and forward-looking approach, from the perspective of progress in medicine, concerns the effectiveness of the ketogenic diet in treating diabetes mellitus.

From the historical point of view, ketogenic diets in T1DM treatment date back to the early twentieth century [8]. The first publication describing the use of a strictly ketogenic diet in type 1 diabetes is the study written by Henwood et al. in 2006 [9]. In terms of T2DM, there is research describing the use of a ketogenic diet as early as 1914–1922 [10]. The diet in question limited the amount of energy from carbohydrates to 8%. Therefore, it can be considered to be a ketogenic diet. The first major study describing the use of a strictly ketogenic diet in type 2 diabetes patients was published in 1996 [11]. Importantly, in both cases, a number of benefits were observed. This encouraged the researchers to further explore the subject, which resulted in a significant increase in the number of studies on this topic in successive years, as shown in Figure 1. The two publications mentioned were the first examples of studies on the effect of a strictly ketogenic diet in type 1 and type 2 diabetes. It is worth emphasizing, however, that recommendations of a diet with the limitation of carbohydrates and an increase in the energy share from fat in diabetes individuals could have also been encountered earlier, in the form of such diets as the paleo, Atkins, South Beach or Zone diets. However, these are not typical ketogenic diets but they show a growing tendency that affects the current studies on strictly ketogenic diets in the treatment of diabetes [12–14].

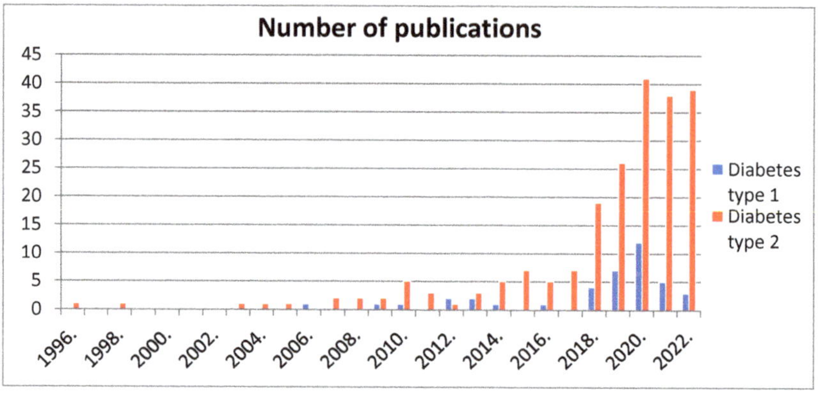

Figure 1. Comparison of the number of publications in the PubMed search engine for type 1 diabetes (phrases: "ketogenic diet and diabetes type 1"; "ketogenic diet and type 1 diabetes"; "ketogenic diet and T1DM"; "ketogenic diet and autoimmune diabetes") and type 2 diabetes (phrases: "ketogenic diet and diabetes type 2"; "ketogenic diet and type 2 diabetes"; "ketogenic diet and T2DM").

On the one hand, available data may provide indirect evidence in terms of the possible prophylactic application of that diet in diabetes (mainly type 2). Success also depends, to some extent, on its treatment. The results are achieved in the first place from the significant influence of the ketogenic diet on glycemia parameters [15–17]. On the other hand, the number of studies suggesting direct benefits in the treatment of diabetes is still insufficient. Other factors, such as the low number of studies on the effect of a ketogenic diet on diabetes, the broad spectrum of available indirect evidence, and the high potential of that diet contribute to the great need for filling the undeveloped scientific space with the development of studies and publications in this area. The multidimensionality and extensiveness of the issue of diabetes, resulting from its various types, also require a clear distinction between studies on the potential use of the ketogenic diet, depending on the type of the disease. Therefore, there is a justified need to conduct research in this area. Taking into account the prevalence of diabetes, such publications could finally lead to the practical implementation of a specific dietetic management plan. This could result in an improvement in the quality of life and the survival of millions of patients struggling with this disease.

2. Ketogenic Diet

The ketogenic diet, according to the most accurate and, at the same time, the most general definition, is a diet leading to the increased formation of ketone bodies (such as β-hydroxybutyrate, acetoacetate, and acetone) in the organism. Its task includes somewhat mimicking a condition of fasting, without the negative consequences of starvation [18]. Despite the plethora of various types of diets, this one undoubtedly distinguishes itself from all others. It leads to a change in the body's preferences concerning its main source of energy supply. While, in other diets, the main source of energy is glucose, in this case, the organism preferentially targets ketone bodies. The organism is then put into the so-called nutritional ketosis state, which, in turn, has many favorable applications [19,20]. It is a diet with a low carbohydrate content, high fat content, and moderate protein content. It can be assumed that the majority of ketogenic diets in practice concern a limitation of the amount of carbohydrates to a maximum of 50 g daily. Moreover, the total supply of carbohydrates can be lowered to 30 g/daily in order to adapt the organism to a more effective use of ketone bodies. However, this is not an elimination diet but only requires reducing to a minimum those products with a higher carbohydrate content. At the same time, the percentage share of fats increases, usually to 70–80%, and protein frequently accounts for about 20% of the energy share. Additionally, the ratio between protein and fat may be more variable and is highly dependent on the specifics of each particular case. This type of theoretical sharing of the macro-components in most cases fits into the frame of the ketogenic diet, but it is worth stressing that, possibly, the least processed products should be used. The products that are most frequently present in such menus include, among other foodstuffs, eggs, meat, and fish (particularly oily fish), plant oils (e.g., olive oil and coconut oil), giblets (e.g., the liver, heart, and kidneys), non-starchy vegetables (all, but primarily the green ones, i.e., broccoli, spinach, lamb's lettuce, arugula, and kale), avocado, olives, and nuts [21].

3. Diabetes Mellitus

Diabetes mellitus is a disease showing a dramatically increasing incidence. The number of diabetes patients worldwide has already exceeded 460 million and is expected to reach 700 million by 2045. This is a very wide field of application, but most frequently, two main diabetes types are distinguished that are characterized by specific mechanisms of development, although the signs are frequently similar [22,23]. Type 1 diabetes mellitus accounts for 5–10% of all cases of the disease and is most frequently diagnosed in children. It is a chronic disease in which the insulin-producing pancreatic beta cells are damaged. Such damage occurs, e.g., as the result of an autoimmune reaction in the body. Because of the absence of a sufficient amount of insulin, the transport of glucose into the cells is impaired and is manifested as elevated serum glucose concentrations [24–26]. The manifestations of type 2 diabetes are the same, although the mechanisms of its development are different. In this case, what occurs is an impairment of the response of cells to insulin rather than a shortage of that hormone. The disease usually develops over many years, since it is diagnosed most frequently in adults [27,28]. Criteria for the diagnosis of diabetes mellitus are fasting plasma glucose$\geq$ 126 mg/dL (7.0 mmol/L) or random plasma glucose $\geq$ 200 mg/dL (11.1 mmol/L) or 2-hour plasma glucose reading with a 75 g oral glucose tolerance test (OGTT) $\geq$ 200 mg/dL (11.1 mmol/L) or glycated hemoglobin (HbA1c) $\geq$ 6.5% (48 mmol/mol) [29].

4. The Effect of the Ketogenic Diet on the Pharmacotherapy of Type 1 and Type 2 Diabetes

A diagnosis of diabetes is associated with specific, individually selected therapeutic management, including pharmacotherapy. The unusually high prevalence of this subject matter demonstrates that this subject is worth discussing in detail. The results of the available studies suggest an interesting influence of the ketogenic diet on pharmacological treatment. They demonstrate, among other findings, a significant reduction in the body's requirements for insulin and oral antidiabetic drug doses [30]. It has been shown that the ketogenic diet can reduce the requirement for insulin in type 1 diabetes patients using insulin pumps by as much as 44.3% [31]. It turns out that this value is close to that observed in type 2 diabetes patients, in whom there is frequently a reduction in the requirement for insulin, reducing it by half, on average [32]. The American Association of Clinical Endocrinology suggests, in turn, that patients using a treatment with sodium-glucose cotransporter 2 (SGLT) inhibitors should stop taking them even before the beginning of a ketogenic diet. This is because of the increased risk of the development of diabetes ketoacidosis [33]. Patients receiving treatment with glucagon-like peptide 1 (GLP-1) receptor agonists during ketogenic diet use should be strictly monitored. The necessity for the complete termination of their administration is also suggested. That need is associated, as in the case of SGLT2 inhibitors, with an increased risk not only of diabetes ketoacidosis but also of hypoglycemic episodes [34,35]. In the case of metformin, no general contraindications have been demonstrated, although each case should be individually considered [36]. In that particular nutrition model, it is thus possible to completely withdraw pharmacotherapy or at least to reduce it. A possibility of remission of the disease has also been suggested [37]. That finding was demonstrated in one study, in which most of the patients struggling with type 2 diabetes could have reduced the doses or completely discontinued their treatment with antidiabetic drugs during the 16 weeks of the study's duration [38]. Taking into account the significant effect of a ketogenic diet on glycemia values and the possibility of reducing drug doses, the authors of the review published in 2021 suggest that the ketogenic diet in type 2 diabetes patients seems to be a promising intervention that can be applied to improve glycemia control [39].

Considering the above-mentioned data, in some patients with pharmacologically treated type 2 diabetes, the prescribed drugs could be completely withdrawn or their doses could be reduced. That would thus contribute to the minimization or avoidance of the potential adverse effects of pharmacotherapy, at least in some patients. Taking into account the high prevalence of the disease, the given proportion of patients would possibly constitute quite a substantial group. It is worth stressing here that the continuous monitoring of the health condition of these patients is extremely important since the subject has not as yet been sufficiently studied. Continuous monitoring and control will enable the modification (or the complete withdrawal) of drug doses in such a way as to prevent episodes of hypoglycemia, diabetes ketoacidosis, and other complications seen in the disease. This is also important from a future perspective when a ketogenic diet could be recommended more frequently in diabetes treatment [30].

5. The Effect of the Ketogenic Diet on the Course of Type 1 Diabetes

Studies assessing the effect of a ketogenic diet on diabetes are relatively scarce, and those concerning strictly type 1 diabetes are particularly scarce. On the other hand, diabetes patients and ketogenic diet enthusiasts want to apply it based on certain assumptions, anecdotal evidence, or their own feelings. This is the result of a number of observed potential advantages to a ketogenic diet, which may motivate such individuals to follow the principles of such a diet in spite of a current lack of unequivocal evidence as to its effectiveness in this regard [40]. Therefore, there is a vast gap in the literature for research concerning the effect of a ketogenic diet on diabetes. One of the main causes of the low number of studies in this respect relates to concerns regarding ketoacidosis in diabetes patients who may be persuaded to implement such a type of nutritional management. Frequently, the physiological condition of nutritional ketosis is regarded as a risk factor for ketoacidosis, which can develop, e.g., as the result of type 1 diabetes complications. These two notions should be clearly distinguished. Ketoacidosis is the simultaneous occurrence of very high serum concentrations of ketone bodies (15–25 mmol/L) and glucose (250 mg/dL and higher). That condition leads to a dangerous decrease in blood pH value to a level below 7.3. The physiological condition of nutritional ketosis is characterized by a low value within the normal range (70–99 mg/dL) of glucose concentration and a slight concentration (compared with ketoacidosis) of ketone bodies (usually within the range of 0.5–3 mmol/L). It causes no reduction in blood pH value [19,21,41–43]. In the case of healthy individuals, and even in those with type 2 diabetes, the concerns regarding ketoacidosis development due to the adoption of a ketogenic diet are completely irrational and devoid of any scientific foundation. The situation is different in the case of patients suffering from type 1 diabetes. In these patients, such a type of anxiety can be fully justified, and this subject is frequently discussed in the publications concerning that problem. Such publications, among other concerns, also mention the risk of hypoglycemia or dyslipidemia development [44,45].

5.1. Possible Mechanisms of Therapeutic Ketogenic Diet Activity in Type 1 Diabetes

There are some circumstances according to which the ketogenic diet can affect type 1 diabetes that is already at the stage of autoimmunization. The autoimmune process, associated with the destruction of pancreatic beta cells, may be related to intestinal homeostasis disorders. These disorders arise, among other factors, from a reduction in the number of intestinal bacteria producing lactate and butyrate, with the simultaneous increase in the number of *Bacteroides* organisms. A protective effect of butyrate is suggested on the autoimmunization process of pancreatic cells. Therefore, it can indirectly prevent the development of type 1 diabetes [46,47]. During ketogenic diet observation, the amount of β-hydroxybutyrate (the main ketone body) increases. On the one hand, it plays a role similar to that of the medium-chain fatty acids produced by intestinal bacteria. On the other hand, it inhibits inflammatory conditions, for example, through the reduction of the number of intestinal proinflammatory Th17 cells [48–50]. The extremely widespread effect of the ketogenic diet on the microbiome means that in the future, it would be worth looking at the potential mechanism of its influence on the immunization process, including its effect on pancreatic beta cells [51]. This can also be of importance when taking into account the character of the ketogenic diet itself, which, in principle, should mimic fasting conditions in the body, and which, conversely, is frequently also used in diets employing so-called metabolic windows (e.g., the 16:8 diet). A study in mice demonstrated that a low-carbohydrate, low-protein, and high-fat diet with a calorie deficit was able to affect pancreatic beta cells. It should be stressed, however, that a low-carbohydrate diet is not tantamount to a ketogenic diet. In the case of the pancreatic islets seen in type 1 diabetes, the periods of fasting decrease the activity of PKA and mTOR and induce Sox2 and Ngn3 expression and insulin production. Thus, they promote the reprogramming of the pancreatic islet cells, inducing an expression of genes similar to that observed during fetal life. It has also been demonstrated that they can reverse insulin deficiency in murine and human cells in type 1 diabetes [52,53]. Such a possibility was also suggested by the authors of a study in 2014, in which they presented a case report on a 19-year-old man who had been diagnosed with type 1 diabetes. After the recommendation of a standard diabetes diet on which he consumed 240 g of carbohydrates daily in six meals, over the period of 20 days of diet observation, his glucose concentrations fluctuated within the 68–267 mg/dL range and the general feeling was that he failed to improve. A paleolithic version of the ketogenic diet was then applied. His glucose concentrations had returned to normal already in the first few days; therefore, the administration of insulin was discontinued. After 10 weeks, the testing was repeated to determine C peptide concentration, which increased from the initial 0.6 ng/mL level to 2.2 ng/mL. The patient's glycated hemoglobin level was 5.5% and his glucose concentration was 88 mg/dL. Although that change could have occurred independently of the diet used, the authors of the study suggest that the evident increase in the C peptide level accompanying insulin withdrawal may indicate the return of insulin production in the pancreas [54]. Many other mechanisms not strictly concerning the autoimmunology process, such as its effect on glucose, insulin, and insulin resistance, are, to a significant degree, common for both types of diabetes. Therefore, the effect on the aspects mentioned is discussed in more detail in Section 6, "Effect of a ketogenic diet on the prevention and treatment of type 2 diabetes mellitus".

5.2. The Ketogenic Diet in the Treatment of Type 1 Diabetes in Children

Based on the conducted studies, the data provided frequently concern individual cases. The available case studies on children with type 1 diabetes who observed a ketogenic diet provide very interesting information. This scenario has been described in, among other papers, the study conducted by McClean et al. [55]. An effective use of the ketogenic diet was demonstrated, among other patients, in a nearly four-year-old boy, suffering from myoclonic-astatic epilepsy and type 1 diabetes, whose glycated hemoglobin values were within the 5.7–6.4% range. No severe hypoglycemia or ketoacidosis episodes were observed. After the institution of the diet, the boy continued to follow it for six years [56]. Another paper presents the case of a 14-year-old girl with type 1 diabetes who observed a direct improvement after beginning a ketogenic diet, an improvement that was also noted in her glucose level measurements [57]. In another study, a girl aged 3.5 years who was suffering from epilepsy and type 1 diabetes was followed up for 15 months while on a ketogenic diet. Since the institution of the diet, no clinical signs of epilepsy had been reported and an improvement in the girl's development was recorded. Her glycated hemoglobin (HbA1c) level improved, the control of her glycemia was very good, and no severe adverse effects were observed [58]. Another study described the case of a nine-year-old child with type 1 diabetes, in whom, after following a ketogenic diet, insulin administration could have been discontinued [59]. Another two-year-old girl with epilepsy and type 1 diabetes was put on a ketogenic diet. The recorded value of glycated hemoglobin at the time of making the diagnosis was 7.6% and, after six months, it decreased to 6.8%. Over that time period, no severe episodes of ketoacidosis and only many mild hypoglycemia episodes were noted. The girl was put on the diet for 10 months [60]. Interesting data were also provided by a study conducted on a four-year-old girl with pyruvate dehydrogenase deficiency, static encephalopathy, convulsive disorder, and diabetic ketoacidosis. She was simultaneously treated with a ketogenic diet and exogenous insulin, which, at first, seemed quite risky. However, the follow-up period, lasting 28 months, revealed that the ketogenic diet contributed to an improvement in the activity and development of the girl and to the excellent control of glycemia [9]. All the above-mentioned cases are summarized in Table 1.

In 2018 a study was conducted that listed the potential risks resulting from the application of that nutritional model in children with type 1 diabetes, based on six cases. The first of the cases concerned a boy, aged 13 years old, in whom a significant reduction in glycated hemoglobin levels occurred (from 10.3% to 6.3%), while the authors regarded reaching the cholesterol concentration of 5.5 mmol/L as a potentially negative side effect. The second case was that of a 12-year-old girl, in whom the following of a ketogenic diet resulted in a reduction in HbA1c level (from 7.4% to 6%) but hypoglycemia episodes, a cholesterol concentration of 5 mmol/L, and hypertension were observed. The third case described was of a six-year-old boy, in whom a ketogenic diet was associated with hunger, body weight loss, and poor development. The fourth child studied was a girl of four years old, in whom a significant reduction in the HbA1c value (from 14% to 8.1%), many hypoglycemic episodes, slow growth, and delayed bone age were observed. In the fifth case of a 3.5-year-old boy, an HbA1c reduction (from 6.1% to 5.3%), significant hypoglycemia, slower growth, low growth hormone levels, and high cholesterol concentrations were recorded. In the last case concerning a 3.5-year-old girl, unfavorable signs were observed, taking the form of poor body mass gain and disturbed growth [61]. The publication frequently refers to increased total cholesterol levels, which, according to current knowledge, are, rather, of non-significant importance within such a range. The frequent hypoglycemia episodes observed in the study may be an actual problem that is worth keeping in mind; the condition of such patients should be adequately monitored. The authors also refer to the children's unmet requirements of calcium, in particular, as well as magnesium and phosphorus. These unmet needs, however, are not the effects of the ketogenic diet itself but are rather from an inadequately composed ketogenic diet. Frequently, the diet compensates for the inadequate balance and incorrect nutritional value of the menus that are presented in the studies.

Table 1. Effect of ketogenic diets as prescribed for children with type 1 diabetes.

Gender and Age	Disease	Dietary Intervention	Benefits	Adverse Effects	References
Boy, 4 years	Type 1 diabetes and myoclonic-astatic epilepsy	Ketogenic diet	1. Acceptable control of epileptic seizures 2. Improvement of cognitive functions 3. Maintenance of the target glycemia values	No severe episodes of hypoglycemia or ketoacidosis were observed	[56]
Girl, 14 years	Type 1 diabetes	Ketogenic diet	1. Significant improvement of subjective sensations 2. Significant improvement in glycemia	Not observed	[57]
Girl, 3.5 years	Type 1 diabetes, right hemiparesis, epilepsy	Ketogenic diet	1. Absence of epileptic seizures 2. Improvement of development, motor functions, and activity 3. Proper glycemic control and improvement in the HbA1c value	1 episode of ketoacidosis, apart from which no adverse effects were observed	[58]
A 9-year-old child (no information on gender)	Type 1 diabetes	Paleolithic ketogenic diet	1. Improvement of insulin and glucose levels 2. Discontinuation of insulin treatment	Not observed	[59]
Girl, 2 years	Type 1 diabetes and epilepsy	Ketogenic diet	1. No episodes of epileptic seizures 2. Improvement (reduction) of glycated hemoglobin level 3. No new episodes of diabetic ketoacidosis	Mild hypoglycemia episodes	[60]
Girl, 4 years	Pyruvate dehydrogenase deficiency, diabetic ketoacidosis, static encephalopathy, convulsive disorders	Ketogenic diet	1. Proper glycemic control 2. Improvement of activity level 3. Significant developmental achievement 4. Compensation of linear growth from < 5th to the 50th percentile	No major adverse effects were observed	[9]

5.3. The Ketogenic Diet in the Treatment of Type 1 Diabetes in Adults

The results of the available studies in adults are also promising. In one of these studies on 22 adults, suffering from type 1 diabetes, who decided to go on a ketogenic diet (or, at least, a very low-carbohydrate diet, i.e., 70–90 g of carbohydrates daily), some specific benefits were demonstrated. The joint duration of the follow-up was 12 months; during this period, a significant reduction was observed in hypoglycemic episodes, glycated hemoglobin level (from 7.5% to 6.4%), and the postprandial requirement for insulin by almost half (from 21.1 IU daily to 12.4 IU daily). The total cholesterol and HDL concentration values were not significantly changed, while the concentration of triglycerides alone decreased by 16%, on average [62]. However, it is worth emphasizing, however, that a low-carbohydrate diet was being followed and it was not known whether such an amount was sufficient to induce ketosis, since the level of the ketone bodies was not measured. In the conclusions drawn from another study conducted in 11 adult type 1 diabetes patients who had been on a ketogenic diet for various time periods (2.6 years, on average), the authors demonstrated that a ketogenic diet in patients with type 1 diabetes was, as a rule, associated with normal glycated hemoglobin (HbA1c) concentrations and slight glycemic fluctuations (1.5 ± 0.7 mmol/L). However, they also mentioned that this could have been associated with the large number of hypoglycemia episodes and also with dyslipidemia [63]. Another randomized controlled study that was conducted on adults with type 1 diabetes compared the effect of a low-carbohydrate diet with that of a standard carbohydrate diet over a period of 12 weeks. While, in the carbohydrate group, no changes were observed, in the low-carbohydrate group, a significant reduction in glycated hemoglobin level (from 8.9% to 8.2%), and a reduction in the daily insulin dose (from 64.4 IU to 44.2 IU) and body weight (from 83.2 kg to 78 kg) occurred. Therefore, the authors suggested a certain effectiveness of the diet in type 1 diabetes patients [64]. However, a low-carbohydrate diet is not tantamount to a ketogenic diet, so the results of that study do not necessarily need to be the same as those observed from a ketogenic diet. In 2018, a questionnaire-based survey was distributed to a larger group, i.e., 316 type 1 diabetes patients, with the subjects being on very low-carbohydrate diets (36 g carbohydrate daily on average) for a mean period of 2.2 years. Very low amounts of carbohydrates, on average, are found in the ketogenic diet. It should, however, be kept in mind that these are averaged values. Taking that into account, it is possible that not all study participants entered a state of ketosis. The mean glycated hemoglobin value reported in these patients was 5.67%. Only seven patients gave a history of diabetes-related hospitalizations in the last year: four were due to ketoacidosis, and only two were due to hypoglycemia. The authors concluded that a ketogenic diet affected glycemia control in type 1 diabetes, with a low rate of adverse effects, in both adults and children on a very low-carbohydrate diet (the mean value of 36 g of carbohydrates meets one criterion of a ketogenic diet) [65]. One case study described a 37-year-old male with type 1 diabetes who, in 20 days, had covered a distance of 4011 km on a bicycle in Australia, while being on a ketogenic diet. Continuous glycemia monitoring during the ride revealed an unusual glycemic stability (mean value 6.1 mmol/L) and only one episode of major hypoglycemia. The authors have concluded that this observed glycemic stability suggests that the body's adaptation to using fats can alleviate glycemia fluctuations. In addition, it can lead to a lower dependence on carbohydrate consumption in order to maintain normal glycemia values during physical exercise [66]. It is worth keeping in mind, however, that an inadequately applied ketogenic diet can be fraught with adverse effects, as was observed in the case of a 22-year-old woman. She was not aware of her type 1 diabetes, and she started to follow a ketogenic diet, which led to diabetic ketoacidosis. That particular case was described in a paper published in 2021 [67]. All the above cases are summarized in Table 2.

Table 2. Effect of a ketogenic diet in adults with type 1 diabetes.

Gender and Age	Disease	Dietary Intervention	Benefits	Adverse Effects	References
Group of 17 women and 7 men. Mean age 51 ± 10 years	Type 1 diabetes	Low-carbohydrate diet (70–90 g carbohydrates)	1. Significant reduction in hypoglycemia episodes and glycated hemoglobin concentration (from 7.5% to 6.4%), 2. Significant reduction in the postprandial requirement for insulin (from 21.1 IU daily to 12.4 IU daily) 3. Reduction of triglyceride concentration by 16% on average	Diabetic gastroparesis was observed in 6 individuals	[62]
Group of 7 men and 4 women. Mean age 36.1 ± 6.8 years	Type 1 diabetes	Ketogenic diet	1. Normal HbA1c concentration (The mean HbA1c levels were 35 ± 4 mmol/mol (5.3 ± 0.4%)) 2. Slight changes in glycemia values (little daily glycemic variability (1.5 ± 0.7 mmol/L)	Hypoglycemia episodes and dyslipidemia were observed	[63]
Group of 4 men and 1 woman. Mean age 44.5 ± 10.4 years (in the normal carbohydrate amount group, there were an additional 3 men and 2 women, mean age 44.8 ± 8.3 years)	Type 1 diabetes	Target: ketogenic diet, 50–75 g carbohydrates as a result: low-carbohydrate diet, up to 100 g of carbohydrates	1. Improved glycemic control (significant reductions in HbA1c (63 to 55 mmol/mol)) 2. Reduction of insulin doses (significant reductions in daily insulin use (64.4 to 44.2 units/day)) 3. Body mass reduction (83.2 to 78.0 kg)	One participant reported a higher irritability and another one a greater number of minor diseases	[64]
Group of 316 women and men. Mean age 27 ± 19	Type 1 diabetes	Carbohydrates of 36 g on average, i.e., a ketogenic diet	1. Improved glycemic control (mean HbA1c was 5.67% ± 0.66%) 2. Lower requirement for insulin	A group of 7 individuals in a year, who reported a hospitalization associated with diabetes (including ketoacidosis and hypoglycemia); in the remaining majority of cases, there were no adverse effects	[65]

Table 2. *Cont.*

Gender and Age	Disease	Dietary Intervention	Benefits	Adverse Effects	References
Man. 37 years	Type 1 diabetes	Ketogenic diet	1. Maintenance of glycemic stability (average interstitial glucose 6.1 mmol/L) and 80.4% of the time spent within a range of 3.9–10 mmol/L. Interstitial glucose was < 3 mmol/L for 2.1% of this time 2. No problems with riding 4011 km on a bicycle over 20 days	1 episode of major hypoglycemia	[66]
Woman. 22 years	Undiagnosed type 1 diabetes	Ketogenic diet	None	Development of diabetic ketoacidosis in 4 days	[67]

6. The Effect of the Ketogenic Diet on the Prevention and Treatment of Type 2 Diabetes

In type 2 diabetes, the significantly greater popularity of the ketogenic diet is observed, compared with that in type 1 diabetes. In this case, the occurrence of the possible adverse effects is much less of a concern. Despite that finding, clinicians and scientific societies still cannot find any possible application for it. The favorable mechanisms of a ketogenic diet that have an effect on the prophylaxis and treatment of type 2 diabetes may be multifaceted. Based on the available studies, a conclusion can be drawn that the most important aspect of the issue is the significant effect of this diet on glycemia values.

6.1. The Effect of the Ketogenic Diet in the Therapy of Type 2 Diabetes—Meta-Analyses and Systematic Reviews

A number of meta-analyses and systematic reviews on the topic of a ketogenic diet are already available. We considered only those meta-analyses and systematic reviews that were published in English from 2012 to 2022. Search terms included the keywords: "Diabetes" AND Ketogenic Diet". The literature search was completed on 29 November 2022. This study was performed according to PRISM (preferred reporting items for systemic review) guidelines (Figure 2). The eligibility of studies was assessed by applying clear inclusion/exclusion criteria. The studies included comprised full-text English meta-analyses and systematic reviews on the use of a ketogenic diet by patients with T2DM. Studies excluded were those involving animal studies and those that were focused on other diseases.

A comprehensive meta-analysis of randomized controlled studies demonstrated that the ketogenic diet was more effective than the low-fat diet in terms of the improvement of glycemia parameters, body mass, and lipid profile, particularly in patients with previously diagnosed diabetes who were also overweight [68]. Another meta-analysis concerned the effect of a ketogenic diet on glycemia control, insulin resistance, and lipid metabolism in patients with type 2 diabetes. It was demonstrated that the diet caused an average reduction in glucose concentration by 1.29 mmol/L and a reduction in glycated hemoglobin (% HbA1c) by 1.07% and in the concentrations of total cholesterol, LDL fraction, and triglycerides, as well as an increase in HDL cholesterol concentrations. A reduction in body mass by 8.66 kg, on average, in waist circumference by 9.17 cm, and in BMI by 3.13 kg/m^2 were also observed. In the conclusion of the study, it was said that the ketogenic diet favorably affected the control of the glycemia and lipid profile in type 2 diabetes patients and, moreover, significantly contributed to body mass reduction [69]. A meta-analysis of randomized controlled trials (RCT) in 2022 also provides important information and presents the benefits of that diet when used for type 2 diabetes. The authors strove for an assessment of its effect on the control of glycemia and body mass in type 2 diabetes patients who were overweight. They found that a ketogenic diet exerted a significant favorable effect on body mass loss (by 5.63 kg, on average), a reduction in waist circumference (by 2.32 cm, on average), a reduction in the concentrations of glycated hemoglobin (by 0.38 HbA1c, on average) and triglycerides (by 0.36 mmol/L, on average) and an increase in HDL cholesterol levels (by 0.28 mmol/L, on average). The authors suggest that the diet can be recommended for type 2 diabetes [70]. Another meta-analysis in 2022 compared the effect of a ketogenic diet with that of a standard diet recommended for diabetes patients. The scope of the study was only limited, however, to an assessment of drug metabolism. The greatest changes were observed in terms of triglyceride concentration reduction in patients on a ketogenic diet. Compared with the control group, an average triglyceride concentration reduction was found in the 3rd, 6th, and 12th months of the treatment, by −0.49 mmol/L, −0.82 mmol/L, and −0.18 mmol/L, respectively. The greatest change was found in the 3rd month. Importantly, however, no significant differences were demonstrated in the total cholesterol, HDL, and LDL concentrations. The authors stressed that type 2 diabetes patients on a ketogenic diet had no elevated total cholesterol and LDL concentrations. Similarly, in the studies included in the meta-analysis, no reduced cholesterol concentrations were observed in the patients [15]. In 2022, a systematic review with a meta-analysis was conducted, in order to estimate the effect of a ketogenic diet in patients with type 2 diabetes and in those with a pre-diabetic condition. The control group in the studies mentioned consisted of individuals following a diet with a higher content of carbohydrates than a ketogenic diet. It was shown that in the patients on a ketogenic diet, compared with the control group, the concentration of triglycerides decreased by −0.28 mmol/L, while that of HDL cholesterol increased by 0.04 mmol/L. In four studies, the changes in glycated hemoglobin concentration after 12 months, with the estimation of permanent effects (very low carbohydrate/ketogenic diets (VLC/KD) minus the control group), were 0.01% (−0.22 to 0.25). In two studies, an HbA1c change of −0.65% (−0.991–0.31) was noted in relation to the initial value. The authors concluded that a ketogenic diet could effectively reduce HbA1c and triglyceride concentrations in individuals with type 2 diabetes and those in a pre-diabetic condition. They stress, however, that still more well-planned studies are needed [71]. The benefits resulting from a ketogenic diet were also confirmed by another meta-analysis conducted in 2022. The authors compared the effect of the ketogenic diet with that of the standard diets recommended for diabetes patients. Compared with the diets in the control groups, in patients on a ketogenic diet, the following were observed: an HbA1c concentration reduction after three months (by 6.7 mmol/mol on average) and after six months (by 6.3 mmol/mol on average), a body mass reduction after three months (by 2.91 kg, on average) and after six months (by 2.84 kg, on average). Moreover, for 12 months, the diet demonstrated an advantage over the control diets in respect of triglyceride level reduction, HDL cholesterol concentration increase, and a reduction in the administration of antidiabetic drugs. The authors clearly stated, however, that the quality of the currently available evidence was still insufficient to allow for the standard recommendation of a ketogenic diet in patients with type 2 diabetes [72]. In a systematic review with a meta-analysis published in 2022, the advantages of a ketogenic diet were also demonstrated in type 2 diabetes patients. Compared with the control diets, the individuals following a ketogenic diet decreased their glycated hemoglobin values by 1.45% HbA1c on average and reduced their body mass by 2.67 kg on average [73]. This finding was confirmed by another systematic review [39]. It demonstrated that a ketogenic diet improved HbA1c concentration in diabetes patients already showing changes after three weeks; the effect was maintained for at least a year. That change was also associated with a reduction in administered antidiabetic drugs and the sustained persistence of a reduced body mass. In a systematic review of 2022, the effect was studied of a ketogenic diet on insulin sensitivity in type 2 diabetes patients. Once again, its favorable effect on glycemia was confirmed [74]. Another review also suggested its benefits when treating patients with metabolic diseases, such as type 2 diabetes. The authors demonstrated its advantages in respect of reducing the concentrations of HbA1c, glucose, insulin, total cholesterol, triglycerides, ALT, AST, and the HOMA-IR index. It is worth stressing the finding that no severe adverse effects were noted [75]. A summary of the results of all analyses and systematic reviews is shown in Table 3.

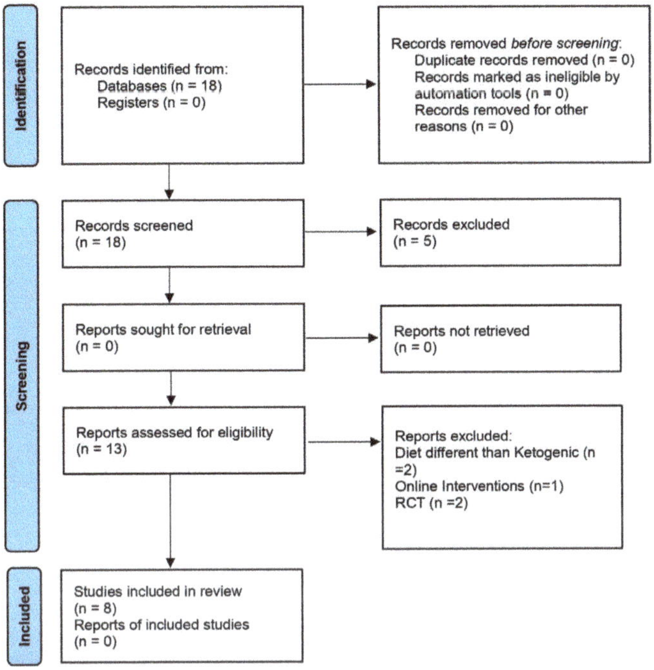

Figure 2. Flowchart presenting the study selection according to PRISMA guidelines.

Table 3. Summary of the meta-analyses and systematic reviews concerning the effect of a ketogenic diet on patients with type 2 diabetes.

Year and Type of Publication	Number of Studies Considered	Diet Type	Publication Aim	Results	References
2020 Meta-analysis	14 RCT	Ketogenic diet	Assessment of the effectiveness of a ketogenic diet in metabolic compensation in patients who are overweight/obese, with and without type 2 diabetes, compared with a low-fat diet	Advantages of a ketogenic diet over a low-fat diet in the control of glycemia (lower HbA1c levels (SMD −0.62), body mass (SMD −0.46) and lipid profile (reduction in triglyceride concentration (mean -0.45, increase in HDL (SMD 0.31) concentration)	[68]
2020 Systematic review and meta-analysis	13	Ketogenic diet	Assessment of the effect of a ketogenic diet on the control of glycemia, insulin resistance, and lipid metabolism in type 2 diabetes patients	Reduction in the concentrations of glucose (by 1.29 mmol/L on average), glycated hemoglobin (by 1.07% HbA1c on average), total cholesterol (by 0.33 mmol/L on average, LDL (by 0.05 mmol/L on average), and the reduction of body mass (by 8.66 kg on average), waist circumference (by 9.17 cm on average), and BMI (by 3.13 kg/m^2 on average) HDL concentration increased (by 0.14 mmol/L, on average)	[69]
2022 Meta-analysis	8 RCT	Ketogenic diet	Studying the role of the ketogenic diet in controlling body mass and glycemia in patients with type 2 diabetes who are overweight	Reduction in body mass (by 5.63 kg on average), waist circumference (by 2.32 cm on average), glycated hemoglobin concentration (by 0.38% HbA1c on average), triglycerides (by 0.36 mmol/L on average) and an increase in HDL cholesterol concentration (by 0.28 mmol/L, on average)	[70]
2022 Meta-analysis	10 RCT	Ketogenic diet	Assessment of the effect of a ketogenic diet on lipid metabolism in patients with type 2 diabetes (compared with the effects of the standard diets)	Advantage of the ketogenic diet in reducing triglyceride concentration, particularly in the 3rd month (compared with the control group, with TG reduction in the 3rd, 6th, and 12th months of treatment, by 0.49 mmol/L, −0.82 mmol/L, and −0.18 mmol/L, on average) No significant differences in total cholesterol, LDL, and HDL concentrations	[76]

Table 3. *Cont.*

Year and Type of Publication	Number of Studies Considered	Diet Type	Publication Aim	Results	References
2022 Systematic review and meta-analysis	8 RCT	Ketogenic diet	Estimation of the effect of a ketogenic diet in type 2 diabetes patients and individuals with pre-diabetic conditions (compared with the diets with a higher content of carbohydrates than in a ketogenic diet)	In individuals on a ketogenic diet, compared with the control group, a reduction of triglyceride concentration by 0.28 mmol/L and an increase in HDL cholesterol level by 0.04 mmol/L. In total, 4 studies demonstrated changes in HbA1c concentration after 12 months, with the estimation of persistent effects (VLC/KD minus control group) at 0.01% level (−0.22 to 0.25). In addition, 2 studies demonstrated a change in HbA1c from the initial value: −0.65% (−0.99; −0.31)	[71]
2022 Meta-analysis	8RCT	Ketogenic diet	Comparison of a ketogenic diet vs. standard diet recommended in patients with type 2 diabetes in the context of parameter changes, i.e.: glycemia, body mass, lipid profile, drug taking, and the discontinuation of drug taking	Compared with the standard recommended diets, in patients on a ketogenic diet, a reduction in HbA1c after 3 and 6 months (by 6.7 mmol/L and 6.3 mmol/L, on average, respectively) and a body mass reduction after 3 and 6 months (by 2.91 kg and 2.84 kg on average, respectively) were observed. In a 12-month period, an advantage of a ketogenic diet over the control diets was seen in respect of triglyceride concentration reduction and a reduction in the requirements for drugs, as well as an increase in HDL concentration	[72]
2022 Systematic review and meta-analysis	15	Ketogenic diet and low-carbohydrate diet	Assessment of the effectiveness of a ketogenic diet (and low-carbohydrate diet) in controlling glycemia and body mass in patients with type 2 diabetes	Patients using a ketogenic diet, compared with control diets, reduced their glycated hemoglobin values by 1.45% HbA1c, on average, and their body mass by 2.67 kg, on average	[73]
2021 Systematic review	14	Ketogenic diet	Assessment of the pleiotropic effect of a ketogenic diet on glycemic control, drug changes, and body mass loss in patients with type 2 diabetes	Improvement of glycated hemoglobin concentrations in patients with type 2 diabetes within three weeks and the persistence of the effect for at least one year. Effectiveness in the long-term maintenance of a reduced body mass	[39]

6.2. The Effect of the Ketogenic Diet in the Therapy of Diabetes Type 2—Randomized Controlled Trials (RCT)

A literature search of the Cochrane Library and PubMed was performed to identify randomized controlled trials. Search terms included the keywords: "Diabetes" AND Ketogenic Diet" in the publications' titles, abstracts, and keywords. Search filters were set as "Trials" and a time range was set up from 2012 until 2022. The initial database search yielded 41 papers (with 17 papers in PubMed and 25 from the Cochrane Library), from which 17 duplicates were excluded. The remaining 24 trials underwent further screening of the titles and abstracts. As a result, 17 articles were excluded for the following reasons: other co-morbidities (5), no RCT (4), an animal-based study (1), and unrelated topics (7). Full-text examination by two independent reviewers narrowed down the number of articles to 7. The available data from randomized controlled trials (RCT) shows a positive prognosis for the ketogenic diet's application in diabetes. One such study demonstrated an advantage of a ketogenic diet over a diet composed according to the dietary recommendations of the American Diabetes Association (ADA). It was conducted for 32 weeks in patients with type 2 diabetes who were overweight, as part of an online intervention. In the group on a ketogenic diet, a greater reduction in glycated hemoglobin concentrations (−0.8%, on average) occurred, compared with the other group (−0.3%, on average). Besides that, in 55% of patients, a reduction in the glycated hemoglobin value below 6.5% was found, which was not observed in any subject in the control group. The improvement of triglyceride concentrations was also more pronounced in the group on a ketogenic diet (−60.1 mg/dL, on average) than in the other group (−6.2 mg/dL, on average). Significant differences were also observed in body mass, the mean loss of which was 12.7 kg in the ketogenic diet group and 3 kg in the control group. Importantly, the individuals on a ketogenic diet showed non-compliance significantly less frequently in further adhering to the recommendations (8% of the individuals) compared with those on the control diet (46% of the subjects) [77,78].

Another randomized controlled study also demonstrated an advantage to the low-calorie ketogenic diet over the standard low-calorie diet in 89 adults with type 2 diabetes. A similar level of safety of use was also observed. However, the values of glycated hemoglobin showed greater improvement in the ketogenic diet group. In addition, a significantly greater bodyweight loss and waist circumference reduction were observed [79]. In 2022m a 12-week-long, randomized controlled study was conducted on 60 adult patients with newly diagnosed diabetes who were overweight. The effect of the usual diabetic diet (30 subjects) was compared with that of a ketogenic diet (30 subjects). It was found that improvements in the values of such parameters as glycated hemoglobin (HbA1c), fasting glucose and insulin levels, lipid profile, body mass, and BMI were more pronounced in those individuals on the ketogenic diet. Thus, it demonstrated its higher effectiveness compared with the standard diets [80]. An RCT from 2022 compared the effects of ketogenic and low-carbohydrate Mediterranean diets, along with their influence on glucose concentration and cardio-metabolic risk factors in T2DM or pre-diabetic patients. Both types of diet had a positive effect on HbA1c. However, the ketogenic diet showed greater benefits by decreasing triglyceride concentrations and body weight and increasing HDL cholesterol concentration. In addition, increases of LDL cholesterol was observed, and the authors pointed out a lower balance in the diet [81]. Another study compared the adherence to both diet types in patients with T2DM and pre-diabetes. The results showed a high similarity in diet adherence in the first 4 weeks (when all food was delivered to the participants). Conversely, when the participants were responsible for providing and preparing their own meals, this similarity was lower. In the context of comparing the ketogenic diet to ADA guidelines, data published in 2014 showed the advantages of this nutrition model. It was demonstrated that a higher percentage of participants (44% vs. 11%) could discontinue one or more oral diabetes medications. Additionally, HbA1c and body weight decreased on a ketogenic diet. [82]. Moreover, in a 12-month timeframe, participants on a ketogenic diet showed greater reductions than participants on a low-fat diet. Furthermore, greater reductions in glycated hemoglobin j (−0.5% vs. −0.2%) and weight (−7.9 kg vs. −1.7 kg) were observed. Reductions in diabetes-related medications were only observed for the ketogenic diet [83]. Details of the described RCT are shown in Table 4.

The most promising results were obtained in the RCT conducted by Myette-Côté et al., although considering the fact that this trial did not focus on the ketogenic diet (but on restricting energy from carbohydrates to 10%), it was excluded from Table 4. The aforementioned study compared the short-term influence of three different types of intervention: a low-fat low-glycemic index diet (GL), a low-carbohydrate high-fat diet (LC), and an LC with 15-minute post-meal walks (LC+Ex). The best results were observed for the LC+Ex diet (a significant reduction in glucose concentration was observed), and the worst results were presented for the GL diet. Moreover, a significant reduction in circulating proinsulin was observed for the LC and LC+Ex diets. Additionally, it was shown that the LC diet improved 4-day glycemic control and fasting proinsulin levels, compared to the GL diet. An additional 15-minute post-meal walk was beneficial with the LC diet [85].

Table 4. Summary of the randomized controlled trials (RCT), listing the effect of a ketogenic diet on patients with type 2 diabetes.

Year and Type of Publication	Number of Patients and Duration	Diet Type	Publication Aim	Main Outcome(for Ketogenic Diet)	Percentage of Drop-Outs	References
2017, RCT	25 (12 in the intervention, 13 control). 32 weeks	Ketogenic diet ad libitum vs. a diet program based on the American Diabetes Association's "Create Your Plate" website	Comparison of the online intervention of a ketogenic diet ad libitum vs. an online diet program based on the American Diabetes Association's "Create Your Plate" website on glycemic control and other health outcomes among overweight individuals with type 2 diabetes	1. Greater HbA1c decrease (−0.8% vs. −0.6%), 2. Lowering HbA1c to less than 6.5% (55% of participants vs. 0% of participants), 3. Greater weight loss (−12.7 kg vs. −3.0 kg), 4. A greater percentage of participants lost at least 5% of their body weight (90% of participants vs. 29% of participants) 5. Greater reduction in triglyceride levels (−60,1mg/dl vs. −28,9mg/dl) 6. Lower numbers of dropouts (8% of participants vs. 46% of participants)	Intervention group: 8% (1/12) and control group: 46% (6/13)	[78]
2016, RCT	89 (45 in the intervention, 44 control). 4 months	Low-calorie ketogenic diet (VLC/KD) vs. a standard low-calorie diet	Evaluating the short-term safety and tolerability of a very low-calorie-ketogenic diet (VLCKD) (< 50 g of carbohydrate daily) in an interventional weight loss program including lifestyle and behavioral modification support (Diaprokal method) in subjects with T2DM.	1. Greater body mass reduction (−14.7 kg vs. −5.05 kg) 2. Greater percentage of participants lost more than 5% and 10% of their body weight (97.6% vs. 50% and 85.4% vs. 16.7%, respectively) 3. Greater reduction in waist circumference (−12 cm vs. −5.4 cm) 4. Greater reduction in HbA1c levels (−0.9% vs. −0.4%) 5. Greater reduction in oral diabetes medication (from 73.3% to 50% of participants vs. from 86.4% to 83.3% of participants)	VLCKD: 11.1% (5/45), LC diet: 18.2% (8/44)	[79]
2022, RCT	60 (30 intervention, 30 control). 12 weeks	Ketogenic diet vs. the routine diet for diabetes	Observation of a periodic ketogenic diet for its effect on overweight or obese patients newly diagnosed with T2DM, with a comparison with the routine diet for diabetes	1. Greater reduction in HbA1c levels (−0.92% vs.−0.27%) 2. Greater reduction in fasting glucose concentration (−1.39 mmol/L vs. −0.56 mmol/L) 3. Greater reduction in fasting insulin concentration (−48.23 pmol/L vs. −3.7 pmol/L) 4. Greater body mass reduction (−8.06 kg vs. −0.61 kg) 5. Greater reduction in waist circumference (−9.29 cm vs. −0.77 cm) 6. Increase HDL concentration (+0.13 mmol/L vs. + 0.03 mmol/L) and decrease of LDL concentration (−0.41 mmol/L vs. −0.18 mmol/L)	Ketogenic diet group: 20% (6/30), Control group: 3.3% (1/30)	[80]
2022, RCT	40 (20 + 20) (33 after drop-outs (16 on the ketogenic diet vs. 17 on the Mediterranean diet)). 12 weeks	Ketogenic diet vs. a low-carb Mediterranean diet	Comparison of 2 low-carbohydrate diets with 3 key similarities (incorporating nonstarchy vegetables and avoiding added sugars and refined grains) and 3 key differences (incorporating, compared with avoiding, legumes, fruits, and whole, intact grains) for their effects on glucose control and cardiometabolic risk factors in individuals with pre-diabetes and T2DM	1. Decrease in triglyceride concentration (−16% vs. −5%) 2. Increase in HDL concentration (+11% vs. +7%) 3. Greater reduction in body mass (−8% vs. −7%) 4. Increase in LDL concentration (+10% vs. −5%)	17.5% (7/40)	[81]

Table 4. *Cont.*

Year and Type of Publication	Number of Patients and Duration	Diet Type	Publication Aim	Main Outcome(for Ketogenic Diet)	Percentage of Drop-Outs	References
2021, RCT	40 (20 + 20) 24 weeks (12 + 12)	Ketogenic diet vs. a low-carb Mediterranean diet	Detailed examination and comparison of the adherence to the two study diets (well-formulated ketogenic diet (WFKD) and Mediterranean Plus (Med-Plus) under two conditions: all food being provided (delivered) and all food being obtained by individual participants (self-provided)	1. Higher adherence to the WFKD diet in the food delivery phase (7.6 vs. 7.3) and self-provided food (5.7 vs. 5.4) on a 10-point scale. 2. After study completion, a clear relationship with diet preference was observed—participants preferred the diet that they were assigned first	12.5% (5/40)	[84]
2014, RCT	34 (16 ketogenic diet + 18 standard ADA diet). 3 months	Ketogenic diet vs. a diet consistent with guidelines from the American Diabetes Association	Comparison of the effects of each diet on glycemic control, medication use, and weight loss among overweight or obese individuals with type 2 diabetes mellitus or prediabetes + testing the feasibility of research design for conducting a larger scale	1. Decrease in HbA1c levels (−0.6% vs. 0%) 2. Higher percentage of participants discontinued one or more diabetes medications (44% vs. 11%) 3. Higher percentage of participants discontinued sulfonylureas (31% vs. 5%) 4. Greater body mass reduction (−5.5 kg vs. −2.6 kg)	LCK group: 6.25% (1/16), MCCR group: 5.55% (1/18)	[82]
2017, RCT	34 (16 ketogenic diet vs. 18 standard diet). 12 months	Ketogenic diet vs. a low fat, moderate carbohydrate, calorie-restricted diet	Comparison of the effects of each diet on glycemic control, medication use, and weight loss among overweight or obese individuals with type 2 diabetes mellitus or pre-diabetes	1. Greater reduction in HbA1c levels (−0.5% vs. −0.2%) 2. Greater body mass reduction (−7.9 kg vs. −1.7 kg) 3. Discontinued metformin medication (30% of participants vs. 0% of participants) 4. Discontinued sulfonylureas or dipeptidyl peptidase-4 inhibitor medications (60% of participants vs. 0% of participants)	LCK group: 12.5% (2/16), MCCR group: 16.7% (3/18)	[83]

6.3. The Effect of the Ketogenic Diet in the Therapy of Diabetes Type 2—Additional Studies

Two systematic reviews were not included in the PRISMA analysis due to their not being identified as a meta-analysis or systematic review by the PubMed filter [74] or not being listed in PubMed [75]. First, one systematic review from 2022 analyzed the influence of the ketogenic diet on insulin sensitivity in T2DM. It showed an improvement in glycemic control and the beneficial influence of the ketogenic diet on insulin sensitivity in type 2 diabetes [74]. The second article investigated the influence of the ketogenic diet on patients with metabolic disorders. Considering the parameters connected with diabetes, they reported an improvement in the concentrations of glycated hemoglobin, glucose, and insulin, and an improvement in HOMA-IR. Moreover, improvements in cholesterol, ALT, and AST concentrations were reported. [75] In 2022, an extensive meta-analysis was conducted, including 50 studies that involved 4291 patients, concerning the effect of a low-carbohydrate diet, in relation to the number of carbohydrates, in patients with type 2 diabetes. It should be stressed that the meta-analysis did not strictly concern the ketogenic diet. For that reason, the importance of that publication in the context of the ketogenic diet is not as reliable as that of the earlier-described meta-analyses of ketogenic diets. Taking into account the widely presented data in that meta-analysis, it is worth mentioning its results as a complement to the subject discussed herein. It was shown that over a period of six months, together with a reduction in the share of energy from carbohydrates from 55–65% to 10%, a linear reduction occurred in the glycated hemoglobin values (by 0.20 HBA1c %, on average), fasting glycemia (by 0.34 mmol/L, on average), body mass (by 1.44 kg, on average), triglyceride concentration (by 0.12 mmol, on average) and systolic blood pressure (by 1.79 mmHg, on average). From the 12-month perspective (from the 6th to the 12th month), the linear tendency was still observed in the glycated hemoglobin values (a reduction by 0.11 HbA1c %, on average) and in triglyceride concentration (reduction by 0.12 mmol, on average). A "U"-shaped effect after 12 months was noted in the case of body mass (the highest reduction being from a limitation of the carbohydrate energy share to 35%) and after 6 months in the case of total cholesterol and LDL fraction levels (the highest reduction was seen at a carbohydrate amount corresponding to 40% of the energy share) [86].

A one-year study of type 2 diabetes patients showed that owing to the ketogenic diet, fasting insulin values decreased by 43%, on average. Most of that effect had been noted already at the initial stage of 70 days [87]. Such benefits are not surprising when taking into account the character of the ketogenic diet. Not infrequently, its significant effect was demonstrated on the glycemic parameters, including the HOMA-IR (homeostasis model assessment of insulin resistance) index, which was established based on fasting glucose and insulin concentrations. In one study, it was already clear after 12 weeks that the index value was reduced from 3.73 to 1.4, on average [17]. The benefits resulting from diets, even with a moderate carbohydrate content (26–45% of the energy share from carbohydrates), low-carbohydrate content (11–25% of the energy share from carbohydrates), and very-low-carbohydrate content (<10% of the energy share from carbohydrates), have been shown in a significant number of research studies. Their effects in reducing the concentrations of glycated hemoglobin and triglycerides, blood pressure, body mass, and increasing HDL cholesterol concentrations have been shown [88–93]. Considering the presented data, it cannot be said that there is no evidence for the ketogenic diet's effectiveness in type 2 diabetes. It is worth clearly stating that not all these effects were observed from the ketogenic diet. Some of them were related to the total amount of carbohydrates and not to the condition of ketosis. Although the number of significant clinical studies is not high, it is sufficient to conclude that the ketogenic diet can be extremely helpful. From the perspective of the currently available data, the potential benefits cannot be ignored. The topic definitely deserves attention and further development of the research in this field.

7. The Ketogenic Diet and Standard Recommended Diabetes Diets

Although no ideal percentage proportions have been established for the calorie share from carbohydrates, fats, and protein to prevent diabetes, the recommended standard diets in patients with existing diabetes conditions are mainly based on the energy supplied by carbohydrates. The suggested amount is a 45–65% share of the energy supply. The recommended amount of fats is from 25% to 40% of the energy share, while that of protein is 15–20%. The maximal amount of cholesterol is 300 mg daily (in the case of dyslipidemia, it is <200 mg) [29,94]. Thus, the recommendations remain in contradiction to the principles of the ketogenic diet (the differences are illustrated in Figure 3). On the one hand, the observed common, although decreasing, reluctance toward employing the ketogenic nutritional model among some doctors and dieticians is not surprising. On the other hand, ignoring the current scientific evidence and potential benefits of such a diet and sticking to the old paradigm may be surprising. Moreover, when using that nutrition model, it is easier to control the glycemia level since the postprandial glucose and insulin levels are usually similar to the fasting ones, thus undergoing almost no changes [95]. Comparing studies on the ketogenic diet with the recommended standard diabetes diet model, it is clear that the ketogenic diet can produce even better effects, as has been described in the earlier parts of the current paper. In view of the literature, there is a chance that, over the years, the current dietary diabetic recommendations will possibly be modified.

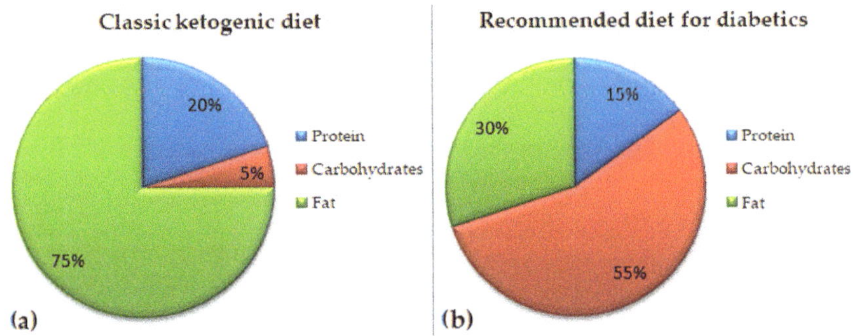

Figure 3. Comparison of an exemplary macronutrient ratio in the classic ketogenic diet (**a**) with that in an exemplary diabetic diet (**b**).

8. The Ketogenic Diet in Practice

The implementation of the ketogenic diet should always be performed under the supervision of a dietician and medical doctor. Additionally, considering the high-fat and low-carbohydrate character of this diet, there is a limited base of usable products. The diet should be prepared with the elimination of starchy products (e.g., rice, groats, bread, and potatoes) and products with high amounts of simple sugars (e.g., sweets and fruits). Small amounts of fruits with low amounts of simple sugars (e.g., strawberries, raspberries, and blueberries) are allowed. Green leafy vegetables (e.g., arugula, spinach, kale, corn salad, and salad) are of high importance. The majority of vegetables, except scratch vegetables, can be put into the everyday menu. Meat, fish, eggs, and dairy products (to a lesser extent) are the source of protein, while avocado, olives, nuts, olive oil, cream, or butter will be good sources of fat. A proper selection of products rich in fatty acids (e.g., Omega-3) will provide an anti-inflammatory effect. Additionally, the level of hydration is significant. Special attention should be given to the number of meals planned for a diabetic patient, due to glucose fluctuations. The source of fluid should be water, but it can also include tea or coffee.

Due to the possible occurrence of the symptoms of "keto-flu", which arise as a result of increased water excretion, in the initial stage of the diet, it is beneficial to pay special attention to hydration and electrolyte (magnesium, sodium, and potassium) supplementation. Moreover, patients should regularly measure the glucose concentration in their blood and the ketone bodies in serum. In the initial stages of a ketogenic diet, diabetic patients should limit their physical activity to prevent potential hypoglycemic episodes. Patients taking medications and insulin should discuss their diet with their medical specialist, with further pharmacological management. It is possible to reduce the doses of drugs or to discontinue them altogether (e.g., insulin). It is worth mentioning that coming back to a high-carbohydrate diet may result in an increase in glycemia. Coming back to a high-carbohydrate diet should be gradual and should always be performed under the supervision of an experienced dietician and medical doctor.

9. Conclusions

Considering all the above-mentioned information concerning the effect of a ketogenic diet on the prophylaxis and treatment of diabetes, it can definitely be said that this is a promising direction for future research into the results derived from the low-carbohydrate character of a ketogenic diet, which induces a condition of ketosis in the body. The results of the current study can inspire optimism since the available data offer evidence of the favorable effect of a ketogenic diet on the prophylaxis and treatment of T2DM disease.

The perspective of drug-dose reduction or its complete withdrawal only renders the topic more interesting and calls for further exploration. Although carbohydrate reduction in the diet is beneficial on its own, it seems that the attainment of a ketosis state is necessary to obtain its therapeutic effects. This observation applies to both types of diabetes; however, the majority of data refers to T2DM.

Based on the current literature, the use of a ketogenic diet in T2DM treatment has its justifications.

The studies mentioned herein suggest an advantage of ketogenic diets over the standard diets recommended for patients with type 2 diabetes. The multifaceted benefits may outweigh those resulting from the standard dietary recommendations for diabetes patients. The observed beneficial effects of a ketogenic diet in T2DM are the reduction and stabilization of glucose and insulin concentrations in serum, the reduction of glycated hemoglobin concentration, and the reduction of the HOMAR-IR indicator, insulin resistance, and body mass. Additionally, the ketogenic diet shows anti-inflammatory effects and changes to the lipid parameters. It is worth mentioning that this effect is more visible for a ketogenic diet when combined with a caloric deficit. The available literature shows that the fears of potential adverse effects of using a ketogenic diet in T2DM patients are, in practice, disproportionate to the number of cases observed. With the correct monitoring of the patient's condition, a skillfully adjusted diet poses no significant risk.

Based on the available data, it can be concluded that it may be reasonable to apply the ketogenic diet in type 1 diabetes (T1D). Although the literature evidence is limited, the beneficial effect of a ketogenic diet in T1D may arise as a consequence of its anti-inflammatory capabilities, glycemic stabilization, and potential effects on the pancreas.

According to the published studies, the occurrence of diabetic ketoacidosis or episodes of severe hypoglycemia is not frequent, although it should be taken into consideration, and the patient should be under medical supervision. It should be made clear, however, that further studies in this field are definitely needed. Although the available data bode well for the future, it seems that it is still too early to "give the green light" to the use of the ketogenic diet; this is caused by, e.g., the unavailability of long follow-up studies or a lack of final results.

In conclusion, the ketogenic diet may be advantageous in T2DM patients. In the case of T1D patients, the observed results are beneficial, but available data are limited and it is difficult to make a final assessment. The topic definitely deserves continued research and clinical follow-up. The results are promising enough to contribute not only to the development of science but also to a potential change in the recommendations for diabetes patients. Thus, it could improve the quality of life of many millions of individuals worldwide.

Author Contributions: D.D., K.K., F.A. and A.P. reviewed the literature; D.D. and F.A. drafted the manuscript; K.K., F.A. and A.P. supervised and revised the manuscript; F.A. and A.P. performed the PRISMA and Cochran analyses; D.D. prepared the visualization; K.K. and A.P. completed funding acquisition. All authors have read and agreed to the published version of the manuscript.

Funding: The publication was financed by the Institute of Health Sciences in the Faculty of Medical and Health Sciences, Siedlce University of Natural Sciences and Humanities. F.A. was funded by the European Union's Horizon 2020 research and innovation program, grant no. 856620.

Institutional Review Board Statement: Not applicable.

Informed Consent Statement: Not applicable.

Data Availability Statement: Not applicable.

Acknowledgments: The authors would like to thank Piotr Słomski for translating and proof-reading the publication and for his scientific commitment.

Conflicts of Interest: The authors declare no conflict of interest.

References

1. Wilder, R.M.A. The effects of ketonemia on the course of epilepsy. *Mayo Clin. Bull.* **1921**, *2*, 307.
2. Ko, A.; Kwon, H.E.; Kim, H.D. Updates on the ketogenic diet therapy for pediatric epilepsy. *Biomed. J.* **2021**. *Epub ahead of print*. [CrossRef] [PubMed]

3. Ruan, Y.; Chen, L.; She, D.; Chung, Y.; Ge, L.; Han, L. Ketogenic diet for epilepsy: An overview of systematic review and meta-analysis. *Eur. J. Clin. Nutr.* **2022**. *Epub ahead of print*. [CrossRef]
4. Pizzo, F.; Collotta, A.D.; Di Nora, A.; Costanza, G.; Ruggieri, M.; Falsaperla, R. Ketogenic diet in pediatric seizures: A randomized controlled trial review and meta-analysis. *Expert Rev. Neurother.* **2022**, *22*, 169–177. [CrossRef]
5. Weber, D.D.; Aminazdeh-Gohari, S.; Kofler, B. Ketogenic diet in cancer therapy. *Aging* **2018**, *10*, 164–165. [CrossRef]
6. Phillips, M.C.L.; Deprez, L.M.; Mortimer, G.M.N.; Murtagh, D.K.J.; McCoy, S.; Mylchreest, R.; Gilbertson, L.J.; Clark, K.M.; Simpson, P.V.; McManus, E.J.; et al. Randomized crossover trial of a modified ketogenic diet in Alzheimer's disease. *Alzheimer's Res. Ther.* **2021**, *13*, 51. [CrossRef]
7. Dowis, K.; Banga, S. The Potential Health Benefits of the Ketogenic Diet: A Narrative Review. *Nutrients* **2021**, *13*, 1654. [CrossRef] [PubMed]
8. Joslin, E.P. The Treatment of Diabetes Mellitus. *Can. Med. Assoc. J.* **1916**, *6*, 673–684.
9. Henwood, M.J.; Thornton, P.S.; Preis, C.M.; Chee, C.; Grimberg, A. Reconciling diabetes management and the ketogenic diet in a child with pyruvate dehydrogenase deficiency. *J. Child Neurol.* **2006**, *21*, 436–439. [CrossRef] [PubMed]
10. Westman, E.C.; Yancy, W.S., Jr.; Humphreys, M. Dietary treatment of diabetes mellitus in the pre-insulin era (1914–1922). *Perspect. Biol. Med.* **2006**, *49*, 77–83. [CrossRef]
11. Gumbiner, B.; Wendel, J.A.; McDermott, M.P. Effects of diet composition and ketosis on glycemia during very-low-energy-diet therapy in obese patients with non-insulin-dependent diabetes mellitus. *Am. J. Clin. Nutr.* **1996**, *63*, 110–115. [CrossRef]
12. Bolla, A.M.; Caretto, A.; Laurenzi, A.; Scavini, M.; Piemonti, L. Low-Carb and Ketogenic Diets in Type 1 and Type 2 Diabetes. *Nutrients* **2019**, *11*, 962. [CrossRef] [PubMed]
13. Hussain, T.A.; Mathew, T.C.; Dashti, A.A.; Asfar, S.; Al-Zaid, N.; Dashti, H.M. Effect of low-calorie versus low-carbohydrate ketogenic diet in type 2 diabetes. *Nutrition* **2012**, *28*, 1016–1021. [CrossRef]
14. Brouns, F. Overweight and diabetes prevention: Is a low-carbohydrate–high-fat diet recommendable? *Eur. J. Nutr.* **2018**, *57*, 1301–1312. [CrossRef]
15. Schiavo, L.; Pilone, V.; Rossetti, G.; Barbarisi, A.; Cesaretti, M.; Iannelli, A. A 4-Week Preoperative Ketogenic Micronutrient Enriched Diet Is Effective in Reducing Body Weight, Left Hepatic Lobe Volume, and Micronutrient Deficiencies in Patients Undergoing Bariatric Surgery: A Prospective Pilot Study. *Obes. Surg.* **2018**, *28*, 2215–2224. [CrossRef]
16. Alarim, R.A.; Alasmre, F.A.; Alotaibi, H.A.; Alshehri, M.A.; Hussain, S.A. Effects of the Ketogenic Diet on Glycemic Control in Diabetic Patients: Meta-Analysis of Clinical Trials. *Cureus* **2020**, *12*, e10796. [CrossRef]
17. Michalczyk, M.M.; Klonek, G.; Maszczyk, A.; Zajac, A. The Effects of a Low Calorie Ketogenic Diet on Glycaemic Control Variables in Hyperinsulinemic Overweight/Obese Females. *Nutrients* **2020**, *12*, 1854. [CrossRef]
18. Pondel, N.; Liśkiewicz, D.; Liśkiewicz, A. Dieta ketogeniczna–mechanizm działania i perspektywy zastosowania w terapii: Dane z badań klinicznych. *Postępy Biochem.* **2020**, *66*, 270–286. [CrossRef] [PubMed]
19. Puchalska, P.; Crawford, P.A. Multi-dimensional Roles of Ketone Bodies in Fuel Metabolism, Signaling, and Therapeutics. *Cell Metab.* **2017**, *25*, 262–284. [CrossRef]
20. Alharbi, A.; Al-Sowayan, N. The Effect of Ketogenic-Diet on Health. *Jan. Food Nutr. Sci.* **2020**, *11*, 301–313. [CrossRef]
21. Wilson, J.; Lowery, R. *The Ketogenic Bible*; Victory Belt Publishing Inc.: Las Vegas, NV, USA, 2017; ISBN 13:978-1-628601-04-6.
22. Saeedi, P.; Petersohn, I.; Salpea, P.; Malanda, B.; Karuranga, S.; Unwin, N.; Colagiuri, S.; Guariguata, L.; Motala, A.A.; Ogurtsova, K.; et al. Global and regional diabetes prevalence estimates for 2019 and projections for 2030 and 2045: Results from the International Diabetes Federation Diabetes Atlas, 9th edition. *Diabetes Res. Clin. Pract.* **2019**, *157*, 1657. [CrossRef] [PubMed]
23. Cloete, L. Diabetes mellitus: An overview of the types, symptoms, complications and management. *Nurs. Stand.* **2022**, *37*, 61–66. [CrossRef] [PubMed]
24. Katsarou, A.; Gudbjörnsdottir, S.; Rawshani, A.; Dabelea, D.; Bonifacio, E.; Anderson, B.J.; Jacobsen, L.M.; Schatz, D.A.; Lernmark, Å. Type 1 diabetes mellitus. *Nat. Rev. Dis. Primers* **2017**, *3*, 17016. [CrossRef] [PubMed]
25. Available online: https://www.cdc.gov/diabetes/basics/what-is-type-1-diabetes.html (accessed on 14 January 2023).
26. Divers, J.; Mayer-Davis, E.J.; Lawrence, J.; Isom, S.; Dabelea, D.; Dolan, L.; Imperatore, G.; Marcovina, S.; Pettitt, D.J.; Pihoker, C.; et al. Trends in Incidence of Type 1 and Type 2 Diabetes Among Youths-Selected Counties and Indian Reservations, United States, 2002–2015. *MMWR Morb. Mortal. Wkly. Rep.* **2020**, *69*, 161–165. [CrossRef] [PubMed]
27. Kotwas, A.; Karakiewicz, B.; Zabielska, P.; Wieder-Huszla, S.; Jurczak, A. Epidemiological factors for type 2 diabetes mellitus: Evidence from the Global Burden of Disease. *Arch. Public Health* **2021**, *79*, 110. [CrossRef]
28. Goyal, R.; Jialal, I. Diabetes Mellitus Type 2. In *StatPearls [Internet]*; [Updated 28 September 2021]; StatPearls Publishing: Treasure Island, FL, USA, 2022. Available online: https://www.ncbi.nlm.nih.gov/books/NBK513253/ (accessed on 14 January 2023).
29. American Diabetes Association. Standards of Medical Care in Diabetes—2022 Abridged for Primary Care Providers. *Clin. Diabetes* **2022**, *40*, 10–38. [CrossRef]
30. Dashti, H.M.; Mathew, T.C.; Al-Zaid, N.S. Efficacy of Low-Carbohydrate Ketogenic Diet in the Treatment of Type 2 Diabetes. *Med. Princ. Pract.* **2021**, *30*, 223–235. [CrossRef]
31. Ranjan, A.; Schmidt, S.; Damm-Frydenberg, C.; Holst, J.J.; Madsbad, S.; Nørgaard, K. Short-term effects of a low carbohydrate diet on glycaemic variables and cardiovascular risk markers in patients with type 1 diabetes: A randomized open-label crossover trial. *Diabetes Obes. Metab.* **2017**, *19*, 1479–1484. [CrossRef]

32. Westman, E.C.; Yancy, W.S.; Mavropoulos, J.C.; Marquart, M.; McDuffie, J.R. The effect of a low-carbohydrate, ketogenic diet versus a lowglycemic index diet on glycemic control in type 2 diabetes mellitus. *Nutr. Metab.* **2008**, *5*, 36. [CrossRef]
33. Handelsman, Y.; Henry, R.R.; Bloomgarden, Z.T.; Dagogo-Jack, S.; DeFronzo, R.A.; Einhorn, D.; Ferrannini, E.; Fonseca, V.A.; Garber, A.J.; Grunberger, G.; et al. American Association of Clinical Endocrinologists and American College of Endocrinology Position Statement on the association of SGLT-2 inhibitors and diabetic ketoacidosis. *Endocr. Pract.* **2016**, *22*, 753–762. [CrossRef]
34. Ahrén, B.; Hirsch, I.B.; Pieber, T.R.; Mathieu, C.; Gómez-Peralta, F.; Hansen, T.K.; Philotheou, A.; Birch, S.; Christiansen, E.; Jensen, T.J.; et al. ADJUNCT TWO Investigators. Efficacy and safety of liraglutide added to capped insulin treatment in subjects with type 1 diabetes: The ADJUNCT TWO randomized trial. *Diabetes Care* **2016**, *39*, 1693–1701. [CrossRef]
35. Mathieu, C.; Zinman, B.; Hemmingsson, J.U.; Woo, V.; Colman, P.; Christiansen, E.; Linder, M.; Bode, B. ADJUNCT ONE Investigators. Efficacy and safety of liraglutide added to insulin treatment in type 1 diabetes: The ADJUNCT ONE Treat-To-Target randomized trial. *Diabetes Care* **2016**, *39*, 1702–1710. [CrossRef]
36. Moriconi, E.; Camajani, E.; Fabbri, A.; Lenzi, A.; Caprio, M. Very-low-calorie ketogenic diet as a safe and valuable tool for long-term glycemic management in patients with obesity and type 2 diabetes. *Nutrients* **2021**, *13*, 758. [CrossRef] [PubMed]
37. Krejčí, H.; Vyjídák, J.; Kohutiar, M. Low-carbohydrate diet in diabetes mellitus treatment. *Vnitr. Lek.* **2018**, *64*, 742–752. (In English) [CrossRef] [PubMed]
38. Yancy, W.S., Jr.; Foy, M.; Chalecki, A.M.; Vernon, M.C.; Westman, E.C. A low-carbohydrate, ketogenic diet to treat type 2 diabetes. *Nutr. Metab.* **2005**, *2*, 34. [CrossRef]
39. Tinguely, D.; Gross, J.; Kosinski, C. Efficacy of Ketogenic Diets on Type 2 Diabetes: A Systematic Review. *Curr. Diab. Rep.* **2021**, *21*, 32. [CrossRef]
40. Wong, K.; Raffray, M.; Roy-Fleming, A.; Blunden, S.; Brazeau, A.S. Ketogenic Diet as a Normal Way of Eating in Adults With Type 1 and Type 2 Diabetes: A Qualitative Study. *Can. J. Diabetes* **2021**, *45*, 137–143. [CrossRef]
41. Krebs, H.A. The regulation of the release of ketone bodies by the liver. *Adv. Enzym. Regul.* **1966**, *4*, 339–354. [CrossRef]
42. Volek, J.S.; Phinney, S.D. *The Art and Science of Low Carbohydrate Performance*; Beyond Obesity LLC: London, UK, 2012.
43. Oh, R.; Gilani, B.; Uppaluri, K.R. Low Carbohydrate Diet. In *StatPearls*; StatPearls Publishing: Treasure Island, FL, USA, 2020.
44. Sukkari, M.A.; Cotten, L.; Alam, M.; Temponi, E.; John, P.D.; Davis, G.; Vellanki, P. Ketogenic Diet in a Patient With Type 1 Diabetes Mellitus With Hypoglycemia Unawareness. *J. Endocr. Soc.* **2021**, *5*, A460. [CrossRef]
45. Buehler, L.A.; Noe, D.; Knapp, S.; Isaacs, D.; Pantalone, K.M. Ketogenic diets in the management of type 1 diabetes: Safe or safety concern? *Clevel. Clin. J. Med.* **2021**, *88*, 547–555. [CrossRef]
46. De Goffau, M.C.; Luopajärvi, K.; Knip, M.; Ilonen, J.; Ruohtula, T.; Härkönen, T.; Orivuori, L.; Hakala, S.; Welling, G.W.; Harmsen, H.J.; et al. Fecal microbiota composition differs between children with β-cell autoimmunity and those without. *Diabetes* **2013**, *62*, 1238–1244. [CrossRef]
47. Endesfelder, D.; Engel, M.; Davis-Richardson, A.G.; Ardissone, A.N.; Achenbach, P.; Hummel, S.; Winkler, C.; Atkinson, M.; Schatz, D.; Triplett, E.; et al. Towards a functional hypothesis relating anti-islet cell autoimmunity to the dietary impact on microbial communities and butyrate production. *Microbiome* **2016**, *4*, 17. [CrossRef]
48. Ang, Q.Y.; Alexander, M.; Newman, J.C.; Tian, Y.; Cai, J.; Upadhyay, V.; Turnbaugh, J.A.; Verdin, E.; Hall, K.D.; Leibel, R.L.; et al. Ketogenic Diets Alter the Gut Microbiome Resulting in Decreased Intestinal Th17 Cells. *Cell* **2020**, *181*, 1263–1275. [CrossRef] [PubMed]
49. Sasaki, K.; Sasaki, D.; Hannya, A.; Tsubota, J.; Kondo, A. In vitro human colonic microbiota utilises D-β-hydroxybutyrate to increase butyrogenesis. *Sci. Rep.* **2020**, *10*, 8516. [CrossRef]
50. Olson, C.A.; Lum, G.R.; Hsiao, E.Y. Ketone Bodies Exert Ester-Ordinary Suppression of Bifidobacteria and Th17 Cells. *Cell Metab.* **2020**, *31*, 1049–1051. [CrossRef] [PubMed]
51. Paoli, A.; Mancin, L.; Bianco, A.; Thomas, E.; Mota, J.F.; Piccini, F. Ketogenic Diet and Microbiota: Friends or Enemies? *Genes* **2019**, *10*, 534. [CrossRef] [PubMed]
52. Choi, I.Y.; Lee, C.; Longo, V.D. Nutrition and fasting mimicking diets in the prevention and treatment of autoimmune diseases and immunosenescence. *Mol. Cell. Endocrinol.* **2017**, *455*, 4–12. [CrossRef] [PubMed]
53. Cheng, C.-W.; Villani, V.; Buono, R.; Wei, M.; Kumar, S.; Yilmaz, O.H.; Cohen, P.; Sneddon, J.B.; Perin, L.; Longo, V.D. Fasting-mimicking diet promotes ngn3-driven β-cell regeneration to reverse diabetes. *Cell* **2017**, *168*, 775–788. [CrossRef]
54. Tóth, C.; Clemens, Z. Type 1 diabetes mellitus successfully managed with the paleolithic ketogenic diet. *Int. J. Case Rep. Images* **2014**, *5*, 699–703. [CrossRef]
55. McClean, A.M.; Montorio, L.; McLaughlin, D.; McGovern, S.; Flanagan, N. Can a ketogenic diet be safely used to improve glycaemic control in a child with type 1 diabetes? *Arch. Dis. Child.* **2019**, *104*, 501–504. [CrossRef]
56. Aylward, N.M.; Shah, N.; Sellers, E.A. The ketogenic diet for the treatment of myoclonic astatic epilepsy in a child with type 1 diabetes mellitus. *Can. J. Diabetes* **2014**, *38*, 223–224. [CrossRef]
57. Cogen, F.R. Incorporation of the Ketogenic Diet in a Youth With Type 1 Diabetes. *Clin. Diabetes* **2020**, *38*, 412–415. [CrossRef] [PubMed]
58. Dressler, A.; Reithofer, E.; Trimmel-Schwahofer, P.; Klebermasz, K.; Prayer, D.; Kasprian, G.; Rami, B.; Schober, E.; Feucht, M. Type 1 diabetes and epilepsy: Efficacy and safety of the ketogenic diet. *Epilepsia* **2010**, *51*, 1086–1089. [CrossRef] [PubMed]
59. Tóth, C.; Clemens, Z. A child with type 1 diabetes mellitus (T1DM) successfully treated with the Paleolithic ketogenic diet: A 19-month insulin freedom. *Int. J. Case Rep. Images* **2015**, *6*, 753–758. [CrossRef]

60. Castaneda, R.L.A.; Mack, K.J.; Lteif, A. Successful treatment of type 1 diabetes and seizures with combined ketogenic diet and insulin. *Pediatrics* **2012**, *129*, e511–e514. [CrossRef]
61. de Bock, M.; Lobley, K.; Anderson, D.; Davis, E.; Donaghue, K.; Pappas, M.; Siafarikas, A.; Cho, Y.H.; Jones, T.; Smart, C. Endocrine and metabolic consequences due to restrictive carbohydrate diets in children with type 1 diabetes: An illustrative case series *Pediatr. Diabetes* **2018**, *19*, 129–137. [CrossRef]
62. Nielsen, J.V.; Jönsson, E.; Ivarsson, A. A low carbohydrate diet in type 1 diabetes: Clinical experience—A brief report. *Ups. J. Med. Sci.* **2005**, *110*, 267–273. [CrossRef] [PubMed]
63. Leow, Z.Z.X.; Guelfi, K.J.; Davis, E.A.; Jones, T.W.; Fournier, P.A. The glycaemic benefits of a very-low-carbohydrate ketogenic diet in adults with Type 1 diabetes mellitus may be opposed by increased hypoglycaemia risk and dyslipidaemia. *Diabet. Med.* **2018**. *Epub ahead of print*. [CrossRef]
64. Krebs, J.D.; Strong, A.P.; Cresswell, P.; Reynolds, A.N.; Hanna, A.; Haeusler, S. A randomised trial of the feasibility of a low carbohydrate diet vs standard carbohydrate counting in adults with type 1 diabetes taking body weight into account. *Asia Pac. J. Clin. Nutr.* **2016**, *25*, 78–84. [CrossRef]
65. Lennerz, B.S.; Barton, A.; Bernstein, R.K.; Dikeman, R.D.; Diulus, C.; Hallberg, S.; Rhodes, E.T.; Ebbeling, C.B.; Westman, E.C.; Yancy, W.S., Jr.; et al. Management of Type 1 Diabetes with a Very Low-Carbohydrate Diet. *Pediatrics* **2018**, *141*, e20173349. [CrossRef]
66. Nolan, J.; Rush, A.; Kaye, J. Glycaemic stability of a cyclist with Type 1 diabetes: 4011 km in 20 days on a ketogenic diet. *Diabet. Med.* **2019**, *36*, 1503–1507. [CrossRef]
67. Charoensri, S.; Sothornwit, J.; Trirattanapikul, A.; Pongchaiyakul, C. Ketogenic Diet-Induced Diabetic Ketoacidosis in a Young Adult with Unrecognized Type 1 Diabetes. *Case Rep. Endocrinol.* **2021**, *2021*, 6620832. [CrossRef] [PubMed]
68. Choi, Y.J.; Jeon, S.M.; Shin, S. Impact of a Ketogenic Diet on Metabolic Parameters in Patients with Obesity or Overweight and with or without Type 2 Diabetes: A Meta-Analysis of Randomized Controlled Trials. *Nutrients* **2020**, *12*, 2005. [CrossRef] [PubMed]
69. Yuan, X.; Wang, J.; Yang, S.; Gao, M.; Cao, L.; Li, X.; Hong, D.; Tian, S.; Sun, C. Effect of the ketogenic diet on glycemic control, insulin resistance, and lipid metabolism in patients with T2DM: A systematic review and meta-analysis. *Nutr. Diabetes* **2020**, *10*, 38. [CrossRef]
70. Zhou, C.; Wang, M.; Liang, J.; He, G.; Chen, N. Ketogenic Diet Benefits to Weight Loss, Glycemic Control, and Lipid Profiles in Overweight Patients with Type 2 Diabetes Mellitus: A Meta-Analysis of Randomized Controlled Trails. *Int. J. Environ. Res. Public Health* **2022**, *19*, 10429. [CrossRef]
71. Parry-Strong, A.; Wright-McNaughton, M.; Weatherall, M.; Hall, R.M.; Coppell, K.J.; Barthow, C.; Krebs, J.D. Very low carbohydrate (ketogenic) diets in type 2 diabetes: A systematic review and meta-analysis of randomized controlled trials. *Diabetes Obes. Metab.* **2022**, *24*, 2431–2442. [CrossRef]
72. Rafiullah, M.; Musambil, M.; David, S.K. Effect of a very low-carbohydrate ketogenic diet vs recommended diets in patients with type 2 diabetes: A meta-analysis. *Nutr. Rev.* **2022**, *80*, 488–502. [CrossRef]
73. Zaki, H.A.; Iftikhar, H.; Bashir, K.; Gad, H.; Fahmy, A.S.; Elmoheen, A. A Comparative Study Evaluating the Effectiveness Between Ketogenic and Low-Carbohydrate Diets on Glycemic and Weight Control in Patients With Type 2 Diabetes Mellitus: A Systematic Review and Meta-Analysis. *Cureus* **2022**, *14*, e25528. [CrossRef]
74. Skow, S.L.; Jha, R.K. A Ketogenic Diet is Effective in Improving Insulin Sensitivity in Individuals with Type 2 Diabetes. *Curr. Diabetes Rev.* **2022**. *Epub ahead of print*. [CrossRef] [PubMed]
75. Charlot, A.; Zoll, J. Beneficial Effects of the Ketogenic Diet in Metabolic Syndrome: A Systematic Review. *Diabetology* **2022**, *3*, 292–309. [CrossRef]
76. Li, M.; Yuan, J. Effects of very low-carbohydrate ketogenic diet on lipid metabolism in patients with type II diabetes mellitus: A meta-analysis. *Nutr. Hosp.* **2022**, *39*, 916–923. (In English) [CrossRef]
77. Wolff, K.; Cavanaugh, K.; Malone, R.; Hawk, V.; Gregory, B.P.; Davis, D.; Wallston, K.; Rothman, R.L. The Diabetes Literacy and Numeracy Education Toolkit (DLNET): Materials to facilitate diabetes education and management in patients with low literacy and numeracy skills. *Diabetes Educ.* **2009**, *35*, 233–245. [CrossRef] [PubMed]
78. Saslow, L.R.; Mason, A.E.; Kim, S.; Goldman, V.; Ploutz-Snyder, R.; Bayandorian, H.; Daubenmier, J.; Hecht, F.M.; Moskowitz, J.T. An Online Intervention Comparing a Very Low-Carbohydrate Ketogenic Diet and Lifestyle Recommendations Versus a Plate Method Diet in Overweight Individuals With Type 2 Diabetes: A Randomized Controlled Trial. *J. Med. Internet Res.* **2017**, *19*, e36. [CrossRef] [PubMed]
79. Goday, A.; Bellido, D.; Sajoux, I.; Crujeiras, A.B.; Burguera, B.; García-Luna, P.P.; Oleaga, A.; Moreno, B.; Casanueva, F.F. Short-term safety, tolerability and efficacy of a very low-calorie-ketogenic diet interventional weight loss program versus hypocaloric diet in patients with type 2 diabetes mellitus. *Nutr. Diabetes* **2016**, *6*, e230. [CrossRef] [PubMed]
80. Li, S.; Lin, G.; Chen, J.; Chen, Z.; Xu, F.; Zhu, F.; Zhang, J.; Yuan, S. The effect of periodic ketogenic diet on newly diagnosed overweight or obese patients with type 2 diabetes. *BMC Endocr. Disord.* **2022**, *22*, 34. [CrossRef]
81. Gardner, C.D.; Landry, M.J.; Perelman, D.; Petlura, C.; Durand, L.R.; Aronica, L.; Crimarco, A.; Cunanan, K.M.; Chang, A.; Dant, C.C.; et al. Effect of a ketogenic diet versus Mediterranean diet on glycated hemoglobin in individuals with prediabetes and type 2 diabetes mellitus: The interventional Keto-Med randomized crossover trial. *Am. J. Clin. Nutr.* **2022**, *116*, 640–652, Erratum in *Am. J. Clin. Nutr.* 2022 Nov. 09. [CrossRef]

82. Saslow, L.R.; Kim, S.; Daubenmier, J.J.; Moskowitz, J.T.; Phinney, S.D.; Goldman, V.; Murphy, E.J.; Cox, R.M.; Moran, P.; Hecht, F.M. A randomized pilot trial of a moderate carbohydrate diet compared to a very low carbohydrate diet in overweight or obese individuals with type 2 diabetes mellitus or prediabetes. *PLoS ONE* **2014**, *9*, e91027. [CrossRef] [PubMed]
83. Saslow, L.R.; Daubenmier, J.J.; Moskowitz, J.T.; Kim, S.; Murphy, E.J.; Phinney, S.D.; Ploutz-Snyder, R.; Goldman, V.; Cox, R.M.; Mason, A.E.; et al. Twelve-month outcomes of a randomized trial of a moderate-carbohydrate versus very low-carbohydrate diet in overweight adults with type 2 diabetes mellitus or prediabetes. *Nutr. Diabetes* **2017**, *7*, 304. [CrossRef]
84. Landry, M.J.; Crimarco, A.; Perelman, D.; Durand, L.R.; Petlura, C.; Aronica, L.; Robinson, J.L.; Kim, S.H.; Gardner, C.D. Adherence to Ketogenic and Mediterranean Study Diets in a Crossover Trial: The Keto-Med Randomized Trial. *Nutrients* **2021**, *13*, 967. [CrossRef]
85. Myette-Côté, É.; Durrer, C.; Neudorf, H.; Bammert, T.D.; Botezelli, J.D.; Johnson, J.D.; DeSouza, C.A.; Little, J.P. The effect of a short-term low-carbohydrate, high-fat diet with or without postmeal walks on glycemic control and inflammation in type 2 diabetes: A randomized trial. *Am. J. Physiol. Regul. Integr. Comp. Physiol.* **2018**, *315*, R1210–R1219. [CrossRef]
86. Jayedi, A.; Zeraattalab-Motlagh, S.; Jabbarzadeh, B.; Hosseini, Y.; Jibril, A.T.; Shahinfar, H.; Mirrafiei, A.; Hosseini, F.; Bidar, S.S. Dose-dependent effect of carbohydrate restriction for type 2 diabetes management: A systematic review and dose-response meta-analysis of randomized controlled trials. *Am. J. Clin. Nutr.* **2022**, *116*, 40–56. [CrossRef] [PubMed]
87. Hallberg, S.J.; McKenzie, A.L.; Williams, P.T.; Bhanpuri, N.H.; Peters, A.L.; Campbell, W.W.; Hazbun, T.L.; Volk, B.M.; McCarter, J.P.; Phinney, S.D.; et al. Effectiveness and Safety of a Novel Care Model for the Management of Type 2 Diabetes at 1 Year: An Open-Label, Non-Randomized, Controlled Study. *Diabetes Ther.* **2018**, *9*, 583–612. [CrossRef] [PubMed]
88. Johnstone, A.M.; Horgan, G.W.; Murison, S.D.; Bremner, D.M.; Lobley, G.E. Effects of a high-protein ketogenic diet on hunger, appetite, and weight loss in obese men feeding ad libitum. *Am. J. Clin. Nutr.* **2008**, *87*, 44–55. [CrossRef] [PubMed]
89. Kodama, S.; Saito, K.; Tanaka, S.; Maki, M.; Yachi, Y.; Sato, M.; Sugawara, A.; Totsuka, K.; Shimano, H.; Ohashi, Y. Influence of fat and carbohydrate proportions on the metabolic profile in patients with type 2 diabetes: A meta-analysis. *Diabetes Care* **2009**, *32*, 959–965. [CrossRef]
90. Fan, Y.; Di, H.; Chen, G.; Mao, X.; Liu, C. Effects of low carbohydrate diets in individuals with type 2 diabetes: Systematic review and meta-analysis. *Int. J. Clin. Exp. Med.* **2016**, *9*, 11166–11174.
91. Snorgaard, O.; Poulsen, G.M.; Andersen, H.K.; Astrup, A. Systematic review and meta-analysis of dietary carbohydrate restriction in patients with type 2 diabetes. *BMJ Open Diabetes Res. Care* **2017**, *5*, e000354. [CrossRef]
92. van Zuuren, E.J.; Fedorowicz, Z.; Kuijpers, T.; Pijl, H. Effects of low-carbohydrate-compared with low-fat-diet interventions on metabolic control in people with type 2 diabetes: A systematic review including GRADE assessments. *Am. J. Clin. Nutr.* **2018**, *108*, 300–331. [CrossRef]
93. McArdle, P.; Greenfield, S.; Rilstone, S.; Narendran, P.; Haque, M.S.; Gill, P. Carbohydrate restriction for glycaemic control in type 2 diabetes: A systematic review and meta-analysis. *Diabet. Med.* **2019**, *36*, 335–348. [CrossRef]
94. Czupryniak, L. Current Topics in Diabetes. *Off. J. Diabetes Pol.* **2022**, *2*, 1–134.
95. Nuttall, F.Q.; Almokayyad, R.M.; Gannon, M.C. Comparison of a carbohydrate-free diet vs. fasting on plasma glucose, insulin and glucagon in type 2 diabetes. *Metabolism* **2015**, *64*, 253–262. [CrossRef]

nutrients MDPI

Article

Real World Practice Study of the Effect of a Specific Oral Nutritional Supplement for Diabetes Mellitus on the Morphofunctional Assessment and Protein Energy Requirements

Juan J. López-Gómez [1,2,*], Cristina Gutiérrez-Lora [2], Olatz Izaola-Jauregui [1,2], David Primo-Martín [1,2], Emilia Gómez-Hoyos [1,2], Rebeca Jiménez-Sahagún [1,2] and Daniel A. De Luis-Román [1,2]

1 Servicio de Endocrinología y Nutrición, Hospital Clínico Universitario de Valladolid, 47003 Valladolid, Spain
2 Centro de Investigación Endocrinología y Nutrición, Universidad de Valladolid, 47002 Valladolid, Spain
* Correspondence: jjlopez161282@hotmail.com; Tel.: +34-983-42-00-00

Citation: López-Gómez, J.J.; Gutiérrez-Lora, C.; Izaola-Jauregui, O.; Primo-Martín, D.; Gómez-Hoyos, E.; Jiménez-Sahagún, R.; De Luis-Román, D.A. Real World Practice Study of the Effect of a Specific Oral Nutritional Supplement for Diabetes Mellitus on the Morphofunctional Assessment and Protein Energy Requirements. *Nutrients* **2022**, *14*, 4802. https://doi.org/10.3390/nu14224802

Academic Editor: Sareen Gropper

Received: 29 October 2022
Accepted: 11 November 2022
Published: 13 November 2022

Abstract: Introduction: The prevalence of malnutrition in patients with diabetes mellitus is high. In these patients, monitoring nutritional intervention is complex. Aims: To evaluate the evolution in the nutritional status in patients with diabetes/prediabetes and malnutrition with a diabetes-specific enteral formula. Methods: Real-life study of one arm in 60 patients with diabetes and prediabetes, performing a dietary adaptation with diabetes-specific oral nutritional supplementation. A morphofunctional assessment was performed, consisting of intake assessment, anthropometry, body composition (bioimpedance and muscle ultrasound), handgrip strength and biochemical markers. The diagnosis of malnutrition was made using the criteria of the Global Leadership Initiative on Malnutrition (GLIM). The variables were measured at baseline and 3 months after starting the intervention. Results: The mean age was 67.13 (14.9) years. In total, 30 (50%) of the patients were women. Of the total, 60% of the patients had diabetes mellitus and 40% of the patients had prediabetes. The initial body mass index was 24.65 (5.35) kg/m^2. It was observed that 80% of the patients had malnutrition, whereas after the intervention, the prevalence was 51.7% ($p < 0.01$). At the beginning of the study, 20% of the patients suffered from sarcopenia and after the intervention it was 16.7% ($p = 0.19$). Conclusions: Medical Nutrition Therapy with an adapted oral diet associated with diabetes-specific oral nutritional supplementation reduces malnutrition in patients at nutritional risk and disturbances of carbohydrate metabolism.

Keywords: diabetes; prediabetes; oral nutritional supplement; enteral nutrition; morphofunctional assessment

1. Introduction

Disease-related malnutrition (DRM) is a pathology with a high prevalence, reaching up to 60% in hospitalized patients with chronic diseases [1]. This malnutrition is more striking in elderly patients and is closely related to sarcopenia, another highly prevalent disease in elderly patients.

Recently, it is being postulated that diabetes mellitus may be a factor favoring malnutrition and sarcopenia. In fact, in institutionalized diabetic patients over 65 years, it has been observed that 21.2% are malnourished and that 39.1% are at risk of malnutrition [2]. This can be related to two situations that occur in diabetic patients: First, there is a sustained metabolic alteration that makes it difficult to manage energy properly, especially carbohydrates. This circumstance promotes a prooxidative state that increases the risk of chronic complications and produces a deterioration in muscle mass and a worser nutritional status. Furthermore, the use of nutrient-restrictive diets is a risk factor for causing imbalances in energy balance. In fact, in the study carried out by Serrano-Valles et al., it was observed that patients with type 2 diabetes mellitus had a worser nutritional situation than patients without diabetes, and this situation was associated with a longer hospital stay [3]. On the other hand, an increase in sarcopenia [4] and sarcopenic obesity [5] has been observed in patients with diabetes mellitus, which conditions a decline in muscle strength and functionality and is associated with a worsening of quality of life of the patient and an increase in mortality [6]. Moreover, glycemic control correlates with muscle mass and function [7].

The diagnosis of malnutrition is difficult because it does not depend only on the weight at a given time, but also on its evolution and the underlying pathological situations [8]. Classically, body mass index has been used as a measure of the patient's nutritional status, but this measure is not the most appropriate and has evident limitations in different pathologies that can make it possible to maintain an adequate weight with a deterioration of the "metabolically active mass" [9]. These pathologies can produce an increase in fat mass (obesity) or body water (heart failure, liver failure, kidney failure) [10]. Therefore, the clinical use of body composition measurements is essential for adequate assessment of this malnutrition, especially in the evaluation of muscle mass and function.

In this context, nutritional assessment can no longer be based on the determination of anthropometric measurements. The concept of morphofunctional assessment postulates that the diagnosis and monitoring of nutritional status must be carried out using techniques that determine the evaluation of intake, anthropometry, body composition, muscle strength and function. This new concept of nutritional evaluation should be implemented in the clinical management of the patient and in the determination of variables in clinical research in nutrition [11,12].

The prevalence of malnutrition in patients with diabetes mellitus and the difficulty in assessing it is high due to the limitations of classical techniques such as body mass index. On the other hand, monitoring nutritional improvement in these patients is complex in relation to the above-mentioned reasons. In this context, we need to deploy a comprehensive assessment of nutritional status. This type of assessment combines methods of determining body composition and muscle strength according to the concept of the Morphofunctional Assessment of Disease-Related Malnutrition. This new concept can allow us to obtain more valuable information in the diagnosis and evolution of the patient in Medical Nutrition Therapy [10].

Intervention studies with specific nutritional oral supplementation in diabetes are scarce and conducting clinical trials of intervention is difficult due to ethical problems in the comparative arm. This means that real life studies can provide us with additional information. These studies could generate evidence obtained from routine clinical practice data. This is the principal value from data obtained outside the context of randomized controlled trials [13].

In patients with diabetes and malnutrition, nutritional intervention for caloric increase can be associated with a worsening of glycemic control. This is the reason to use diabetes-specific oral nutritional supplements. These formulas can have the following characteristics: (a) the reduction of energy from carbohydrates and replace it with energy from lipids; (b) use of carbohydrates with a low glycemic index such as lactose or isomaltulose; or (c) increase the amount of soluble fiber to decrease glucose absorption [14]. These types of formulas usually have a high percentage of protein to improve nutritional status and muscle function. In addition, these formulas are usually enriched in monounsaturated (MUFA) and polyunsatured fatty acids (PUFA) and lead to finding benefits in the lipid profile of these patients [15].

For this reason, a real-life study is proposed to describe the effect of diabetes-specific oral nutrition supplementation in patients with disease-related malnutrition. The main objective of this study was to prove the influence of Medical Nutrition Therapy with a specific oral nutritional supplementation through morphofunctional assessment in patients with malnutrition and diabetes or altered metabolism of carbohydrates.

2. Materials and Methods

2.1. Design

This is an open-label, prospective, interventional study. In this study, the nutritional status of diabetic patients and their evolution was evaluated. The evolution was based on nutritional measurements performed according to medical nutrition therapy (adapted dietary recommendations and diabetes-specific oral nutritional supplementation). It was proposed as a real-world study with data obtained from routine clinical practice.

An exhaustive anamnesis was carried out on affiliation data, personal history, evolution of the disease and nutritional history. Classic anthropometric evaluation, bioelectrical impedanciometry and muscle ultrasound evaluation was performed. Nutritional parameters were measured according to usual clinical practice.

Oral nutritional supplementation was started with a specific normocaloric and hyperproteic formula for diabetic patients in routine clinical practice. The medical and nutritional treatment prescribed at the initial visit and during follow-up was recorded. Records of the evolution of the morphofunctional assessment (anthropometry, nutritional ultrasound, electrical and biochemical bioimpedance analysis) were taken at the beginning and 3 months after the start of nutritional support.

The study was carried out in accordance with the Declaration of Helsinki and all procedures were approved by the Medical Research Ethics Committee (CEIm) of the Hospital Clínico Universitario de Valladolid under code PI 20-1967.

2.2. Study Subjects

The study was developed in patients with malnutrition referred to the Clinical Nutrition consultation of the East Valladolid Area. Patient recruitment was carried out between January 2021 and September 2022.

The patient inclusion criteria were patients with diabetes mellitus or prediabetes at risk of malnutrition, and need for specific oral supplementation of diabetes mellitus and age over 18 years. The exclusion criteria were decompensated liver disease; chronic kidney disease stage IV or higher; inability to walk; and non-signing of the informed consent by the patient.

2.3. Nutritional Intervention

The patients received the following nutritional education and medical nutrition therapy:

- Patients received education on adapted oral diet to increase calories and protein in patients with diabetes or carbohydrate metabolism disorders (prediabetes).
- Patients received nutritional education with a dietitian in adaptation of oral diet to increase protein–energy intake and they received education in consumption of oral nutritional supplementation. The adherence of these diets was assessed every fourteen days with a phone call by a dietitian to improve the calorie restriction and macronutrient distribution. The diet compliance was verified with a telephone nutritional questionnaire every fourteen days and a four-day nutritional questionnaire during face-to-face visits.
- Oral nutritional supplementation with a hyperproteic normocaloric formula specific for diabetes (carbohydrates with a low glycemic index, insoluble fiber) (Nutavant Plus Diabetica®) (Table 1). The amount (1 or 2 bottles) was adjusted according to the nutritional requirements of the patient and the estimation of usual intake [16,17].

Table 1. Composition of Specific Diabetes Formula used as intervention.

Diabetes Specific Formula (250 mL Bottle)	
Caloric Content (kcal)	300
Proteins (g (% TCV [1]))	17 (22.66%)
Lipids (g (% TCV))	11.7 (35.1%)
Saturated (g)	2.6
MCT (g)	1.7
MUFA (g)	5.9
PUFA (g)	2.8
w-3 (g)	0.83
w-6 (g)	1.88
Carbohydrates (g (%TCV))	30 (40%)
Sugars (g)	6.3
Isomaltulose (g)	3
Minerals	
Sodium (mg)	278
Chloride (mg)	113
Potassium (mg)	333
Calcium (mg)	275
Phosphate (mg)	238
Magnesium (mg)	50
Iron (mg)	2.8
Zinc (mg)	2
Copper (mg)	0,20
Iodine (mg)	30
Selenium (mg)	11
Manganese (mg)	0.40
Chrome (mg)	45
Molybdenum (mg)	10.6
Fluoride (mg)	0.58
Vitamins	
Vitamin A (mg)	160
Vitamin D (mg)	1.6
Vitamin K (mg)	15
Vitamin C (mg)	16
Thiamin (mg)	0.22
Riboflavin (mg)	0.28
Vitamin B6 (mg)	0.28
Niacin (mg)	3.3
Folic Acid (mg)	40
Vitamin B12 (mg)	0.50
Pantothenic acid (mg)	1.2
Biotin (mg)	10
Vitamin E (mg)	2.4
Inositol (mg)	38
Choline (mg)	38
Osmolarity (mOsm/L)	315
Fiber (g)	4.5

[1] %TCV: Percentage Total Calorie Value.

2.4. Study Variables

- Clinical variables: Age (years); gender (male/female); systolic and diastolic blood pressure (mmHg); presence of concomitant pathologies.
- Anthropometry: The anthropometric variables measured were weight (kg); height (meters); body mass index (BMI) (weight/height × height) (kg/m^2); arm circumference (AC); and calf circumference (CC). The percentage of weight loss was calculated: Start Weight Loss = ((Usual weight (kg) – Present weight (kg))/Usual weight) × 100; and 3 Months Weight Loss = ((Initial weight (kg) − 3 months weight)/Initial weight) × 100.

- Biochemical variables: They were performed with a Cobas c-711 autoanalyzer (Roche Diagnostics): Glucose (mg/dL); total cholesterol (mg/dL); HDL cholesterol (mg/dL); LDL cholesterol (mg/dL); triglycerides (mg/dL); albumin (g/dL); HbA1c (%), C-Reactive Protein (CRP) (mg/dL), prealbumin (mg/dL); and CRP/prealbumin ratio.
- Energy Expenditure and Nutritional Requirements: The energy expenditure of the patients was determined by means of the Harris–Benedict Equation multiplied by a Stress Factor of 1.3 and the protein requirements were determined by means of the factor 1–1.5 g of protein per kilogram of the patient's adjusted weight. We based the requirements on the patient's clinical situation and comorbidities as the recommendations made by the clinical guidelines of the European Society for Clinical Nutrition and Metabolism in surgery and oncology suggests. This decision was made because most of the patients had underlying oncological and/or surgical pathology [18,19].
- Nutritional questionnaire: All subjects completed a 4-day prospective nutritional questionnaire to assess calorie and macronutrient intake. This questionnaire was conducted before starting the intervention and 3 months after its start. The importance of not modifying dietary habits was insisted on so that it would be representative. All study participants were instructed to record food intake, daily and prospectively, with the help of food scales to facilitate precision in portion sizes. They were also asked about the way of preparing said foods. Records were reviewed by a dietitian and analyzed by a Dietsource® data processing computer system (Nestle, Geneve, Switzerland). Total calorie intake was used as an indicator of nutritional intake. No subject was taking dietary supplements or following any type of diet at the start of the study or in the 6 months prior to the study. Nutritional intake was measured in absolute values (in kilocalories (kcal) or grams (g)) and in percentages of the total caloric value. The nutritional questionnaire assessed the total energy intake, measured in kilocalories, as well as the different macronutrients: proteins, carbohydrates, fats and fiber, all of them measured in grams. The amount of protein ingested per kilogram of body weight was also calculated.
- Muscle functionality variables: Hand dynamometry (JAMAR® dynamometer): non-dominant hand dynamometry was performed with the patient seated and the arm at a right angle to the forearm. Three measurements were made and the average of the three measurements was made. The diagnostic criteria of low muscle strength proposed by the European Working Group on sarcopenia in older people (EWGSOP2) [20] were used. (<27 kg in men and <16 kg in women).
- Corporal Composition:

Bioimpedanciometry (BIA 101 Anniversary; EFG Akern): The BIA was performed between 8:00 and 9:15 h, after an overnight fast and after a time of 15 min in the supine position. The BIA measured the geometrical components of impedance (Z), resistance (R) and the capacitance component (X). The PhA is derived for the following equation PhA = $(X/R) \times (180°/\pi)$. The BIA provided data regarding fat mass (FM), fat-free mass (FFM), skeletal muscle mass (SMM), fat free mass index (FFMI) and percentage of skeletal muscle mass (%MM) [14]. All these data are based on raw electrical data from BIA multifrequency at 50 Hz [14].

Skeletal Appendicular Mass Index (ASMI): ASMI (kg/m^2) was estimated by bioimpedanciometry applying Sergi's formula [19]: $-3.964 + (0.227 \times \text{RI}) + (0.095 \times \text{weight}) + (1.384 \times \text{sex}) + (0.064 \times \text{Xc})$, where RI is Resistivity Index and Xc is reactance (sex: Male = 1; Female = 0).

European Working Group on Sarcopenia in Older People (EWGSOP2) diagnostic criteria of sarcopenia for low muscle mass (ASMI < 7 kg/m^2 in men and ASMI < 5.5 kg/m^2 in women) were used [21].

Muscle ultrasound (Mindray Z60): Muscle ultrasound of the quadriceps rectus femoris (QRF) of the left and right lower extremities with a 10 to 12 MHz probe and a multifrequency linear matrix (Mindray Z60, Madrid, Spain) were performed in all subjects (patient in supine position). The probe was aligned perpendicularly to the longitudinal and transverse axis of the non-dominant QRF. The evaluation was performed without compression at the level of the lower third from the superior pole of the patella and the anterior superior iliac spine, measuring the anteroposterior muscle thickness, circumference and cross-sectional area [17]. The measurements made using this technique were: muscle area (cm^2) (MARA) and the index of the muscle area with respect to height (cm^2/m^2) (MARAI), the X-axis of QRC (cm), Y-axis of QRC (cm) and X/Y index [12].

- Malnutrition and Sarcopenia diagnosis: The diagnosis of malnutrition was made using the Global Leadership Initiative on Malnutrition (GLIM) criteria, using the ASMI estimated by bioimpedance measurement measured by impedance measurement as an evaluation variable for muscle deterioration (ASMI muscle mass reduction < 7 kg/m^2 in men was considered and <5.5 kg/m^2 in women) [8]. On the other hand, the diagnosis of sarcopenia was made according to the revised criteria for sarcopenia of the EWGSOP2, using the ASMI estimated by bioimpedance as a determination of decreased muscle mass with handgrip strength to estimate the function to diagnose sarcopenia [20].

2.5. Data Analysis

The data was stored in a database of the statistical package SPSS 23.0 (SPSS Inc., Chicago Illinois, USA) with an official license from the University of Valladolid. A normality analysis of continuous variables was performed with the Kolmogorov-Smirnov test.

Continuous variables were expressed as mean (standard deviation). The difference in means between parametric variables was analyzed with the unpaired and paired t-Student test, and the non-parametric variables with the Mann–Whitney U-test and the Kruskal–Wallis K-test. An intention-to-treat analysis of patients who consumed supplementation more than a half time of intervention was conducted. A significant difference was considered as a *p*-value of less than 0.05.

3. Results

A total of 75 patients were recruited, of whom 60 (80%) were analyzed (Figure 1). In total, 36 (60%) patients had a diagnosis of diabetes mellitus and 24 (40%) patients suffered from some disorder of carbohydrate metabolism without a diagnosis of diabetes mellitus (altered fasting blood glucose, intolerance to carbohydrates, glycated hemoglobin in prediabetes range (5.7–6.4%).

Of the total number of patients, 30 (50%) patients were women, and 30 (50%) patients were men. The mean age of the patients was 67.13 (14.09) years.

Most of the patients suffered from oncological pathology (68.3%%); the rest of the patients suffered from cardiopulmonary pathology (10%), non-oncological digestive pathology (13.3%), and neurological (1.7%) and other pathologies (6.7%).

3.1. Sample Description

After performing the nutritional assessment, data were obtained from anthropometry, dynamometry, electrical bioimpedance measurement, and muscle ultrasound (Table 2). According to the GLIM criteria, 48 (80%) patients suffered from malnutrition and according to the EWGSOP2 criteria, 12 (20%) patients presented sarcopenia.

After carrying out the initial analysis of the diet, a consumption below the caloric-protein requirements was observed (Table 3).

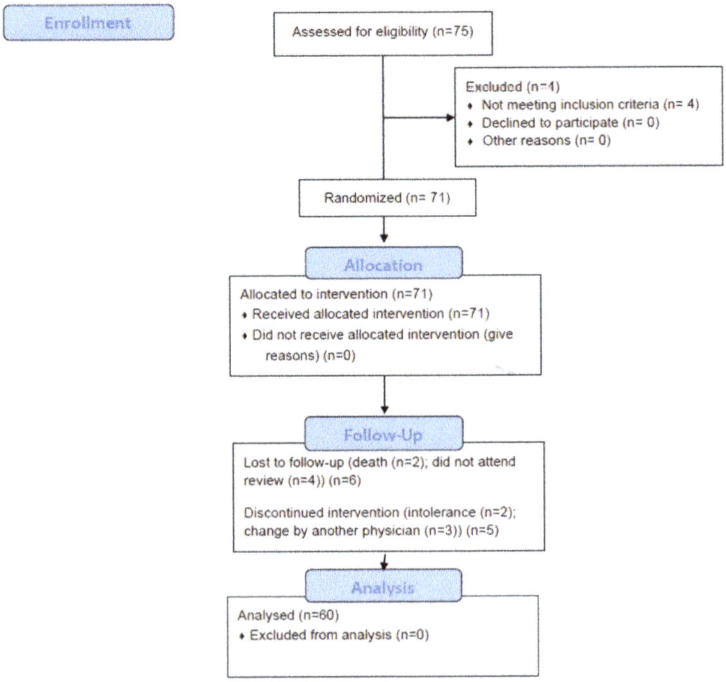

Figure 1. Flow chart.

Table 2. Differences in the Morphofunctional Assessment at the beginning according to sex.

	Total	Men	Women	*p*-Value
Sarcopenia (EWGSOP2)	20%	3.3%	36.7%	<0.01
Malnutrition (GLIM)	80%	86.7%	73.3%	0.19
Diabetes Mellitus	60%	66.7%	53.3%	0.29
Age (years)	67.13 (14.9)	68.70 (12.11)	65.57 (15.89)	0.39
	Anthropometry			
BMI (kg/m^2)	24.65 (5.35)	25.53 (4.30)	23.77 (6.18)	0.20
Braquial circumference (cm)	24.71 (3.52)	25.43 (2.53)	23.99 (4.21)	0,11
Calf Circumference (cm)	31.69 (3.61)	32.75 (3.11)	30.63 (3.81)	0,02
	Handgrip Strength			
Handgrip Strength (kg)	20.60 (8.26)	25.42 (7.65)	15.79 (5.67)	<0.01
	Bioelectrical Impedanciometry			
Resistance (ohm)	545.4 (91.25)	502.53 (75.11)	588.27 (86.59)	<0.01
Reactance (ohm)	46.9 (9.26)	44.7 (9.11)	49.10 (9.01)	0.06
Fase Angle (°)	4.95	5.11 (0.86)	4.78 (0.66)	0.11
ASMI (kg/m^2)	6.43 (1.11)	7.07 (0.91)	5.79 (0.91)	<0.01
FFMI (kg/m^2)	17.46 (3.05)	18.27 (2.70)	16.65 (3.20)	0.04
FMI (kg/m^2)	6.78 (3.32)	6.58 (2.34)	6.98 (4.10)	0.64
BCMI (kg/m^2)	8.29 (1.66)	8.93 (1.69)	7.64 (1.37)	<0.01
%TBW	56.17 (8.90)	58.60 (4.49)	53.75 (11.35)	0.03
	Rectus Femoris Ultrasonography			
RFAI (cm^2/m^2)	1.27 (0.47)	1.36 (0.55)	1.17 (0.35)	0.14
X/Y (cm^2/m^2)	3.59 (1.57)	3.39 (1.56)	3.79 (1.57)	0.34

ASMI: Appendicular Skeletal Muscular Index; FFMI: Fat-Free Mass Index; FMI: Fat Mass Index; BCMI: Body Cell Mass Index; %TBW: Percentage Total Body Water; RFAI: Rectus Femoris Area Index; X/Y: Index Transversal axis (X)/anteroposterior axis (Y).

Table 3. Difference in calorie and protein requirements and consumption between men and women.

	Total	Men	Women	p-Value
Calories Requirement (kcal/day)	1772 (178.12)	1894 (149)	1650 (107)	<0.01
Calories Consumption (kcal/day)	1364 (417)	1333 (455)	1433 (410)	0.40
Calories Consumption (%)	78.76 (16.88)	70.33 (22.83)	86.87 (23.98)	0.01
Protein Requirements (g/day)	79.26 (16.88)	87.82 (12.33)	70 (16.61)	<0.01
Protein Consumption (g/day)	1.15 (0.44)	1.07 (0.41)	1.22 (0.47)	0.23
Protein Consumption (%)	88.58 (34.20)	81.81 (31.27)	94.13 (36.53)	0.23

The metabolic parameters of the patients prior to the intervention were evaluated. An increase in glycated hemoglobin, glucose and triglycerides was observed in patients with diagnosed diabetes mellitus compared to those patients with carbohydrate alterations (Table 4).

Table 4. Difference in metabolic parameters depending on the presence of diabetes mellitus (DIABETES) or alterations in carbohydrates metabolism (NO DIABETES).

	Diabetes	No Diabetes	p-Value
HbA1c (%)	6.86 (1.19)	6.03 (0.58)	<0.01
Glucose (mg/dL)	124.92 (38.19)	94.62 (19.38)	<0.01
Total cholesterol (mg/dL)	153.75 (37.13)	167 (39.95)	0.19
HDL cholesterol (mg/dL)	57.81 (32.50)	63.86 (23.60)	0.45
LDL cholesterol (mg/dL)	79.54 (29,66)	84.40 (23.93)	0.53
Trygliceride s (mg/dL)	108.81 (54.06)	81.79 (31.18)	0.03
Albumin (g/dL)	4.06 (0.56)	4.13 (0.38)	0.59
CRP/prealbumin	0.43 (0.51)	0.60 (0.89)	0.37

HbA1c: Glycated Hemoglobin.

3.2. Nutrional Therapy Intervention

Nutritional supplementation was started with a specific hyperproteic normocaloric diabetes formula. It was prescribed based on the requirements and the calculated dietary intake. A bottle (250 mL) was consumed by 44 patients (73.3%), two bottles by 15 patients (25%) and half a bottle (125 mL) by 1 patient (1.7%). At three months from the beginning of intervention, 56 (93.3%) patients consumed 100% of oral nutritional supplementation prescribed, 1 (1.7%) consumed 50%, 1 (1.7%) consumed 25% and 2 (3.3%) patients consumed no supplementation.

- Influence of the intervention on intake

A significant increase in caloric intake (Baseline: 1364 (417) kcal/day; 3 months: 1666 (519) kcal/day; $p < 0.01$) and of all nutrients (proteins, fats and carbohydrates) was observed, with an increase in the percentage of carbohydrates and a decrease in fat over the total caloric value (Figure 2).

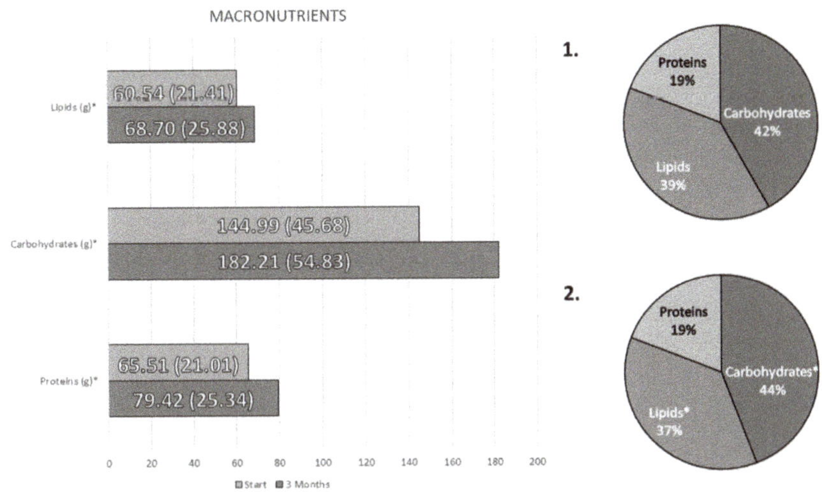

Figure 2. Differences in the consumption of macronutrients between the beginning of the intervention (**1**) and 3 months after it (**2**). * p-value < 0.05.

An improvement in caloric adjustment to the percent of protein and energy requirements was observed. On the other hand, an increase in protein consumption per kg of weight was observed (Initial: 1.14 (0.43) g/kg/day; 3 months: 1.38 (0.49) g/kg/day; p-value < 0.01) (Figure 3).

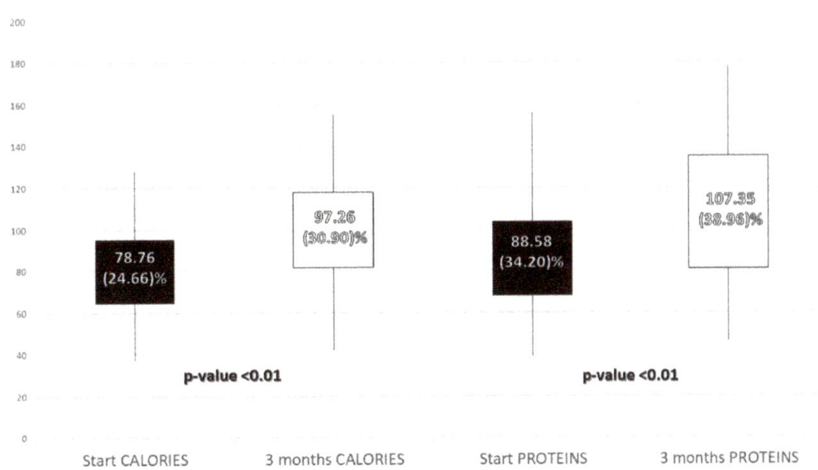

Figure 3. Percent of adjustment to caloric and protein requirements before and 3 months after the intervention.

An improvement in consumption of monounsaturated fatty acids (MUFA) and polyunsaturated fatty acids (PUFA) after the intervention was observed (Table 4). In the same way, an improvement in consumption of minerals except sodium and copper was noted, as was a consumption of vitamins except vitamin A, B1, B3, B12, C and D (Table 5).

Table 5. Changes in macronutrients and micronutrients and their distribution before and 3 months after intervention.

	Start	3 Months	p-Value
Carbohydrates (g)	144.99 (45.68)	182.21 (54.83)	<0.01
Fiber(g)	12.53 (4.76)	17.39 (7.14)	<0.01
Proteins (g)	65.51 (21.01)	70.42 (25.34)	<0.01
Lipids (g)	60.54 (21.41)	68.70 (25.88)	0.03
SFA (g)	17.39 (8.42)	19.77 (9.51)	0.12
SFA (%TCV)	10.37 (8.2–14)	9.69 (7.16–12.48)	0.26
MUFA (g)	23.86 (10.99)	28.94 (12.51)	0.01
MUFA (%TCV)	15.35 (12.38–18.45)	14.28 (12.53–18.45)	0.55
PUFA(g)	6.17 (4.22)	8.44 (3.68)	<0.01
PUFA (%TCV)	3.39 (2.86–4.56)	4.09 (3.61–5.42)	<0.01
EPA (g)	0.08 (0.14)	0.22 (0,56)	0.12
DHA (g)	0.13 (0.20)	0.16 (0.22)	0.52
Cholesterol (mg)	300.85 (157.75)	301.71 (196.96)	0.37
	Minerals		
Phosphorus (mg)	881.10 (338.67)	1116.27 (423.28)	<0.01
Magnesium (mg)	151.65 (60.31)	202.09 (79.43)	<0.01
Calcium (mg)	708.59 (327.23)	982.12 (377.23)	<0.01
Iron (mg)	7.71 (3.09)	10.28 (4.17)	<0.01
Zinc (mg)	6.64 (3.25)	8.09 (3.66)	<0.01
Sodium (mg)	1569.35 (845.39)	1742.61 (824.38)	0.12
Potassium (mg)	1943.22 (687.02)	2205.49 (824.09)	0.04
Iodine (mg)	30.83 (25.16)	60.63 (33.64)	<0.01
Selenium (mg)	35.39 (24.01)	50.66 (27.62)	0.01
Copper (mg)	0.77 (0.54)	0.93 (0.49)	0.08
	Vitamins		
Vitamin A (IU)	1152.26 (1263.39)	1347.76 (973.54)	0.40
Vitamin B1 (mg)	0.89 (0.46)	1.04 (0.72)	0.20
Vitamin B2 (mg)	1.19 (0.59)	1.51 (0.69)	<0.01
Niacin (mg)	11.96 (6.85)	13.84 (6.77)	0.10
Vitamin B5 (mg)	0.14 (0.42)	1.22 (1.00)	<0.01
Vitamin B6 (mg)	1.17 (0.61)	1.43 (0.68)	0.01
Folic Acid (mg)	138.75 (75.32)	174.29 (88.33)	0.01
Vitamin B12 (mg)	5.54 (7.88)	5.26 (4.50)	0.83
Vitamin C (mg)	96.55 (67.59)	109.29 (71.45)	0.33
Vitamin D (mg)	4.02 (6.41)	4.96 (5.57)	0.40
Vitamin E (mg)	5.77 (3.33)	7.62 (3.89)	<0.01
Vitamin K (mg)	1.52 (5.97)	16.93 (15.81)	<0.01

SFA: Saturated Fatty Acids; MUFA: Monounsaturated Fatty acids; PUFA: Polyunsaturated Fatty Acids; EPA: eicosapentaenoic acid; DHA: docasahexaenoic acid; %TCV: percentage total caloric value.

- Influence of the intervention on body composition

A decrease in the percentage of weight loss was observed at 3 months in the total sample and stratified according to sex. No deterioration in anthropometric parameters (weight, muscle circumference, calf circumference), body composition (bioimpedanciometry and muscle ultrasound) or muscle function (hand dynamometry) was observed after the start of oral nutritional supplementation. No difference was observed according to sex (Table 6).

Table 6. Changes in anthropometry, body composition and muscle function according to gender at baseline and 3 months after nutritional intervention.

	Men			Women		
	Anthropometry					
	Baseline	3 Months	*p*-Value	Baseline	3 Months	*p*-Value
%Weight Loss	10.05 (7.03)	−0.25(5.57)	<0.01	12.84 (13.04)	−0.72 (4.95)	<0.01
BMI (kg/m^2)	25.53 (4.30)	24.72 (4.04)	0.21	23.77 (6.18)	23.11 (5.56)	0.53
Arm circumference (cm)	25.43 (2.53)	25.67 (2.77)	0.32	23.99 (4.21)	24.16 (3.91)	0.65
Calf Circumference (cm)	32.75 (3.11)	33.33 (2.75)	0.16	24.16 (3.91)	30.63 (3.81)	0.18
	Handgrip Strength					
Handgrip Strength (kg)	23.81 (7.61)	24.03 (8.81)	0.44	14.77 (6.66)	15.13 (5.69)	0.95
	Bioelectrical Impedanciometry					
Resistance (ohm)	501 (76)	502 (85)	0.95	588 (86)	586 (87)	0.83
Reactance (ohm)	44.61 (9.43)	45.86 (11.43)	0.51	49.10(9.01)	49.43(11.29)	0.82
Phase Angle (°)	5.11 (0.89)	5.24 (1.15)	0.42	4.78 (0.66)	4.81 (0.79)	0.84
ASMI (kg/m^2)	7.11 (0.91)	7.15 (0.94)	0.71	5.79 (0.91)	5.81 (0.94)	0.71
FFMI (kg/m^2)	18.35 (2.76)	18.34 (2.80)	0.98	16.65 (3.20)	16.29 (2.19)	0.46
FMI (kg/m^2)	6.80 (2.15)	6.72 (2.25)	0.76	6.98 (4.10)	6.85 (4.07)	0.51
BCMI (kg/m^2)	8.97 (1.74)	9.09 (2.00)	0.48	7.64 (1.37)	7.67 (1.40)	0.79
%TBW	58.18 (4.13)	58.28 (4.83)	0.88	53.75 (11.35)	55.91 (7.17)	0.32
	Rectus Femoris Ultrasonography					
RFAI (cm^2/m^2)	1.36 (0.55)	1.31 (0.57)	0.31	1.18 (0.35)	1.14 (0.37)	0.19
X/Y (cm^2/m^2)	3.39 (1.56)	3.55 (1.48)	0.46	3.79 (1.57)	3.56 (1.24)	0.46

ASMI: Appendicular Skeletal Muscular Index; FFMI: Fat-Free Mass Index; FMI: Fat Mass Index; BCMI: Body Cell Mass Index; %TBW: Percentage Total Body Water; RFAI: Rectus Femoris Area Index; X/Y: Index Transversal axis (X)/anteroposterior axis (Y).

A significant decrease in the malnutrition (GLIM criteria) rate was observed 3 months after the intervention. However, a slight, non-significant decrease in the prevalence of sarcopenia (EWGSOP2) was observed in the sample studied (Figure 4).

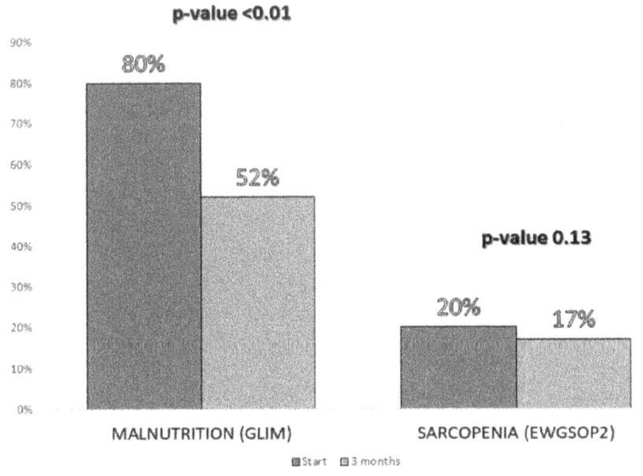

Figure 4. Comparison of the percentages of malnutrition (according to the Global Leadership Initiative on Malnutrition (GLIM) criteria) and sarcopenia (according to the European Working Group on Sarcopenia in Older People (EWGSOP2) criteria) before the intervention and 3 months after the start of the intervention.

In patients with sarcopenia, an improvement in muscle strength measured by dynamometry in the non-dominant hand was observed 3 months after the start of the nutritional intervention (Baseline: 9.83 (5.49); 3 months: 11.33 (6.11), *p*-value = 0.04).

- Influence of the intervention on biochemical parameters

There were no baseline differences in biochemical parameters based on gender. The evolution in the biochemical parameters was analysed according to the presence or not of diabetes mellitus due to the baseline differences of the groups at this level.

A significant increase in glycated haemoglobin was observed in patients with diabetes mellitus, although it was not observed in patients with alterations in carbohydrate metabolism. On the other hand, a significant increase in plasma albumin was observed in patients with diabetes mellitus (Table 7).

Table 7. Difference in the metabolic parameters before and 3 months after the start of the nutritional intervention based on the diagnosis of diabetes (DIABETES) or alterations in the carbohydrate's metabolism (NO DIABETES).

	Diabetes			No Diabetes		
	Baseline	3 Months	*p*-value	Baseline	3 Months	*p*-Value
HbA1c (%)	6.87 (1.24)	7.18 (1.09)	0.02	6.05 (0.60)	6.10 (0.62)	0.30
Glucose (mg/dL)	123.56 (38.79)	131.38 (29.76)	0.10	94.62 (19.39)	89.83 (20.03)	0.55
Total cholesterol (mg/dL)	154 (37.01)	158 (38.99)	0.29	167 (39.95)	172 (46.22)	0.49
HDL cholesterol (mg/dL)	58 (33.48)	60.94 (27.87)	0.25	65.50 (23.94)	63.7 (20.79)	0.34
LDL cholesterol (mg/dL)	79.45 (30.55)	83.62 (31.35)	0.29	88.83 (19.56)	95.83 (43.66)	0.45
Tryglicerides (mg/dL)	109.89 (55.15)	105.34 (50.23)	0.26	81.79 (31.18)	86.92 (29.17)	0.32
Albumin (g/dL)	4.09 (0.53)	4.25 (0.42)	0.02	4.13 (0.38)	4.02 (0.42)	0.14
CRP/prealbumin	0.38 (0.51)	0.70 (2.07)	0.38	0.47 (0.74)	0.26 (0.38)	0.16

4. Discussion

The use of medical nutrition therapy with an adapted oral diet and diabetes-specific supplementation in patients with high nutritional risk and alterations in carborhydrates metabolism (diabetes mellitus and prediabetes) was related to an achievement of nutritional requirements (calorie-protein) in our study. This was associated with cessation of previous weight loss and maintenance of morphofunctional assessment parameters (anthropometry, body composition and muscle strength).

The patients analyzed were referred to the Clinical Nutrition consultation as they were in a situation of nutritional risk. Most of the patients in this sample suffered from cancer. This type of pathology is associated with an increased risk of malnutrition. It has been observed that between 15 and 40% of cancer patients present some degree of malnutrition at diagnosis of the disease [22]. On the other hand, these pathologies are associated with alterations in carbohydrate metabolism in relation to the disease itself or its treatment (chemotherapy, corticosteroids, etc.). This circumstance can be associated with a decreased diagnosis of the state of malnutrition due to the observation of high body mass indexes; and, in addition, it can be associated with a tendency to carry out a dietary restriction to control the metabolic complications of diabetes mellitus [3]. All this can enhance the state of malnutrition by not carrying out adequate medical nutrition therapy.

Most of the patients had malnutrition according to GLIM criteria, and this circumstance can be considered normal given that the patients had been referred for nutritional assessment as they were at high nutritional risk. However, the rate of sarcopenia was quite high (20%) since the population was not an elderly population. Other series of patients with diabetes mellitus have also shown high prevalence of muscle mass deterioration, such as the study by Park et al. which showed a more striking deterioration of skeletal muscle mass in patients with diabetes mellitus [23]. In fact, it has been observed that there are many factors that can negatively influence muscle mass in patients with diabetes, such as poor glycemic control [7], the use of certain treatments for glycemic control that can enhance muscle loss [24] and a sedentary lifestyle, which is a risk factor for diabetes itself [25].

Differences in terms of muscle strength and body composition according to gender have been observed. This factor makes it impossible to analyze the evolution of the total sample so it is necessary to stratify results according to gender.

A decreased caloric and protein intake at baseline was observed with respect to the estimated requirements based on the clinical guidelines. The criteria used to calculate requirements was based on the ESPEN clinical guidelines for surgical and oncological patients, given the characteristics of the patients in the sample [18,19]. This decrease could be due to the underlying pathology and could be related to the high percentage of weight loss observed in our population. After the start of the nutritional intervention, the energy-protein requirements were achieved.

Supplementation was selected in relation to nutritional requirements (protein requirements above energy requirements) and specifically in diabetes mellitus due to the metabolic characteristics of the patients. Use of specific formulas for diabetes in patients with nutrition are widely studied. These types of preparations that change the amount and type of carbohydrates (low glycemic index) and lipids (predominantly monounsaturated fatty acids) have shown an improvement in glycemic control and lipid control [15,26]. However, the evidence regarding its use as a supplementation to an incomplete diet is not as well studied.

The use of this type of supplementation first showed a stabilization of weight loss in our patients. The objective of reaching the caloric-protein requirements was adequate, especially if we consider that most of these patients presented a basic oncological pathology in which there is a tendency for progressive weight loss in relation to the oncological treatment and for those who started from a baseline situation of normal weight. On the other hand, three months after the start of treatment with this type of supplementation, a decrease in the diagnosis of malnutrition was observed in the total sample; therefore, the first objective of supplementation was achieved. These objectives are like those recommended for the use of oral nutritional supplements in patients with malnutrition with oncological or surgical pathology or elderly patients according to the ESPEN recommendations [18,19].

No significant change In body com"osit'on parameters was observed, although there was a trend towards improvement in muscle mass in both men and women. On the other hand, in patients with a diagnosis of sarcopenia at the beginning of the study, it was observed that in women there was a significant improvement in dynamometry. Nutritional intervention in patients with diabetes and sarcopenia can improve this situation, as was shown in the study by Maykish et al. that evaluated the use of different branched-chain amino acids in the management of sarcopenia and their involvement in the modulation of diabetes [27]. On the other hand, patients with diabetes mellitus have a high prevalence of sarcopenic obesity that may require more adapted treatment [5].

The adequate adjustment of the diet in the patient with malnutrition is basic. If the requirements cannot be achieved with an adaptation of the diet, the use of oral nutritional supplementation is necessary to meet these requirements. In our sample, an adequate range of caloric-protein requirements was observed with the nutritional intervention. However, in patients with diabetes, it is also necessary to achieve an adequate glycemic control because it has been observed that poorer glycemic control is associated with a greater decline of muscle mass [7].

A slight increase in glycated hemoglobin and plasma albumin was observed in patients with diagnosed diabetes mellitus. This may be related to the increase in the consumption of carbohydrates after the decrease in the initial intake in relation to the underlying pathology. This data differs from other data observed in studies with diabetes-specific formulas in which an improvement in glycemic parameters was observed, although in these studies, medical nutrition therapy with complete enteral nutrition was usually evaluated [26]. In other studies, in which oral nutritional supplements were used, they were used as a substitute for meals and not in patients with malnutrition, so the results are not comparable with our population [28]. However, in the sample studied, no alterations were observed in basal glycaemia or in lipid parameters (cholesterol and tryglicerides), neither in patients with diabetes nor in patients with prediabetes. This fact could be related to the increase in lipids with MUFA and PUFA consumption despite the increase of carbohydrate consumption [29].

Albumin is an imprecise biomarker that can be interfered with in many situations, such as with inflammation and the hydration state of the patient. In this study, albumin levels were not below the lower limit of normal. The change of this parameter in our sample is unspecific and there is no easy explanation. More specific nutritional biomarkers in the PCR/prealbumin ratio did not show differences but its use is promising to evaluate the prognosis, especially in patients with acute pathologies [10].

The use of diabetes-specific oral nutritional supplementation has shown better postprandial glycemic control after its intake [14,30]. Other studies using a meal-replacement plan during a short period of time have shown an improvement in the glycemic profile [28]. These interventions were not used in patients with disease-related malnutrition. In these studies, the type of prescribed diet, the underlying disease and adherence to oral nutritional supplementation can influence the results.

The main strength of this study was the evaluation of a diabetic-specific formula as an oral nutritional supplement associated with diet in real clinical practice, since there are not many studies that evaluate this method of medical nutrition therapy. This fact allows us to extrapolate our results to generalized clinical practice. On the other hand, the evaluation from a morphofunctional point of view in thess type sof patients allows for a complete assessment and for monitoring the different spheres of nutritional status (evaluation of intake, anthropometry, body composition and muscle function) in order to personalize most appropriate way of treatment.

The limitations of this study are, first, the non-use of a control group that would allow us to evaluate the specific effect of the formula with respect to standard or other specific formulas. In addition, the selected sample has a predominance of cancer patients in whom the effect of nutrition can be variable depending on the stage and treatment of the disease. This situation may interfere with the results and would require a larger sample size to perform adequate stratification.

This study allows us to propose new lines of research on the use of diabetes-specific nutritional supplementation, with a control group and in specific groups of patients at nutritional risk. The use of the different morphofunctional assessment techniques must be basic in all nutritional assessment studies, given that we increasingly have more techniques that can be used in our daily clinical practice, such as bioimpedance measurement and nutritional ultrasonography.

5. Conclusions

Medical Nutrition Therapy with an adapted oral diet and oral diabetes-specific nutritional supplementation reduces malnutrition according to GLIM criteria in patients at nutritional risk with alterations in carbohydrates metabolism. The choice of a diabetes-specific formula produces a slight increase in glycated hemoglobin in patients with diabetes but without a significant alteration in the rest of the metabolic parameters. In patients with BMI in the normal range, this intervention can produce a stabilization in morphofunctional assessment parameters; and, in women with sarcopenia, it shows an improvement in muscle strength measured by hand dynamometry.

Author Contributions: Conceptualization, J.J.L.-G. and D.A.D.L.-R.; Data curation, J.J.L.-G., O.I.-J. and D.P.-M.; Formal analysis, J.J.L.-G. and D.A.D.L.-R.; Funding acquisition, D.A.D.L.-R.; Investigation, J.J.L.-G., O.I.-J., D.P.-M., E.G.-H. and R.J.-S.; Methodology, J.J.L.-G. and D.A.D.L.-R.; Project administration, J.J.L.-G. and D.A.D.L.-R.; Resources, J.J.L.-G., O.I.-J. and D.A.D.L.-R.; Software, J.J.L.-G. and D.A.D.L.-R.; Supervision, J.J.L.-G. and D.A.D.L.-R.; Validation, J.J.L.-G. and D.A.D.L.-R.; Visualization, J.J.L.-G. and D.A.D.L.-R.; Writing—original draft, J.J.L.-G. and D.A.D.L.-R.; Writing—review and editing, J.J.L.-G., C.G.-L. and D.A.D.L.-R. All authors have read and agreed to the published version of the manuscript.

Funding: This research received no external funding.

Institutional Review Board Statement: The study was conducted in accordance with the Declaration of Helsinki, and approved by Ethics Committee of Comité Ética en Investigación con Medicamentos (CEIM) Área Valladolid Este, Spain (PI 20-1967).

Informed Consent Statement: Informed consent was obtained from all subjects involved in the study.

Data Availability Statement: Not applicable.

Conflicts of Interest: The authors declare no conflict of interest.

References

1. León-Sanz, M.; Brosa, M.; Planas, M.; García-de-Lorenzo, A.; Celaya-Pérez, S.; Hernández, J.Á.; Predyces Group Researchers. PREDyCES study: The cost of hospital malnutrition in Spain. *Nutrition* **2015**, *31*, 1096–1102. [CrossRef] [PubMed]
2. Sanz París, A.; García, J.M.; Gómez-Candela, C.; Burgos, R.; Martín, Á.; Matía, P.; Study VIDA group. Malnutrition prevalence in hospitalized elderly diabetic patients. *Nutr. Hosp.* **2013**, *28*, 592–599. [CrossRef]
3. Serrano Valles, C.; López Gómez, J.J.; García Calvo, S.; Jiménez Sahagún, R.; Torres Torres, B.; Gómez Hoyos, E.; Buigues, A.O.; de Luis Román, D. Influencia del estado nutricional sobre la estancia media hospitalaria en el paciente con diabetes mellitus tipo 2. *Endocrinol. Diabetes Nutr.* **2020**, *67*, 617–624. [CrossRef] [PubMed]
4. Kalyani, R.R.; Corriere, M.; Ferrucci, L. Age-related and disease-related muscle loss: The effect of diabetes, obesity, and other diseases. *Lancet Diabetes Endocrinol.* **2014**, *2*, 819–829. [CrossRef]
5. Wang, M.; Tan, Y.; Shi, Y.; Wang, X.; Liao, Z.; Wei, P. Diabetes and Sarcopenic Obesity: Pathogenesis, Diagnosis, and Treatments. *Front. Endocrinol.* **2020**, *11*, 568. [CrossRef]
6. Takahashi, F.; Hashimoto, Y.; Kaji, A.; Sakai, R.; Okamura, T.; Kitagawa, N.; Okada, H.; Nakanishi, N.; Majima, S.; Senmaru, T.; et al. Sarcopenia Is Associated With a Risk of Mortality in People With Type 2 Diabetes Mellitus. *Front. Endocrinol.* **2021**, *12*, 783363. [CrossRef]
7. Sugimoto, K.; Ikegami, H.; Takata, Y.; Katsuya, T.; Fukuda, M.; Akasaka, H.; Tabara, Y.; Osawa, H.; Hiromine, Y.; Rakugi, H. Glycemic Control and Insulin Improve Muscle Mass and Gait Speed in Type 2 Diabetes: The MUSCLES-DM Study. *J. Am. Med. Dir. Assoc.* **2021**, *22*, 834–838.e1. [CrossRef]
8. Cederholm, T.; Jensen, G.L.; Correia, M.I.T.D.; Gonzalez, M.C.; Fukushima, R.; Higashiguchi, T.; Baptista, G.; Barazzoni, R.; Blaauw, R.; Coats, A.; et al. GLIM criteria for the diagnosis of malnutrition—A consensus report from the global clinical nutrition community. *Clin. Nutr.* **2019**, *38*, 1–9. [CrossRef]
9. Romero-Corral, A.; Somers, V.K.; Sierra-Johnson, J.; Thomas, R.J.; Collazo-Clavell, M.L.; Korinek, J.; Allison, T.G.; Batsis, J.A.; Sert-Kuniyoshi, F.H.; Lopez-Jimenez, F. Accuracy of body mass index in diagnosing obesity in the adult general population. *Int. J. Obes.* **2008**, *32*, 959–966. [CrossRef]
10. Okorodudu, D.O.; Jumean, M.F.; Montori, V.M.; Romero-Corral, A.; Somers, V.K.; Erwin, P.J.; Lopez-Jimenez, F. Diagnostic performance of body mass index to identify obesity as defined by body adiposity: A systematic review and meta-analysis. *Int. J. Obes.* **2010**, *34*, 791–799. [CrossRef]
11. García Almeida, J.M.; García García, C.; Vegas Aguilar, I.M.; Bellido Castañeda, V.; Bellido Guerrero, D. Morphofunctional assessment of patient´s nutritional status: A global approach. *Nutr. Hosp.* **2021**, *38*, 592–600. [CrossRef] [PubMed]
12. García-Almeida, J.M.; García-García, C.; Vegas-Aguilar, I.M.; Ballesteros Pomar, M.D.; Cornejo-Pareja, I.M.; Fernández Medina, B.; de Luis Román, D.A.; Guerrero, D.B.; Lesmes, I.B.; Madueño, F.J.T. Nutritional ultrasound®: Conceptualisation, technical considerations and standardisation. *Endocrinol. Diabetes Nutr.* **2022**, S2530016422001471. [CrossRef]
13. Pedrazzoli, P.; Caccialanza, R.; Cotogni, P.; Degli Esposti, L.; Perrone, V.; Sangiorgi, D.; Di Costanzo, F.; Gavazzi, C.; Santoro, A.; Pinto, C. The Advantages of Clinical Nutrition Use in Oncologic Patients in Italy: Real World Insights. *Healthcare* **2020**, *8*, 125. [CrossRef] [PubMed]
14. Kokubo, E.; Morita, S.; Nagashima, H.; Oshio, K.; Iwamoto, H.; Miyaji, K. Blood Glucose Response of a Low-Carbohydrate Oral Nutritional Supplement with Isomaltulose and Soluble Dietary Fiber in Individuals with Prediabetes: A Randomized, Single-Blind Crossover Trial. *Nutrients* **2022**, *14*, 2386. [CrossRef] [PubMed]
15. Sanz-París, A.; Matía-Martín, P.; Martín-Palmero, Á.; Gómez-Candela, C.; Camprubi Robles, M. Diabetes-specific formulas high in monounsaturated fatty acids and metabolic outcomes in patients with diabetes or hyperglycaemia. A systematic review and meta-analysis. *Clin. Nutr.* **2020**, *39*, 3273–3282. [CrossRef]
16. de Souza, M.T.P.; Ozorio, G.A.; de Oliveira, G.N.; López, R.V.M.; Alves-Almeida, M.M.F.; Kulcsar, M.A.V.; Ribeiro, U., Jr.; Singer, P.; Waitzberg, D.L. Effect of age on resting energy expenditure in patients with cancer. *Nutrition* **2022**, *102*, 111740. [CrossRef]
17. Long, C.L.; Schaffel, N.; Geiger, J.W.; Schiller, W.R.; Blakemore, W.S. Metabolic response to injury and illness: Estimation of energy and protein needs from indirect calorimetry and nitrogen balance. *JPEN J. Parenter Enteral. Nutr.* **1979**, *3*, 452–456. [CrossRef]
18. Muscaritoli, M.; Arends, J.; Bachmann, P.; Baracos, V.; Barthelemy, N.; Bertz, H.; Bozzetti, F.; Hütterer, E.; Isenring, E.; Kaasa, S.; et al. ESPEN practical guideline: Clinical Nutrition in cancer. *Clin. Nutr.* **2021**, *40*, 2898–2913. [CrossRef]
19. Weimann, A.; Braga, M.; Carli, F.; Higashiguchi, T.; Hübner, M.; Klek, S.; Laviano, A.; Ljungqvist, O.; Lobo, D.N.; Martindale, R.G.; et al. ESPEN practical guideline: Clinical nutrition in surgery. *Clin. Nutr.* **2021**, *40*, 4745–4761. [CrossRef]
20. Cruz-Jentoft, A.J.; Bahat, G.; Bauer, J.; Boirie, Y.; Bruyère, O.; Cederholm, T.; Cooper, C.; Landi, F.; Rolland, Y.; Sayer, A.A.; et al. Sarcopenia: Revised European consensus on definition and diagnosis. *Age Ageing* **2019**, *48*, 16–31. [CrossRef]
21. Sergi, G.; De Rui, M.; Veronese, N.; Bolzetta, F.; Berton, L.; Carraro, S.; Bano, G.; Coin, A.; Manzato, E.; Perissinotto, E. Assessing appendicular skeletal muscle mass with bioelectrical impedance analysis in free-living Caucasian older adults. *Clin. Nutr.* **2015**, *34*, 667–673. [CrossRef] [PubMed]
22. Gómez-Candela, C.; Luengo, L.M.; Cos, A.I.; Martínez-Roque, V.; Iglesias, C.; Zamora, P.; González-Barón, R. Subjective global assessment in neoplastic patients. *Nutr. Hosp.* **2003**, *18*, 353–357. [PubMed]

23. Park, S.W.; Goodpaster, B.H.; Lee, J.S.; Kuller, L.H.; Boudreau, R.; de Rekeneire, N.; Harris, T.B.; Kritchevsky, S.; Tylavsky, F.A.; Nevitt, M.; et al. Excessive Loss of Skeletal Muscle Mass in Older Adults With Type 2 Diabetes. *Diabetes Care* **2009**, *32*, 1993–1997. [CrossRef] [PubMed]
24. Massimino, E.; Izzo, A.; Riccardi, G.; Della Pepa, G. The Impact of Glucose-Lowering Drugs on Sarcopenia in Type 2 Diabetes: Current Evidence and Underlying Mechanisms. *Cells* **2021**, *10*, 1958. [CrossRef]
25. Eckel, J. Myokines in metabolic homeostasis and diabetes. *Diabetologia* **2019**, *62*, 1523–1528. [CrossRef] [PubMed]
26. Ojo, O.; Weldon, S.M.; Thompson, T.; Crockett, R.; Wang, X.-H. The Effect of Diabetes-Specific Enteral Nutrition Formula on Cardiometabolic Parameters in Patients with Type 2 Diabetes: A Systematic Review and Meta-Analysis of Randomised Controlled Trials. *Nutrients* **2019**, *11*, 1905. [CrossRef]
27. Maykish, A.; Sikalidis, A.K. Utilization of Hydroxyl-Methyl Butyrate, Leucine, Glutamine and Arginine Supplementation in Nutritional Management of Sarcopenia—Implications and Clinical Considerations for Type 2 Diabetes Mellitus Risk Modulation. *J. Pers. Med.* **2020**, *10*, 19. [CrossRef]
28. Mustad, V.A.; Hegazi, R.A.; Hustead, D.S.; Budiman, E.S.; Rueda, R.; Maki, K.; Powers, M.; Mechanick, J.I.; Bergenstal, R.M.; Hamdy, O. Use of a diabetes-specific nutritional shake to replace a daily breakfast and afternoon snack improves glycemic responses assessed by continuous glucose monitoring in people with type 2 diabetes: A randomized clinical pilot study. *BMJ Open Diabetes Res. Care* **2020**, *8*, e001258. [CrossRef]
29. Qian, F.; Korat, A.A.; Malik, V.; Hu, F.B. Metabolic Effects of Monounsaturated Fatty Acid-Enriched Diets Compared With Carbohydrate or Polyunsaturated Fatty Acid-Enriched Diets in Patients With Type 2 Diabetes: A Systematic Review and Meta-analysis of Randomized Controlled Trials. *Diabetes Care* **2016**, *39*, 1448–1457. [CrossRef]
30. Angarita Dávila, L.; Bermúdez, V.; Aparicio, D.; Céspedes, V.; Escobar, M.C.; Durán-Agüero, S.; Cisternas, S.; de Assis Costa, J.; Rojas-Gómez, D.; Reyna, N.; et al. Effect of Oral Nutritional Supplements with Sucromalt and Isomaltulose versus Standard Formula on Glycaemic Index, Entero-Insular Axis Peptides and Subjective Appetite in Patients with Type 2 Diabetes: A Randomised Cross-Over Study. *Nutrients* **2019**, *11*, 1477. [CrossRef]

nutrients

MDPI

Article

Improving Dietary Intake of Essential Nutrients Can Ameliorate Inflammation in Patients with Diabetic Foot Ulcers

Raedeh Basiri [1,2,3,*], Maria Spicer [2], Cathy Levenson [4], Thomas Ledermann [5], Neda Akhavan [2,3] and Bahram Arjmandi [2,3]

1 Department of Nutrition and Food Studies, George Mason University, Fairfax, VA 22030, USA
2 Department of Nutrition and Integrative Physiology, Florida State University, Tallahassee, FL 32306, USA; mspicer@fsu.edu (M.S.); nsa08@my.fsu.edu (N.A.); barjmandi@fsu.edu (B.A.)
3 Center for Advancing Exercise and Nutrition Research on Aging, Florida State University, Tallahassee, FL 32306, USA
4 Department of Biomedical Sciences, College of Medicine, Florida State University, Tallahassee, FL 32306, USA; cathy.levenson@med.fsu.edu
5 Department of Family and Child Sciences, Florida State University, Tallahassee, FL 32306, USA; tledermann@fsu.edu
* Correspondence: rbasiri@gmu.edu

Abstract: Diabetic foot ulcers (DFUs) are classified as chronic wounds and are one of the most common complications of diabetes. In chronic wounds, management of inflammation is a key step in treatment. Nutrition plays an important role in managing and controlling inflammation. This study evaluated the effects of nutrition supplementation and education on inflammatory biomarkers in patients with DFUs. Eligible patients with foot ulcers were randomly assigned to either a treatment (n = 15) or control group (n = 14). Both groups received standard care for wound treatment from the clinic; however, the treatment group was also provided with nutritional supplementation and education. Plasma concentrations of inflammatory biomarkers, namely C-reactive protein (CRP), interleukin 6 (IL6), interleukin 10 (IL10), and tristetraprolin (TTP), were evaluated at baseline and every four weeks, until complete wound closure had occurred or up to 12 weeks. The mean plasma concentration of IL6 significantly decreased in the treatment group (p = 0.001). The interaction between time and group was not statistically significant for the mean plasma concentrations of CRP, IL10, and TTP during the 12 weeks of the study. The results of this study showed the positive effects of nutritional intervention on controlling inflammation in DFU patients. More clinical trials with a larger population and longer duration of time are needed to confirm our results.

Keywords: nutrient supplementation; nutrition education; diabetes; diabetic foot ulcer; chronic wounds; pro-inflammatory cytokines; anti-inflammatory cytokines; nutrition intervention; inflammation; wound healing; cytokines; CRP; IL6; IL10; tristetraprolin; TTP

Citation: Basiri, R.; Spicer, M.; Levenson, C.; Ledermann, T.; Akhavan, N.; Arjmandi, B. Improving Dietary Intake of Essential Nutrients Can Ameliorate Inflammation in Patients with Diabetic Foot Ulcers. *Nutrients* **2022**, *14*, 2393. https://doi.org/10.3390/nu14122393

Academic Editor: Sareen Gropper

Received: 17 May 2022
Accepted: 8 June 2022
Published: 9 June 2022

1. Introduction

Diabetic foot ulcers (DFU) are one of the most common causes of lower extremity amputation in diabetic patients [1]. About a quarter of people with diabetes will develop a foot ulcer, and up to 16% of DFUs will lead to amputation if they go untreated [2]. Wound healing is a complex process, including a series of interactions between various cell types, extracellular matrices, and cytokine mediators. Inflammation is a typical physiologic response to tissue injury, which is essential for disinfecting the wound area and tissue repair processes [3]. In normal wounds, acute inflammatory responses continue for a short duration of time and resolve promptly due to negative feedback mechanisms; however, in chronic wounds such as DFUs, inflammatory responses fail to regulate themselves, which results in chronic inflammation and deterioration of the healing process [4]. In DFU patients, hyperactivity of inflammatory cells results in a higher generation of reactive oxygen species

(ROS), which, along with downregulation of anti-inflammatory factors such as interleukin 10 (IL10), will add to the burden of inflammation [5]. A high concentration of ROS can degrade the extracellular matrix by increasing the expression of matrix metalloproteinase (MMPs) and increasing the production of other free radicals, which worsens the healing process [6,7]. In addition to the high production of ROS in diabetes, there is a diminished ability to remove them due to defective reducing complexes (glutathione), reducing enzymes (glutathione reductase), and reducing amino acids (cysteine) [5]. A combination of hyperglycemia as well as hypoxia, caused by disrupted vasculature in diabetic patients, limits wound healing and inhibits neutrophil and macrophage function which increases the risk of infection [8,9]. Additionally, upregulation of proinflammatory cytokines interleukin 1β (IL1β), interleukin 6 (IL6), and tumor necrosis factor-alpha (TNF-alpha), alongside down-regulation of anti-inflammatory molecules, such as transforming growth factor-beta (TGF-β) and IL10, result in a chronic non-healing wound in diabetic patients [10,11]. Therefore, controlling inflammation and infection within and surrounding the wound site is a major goal in DFU wound care. IL6 and C-reactive protein (CRP) are among the best indicators of inflammation and wound healing in DFU patients. IL6 expression is strongly correlated with both glucose concentration and wound chronicity [12]. A high concentration of IL6 is also a promising predictor of delayed wound healing and infection in DFU patients [13,14]. Additionally, IL6 and CRP are both correlated with the size of the wound in patients with DFUs [15]. A study by Weigelt et al. showed that patients with grade 3 DFUs, according to the University of Texas Wound Classification [16], had significantly higher blood levels of CRP and IL6 compared to those with grade 1 [15]. Decreased expression of anti-inflammatory biomarkers, such as IL10, was also observed in keratinocytes and endothelial cells at the wound margins in DFU patients ($p < 0.05$) [17]. Evidence has shown that a low concentration of IL10 contributes to the development of non-healing wounds in diabetic patients [18]. Inhibition of inflammatory biomarkers increases the expression of anti-inflammatory biomarkers, such as IL10, which is essential for promoting wound healing [19,20]. Various dietary components, including antioxidant vitamins and minerals, have the potential to alleviate chronic inflammation and could play a key role in chronic-wound healing. Vitamins C, A, and E, as well as zinc, manganese, and copper, are strong antioxidants and show potent anti-inflammatory effects [21–28]. Therefore, it is essential to examine if these nutrients can shift diabetic wounds from chronic/non-healing to normal wounds by improving inflammation status. It has been reported that DFU patients have a significantly low intake of the aforementioned nutrients [29,30]. However, evidence of the benefit to wound healing in DFUs through management of inflammation is lacking for a combination of these essential nutrients. Thus, further studies addressing the efficacy of these nutrients in controlling inflammatory conditions linked to DFUs are required. The effectiveness of utilizing educational tools for improving blood glucose and foot self-care behavior in patients with type 2 diabetes has been reported [31,32]; however, to our knowledge the effects of nutrition education in improving inflammatory biomarkers in patients with DFUs have not been evaluated yet. This study provides valuable information about the effects of supplementing with a glucose control formula, consisting of essential nutrients for wound healing, as well as nutrition education on inflammatory biomarkers in patients with DFUs. This study hypothesized that improving dietary intake of antioxidants through nutritional supplementation and nutrition education would improve the inflammation status related to DFUs.

2. Materials and Methods

2.1. Sample Size

The sample size was calculated based on an effect size of Cohen's $d = 0.25$ (standardized mean treatment difference) for the primary outcome measure (reduction in IL6 at 12 weeks), which would be a clinically meaningful change [33]. Using G*Power statistical software, and in order to have 80% power for detecting this difference with an α error of 0.05, we needed at least 24 patients. Anticipating a 10–20% dropout rate, the target sample size was 27–29, with 14–15 patients in each group (control, treatment).

2.2. Screening and Recruitment

This study was approved by the Institutional Review Board (IRB) of both Florida State University and Tallahassee Memorial Hospital (TMH, Tallahassee, FL, USA). The trial is also registered at clinicaltrials.gov(Identifier:NCT04055064). Due to the requirement of IRB regarding confidentiality of personal health information, a point of contact from the clinic (a nurse or one of the medical staff) was identified. They were educated about the study and informed of the inclusion/exclusion criteria. They explained the study and provided flyers to potential participants. Individuals who were willing to participate in the study were then referred to the researcher for further screening. Inclusion criteria were non-pregnant/non-lactating females or males between the ages of 30 and 70 years with type 1 or 2 diabetes. Participants must have been undergoing pharmacological treatment for glycemic control, with at least one foot ulcer of grade 1A based on the University of Texas classification [16].

Patients were excluded from the study if they had: used bioengineered tissue within four weeks prior to baseline; a history of radiation treatment to the ulcer site; hemoglobin A1c (HbA1c) > 12%; known immunosuppression; active malignancy; chronic kidney disease; liver failure/cirrhosis; or heart failure and/or myocardial infarction in the past three months. Additionally, excessive use of alcohol based on the World Health Organization standards, current use of warfarin, or any mental or physiological condition that may interfere with supplement intake warranted exclusion from the study.

After screening, eligible patients were informed about the details of the study and consent was obtained from patients who were interested in participating. This study was a randomized control trial with repeated measures. Participants were randomly assigned to either the treatment or the control group. All participants were consecutively enrolled in the study and followed up until the time that complete wound closure had occurred or up to 12 weeks, whichever came first.

2.3. Intervention/Treatment

All participants, irrespective of their grouping, received standard wound care from the TMH wound care clinic. Additionally, participants in the treatment group were instructed to consume two servings (474 mL) of Boost Glucose Control (Nestle Health Science, Bridgewater Township, NJ, USA), a proprietary produced glucose control nutritional formula (supplement) between meals, preferably one in the morning and one in the afternoon, for 12 weeks or until complete wound closure occurred. Participants in the intervention group were also educated about improving their dietary intake by increasing their consumption of low-fat/high-bioavailable protein sources, vegetables, and high-fiber carbohydrates, as well as decreasing their intake of refined and simple carbohydrates. Nutrition education was conducted in-person by the primary researcher (nutritionist) for 10 min at baseline. Participants were then reminded about the educational materials and were given a chance to ask questions during following visits until complete wound closure had occurred or up to 12 weeks. Consuming two servings of the supplement provided patients in the treatment group with a total energy of 500 calories, 28 g of protein, and essential vitamins and minerals for wound healing. A complete list of the nutrient content of the supplement is shown in Table S1.

We aimed to provide patients with a supplement that could deliver at least 50% of the RDA recommendation for essential nutrients. We anticipated that nutrition education would improve the dietary intake of nutrients and motivate patients to meet the remaining 50% of nutrient recommendations by consuming better food sources. Table 1 shows the recommended daily allowance (RDA) for nutrients involved in wound healing and compares it with the nutrient content of two servings of the supplement.

Table 1. Comparison of content of two servings of supplement vs. RDA recommendations for antioxidants for age and gender.

Nutrient	RDA for Nutrient	Total from Two Supplements/Day	% of RDA Provided; Men vs. Women if RDA Varied
Vitamin C	60	204	304%
Vitamin A	3000 IU	2500 IU	83%
Vitamin E	33.3 IU	66 IU	200%
* Manganese Men/Women	2.3/1.8 mg	0.8 mg	35%/44%
Copper	0.9 mg	0.8 mg	88%
Zinc Men/Women	11/8 mg	6 mg	54%/75%
Protein Men/Women	56 g/46 g	28 g	50%/61%

* No established RDA for Manganese, numbers are showing adequate intake (AI). IU: international unit.

2.4. Data Collection

At baseline, all participants were asked about medical and medication history. Anthropometric measurements, dietary assessments, and blood sample collections were conducted at baseline and repeated every four weeks until complete wound closure had occurred or up to 12 weeks.

2.5. Blood Sampling and Processing

A finger stick blood collection was conducted to assess HbA1c using HbA1c Now + test (Polymer Technology Systems, Indianapolis, IN, USA). Venous blood samples were collected from an antecubital vein using a vacutainer brand collection set. Blood samples were collected for conducting enzyme-linked immunosorbent assay (ELISA) tests on inflammatory biomarkers. Blood samples were obtained at initial assessment/baseline, at 4, 8, and 12 weeks, or at the time of complete wound closure. These were taken at the wound care clinic and were then transported to the Florida State University (FSU) research lab, following the Centers for Disease Control (CDC) guidelines for transporting blood samples. In the FSU research lab, specimens were centrifuged, separated, and aliquoted into labeled tubes, and stored in a −80 °C freezer until needed for analysis.

2.6. Plasma Preparation

After collecting whole blood into anticoagulant tubes, cells were removed from the plasma by centrifugation using an IEC CL31R multispeed refrigerated centrifuge (Thermo Electron Corporation, Waltham, MA, USA) at 1500× g for 10 min [34]. Following centrifugation, plasma was immediately transferred into 0.5 mL aliquots and was stored at −80 °C until analysis.

2.7. Biochemical Analysis

The inflammatory biomarkers IL6, IL10, CRP, and TTP were examined to evaluate the effects of the intervention on inflammation in DFUs. Evaluation of IL6 and CRP was undertaken using a human C-Reactive Protein/CRP Quantikine ELISA Kit and a human IL6 Quantikine ELISA Kit from R&D systems (Biotechne, Minneapolis, MN, USA). For CRP, the mean CV% for intra-assay precision was 5.5% and for inter-assay precision was 6.5%. The reported mean CV% for IL6 for intra-assay precision was 2.6% and for inter-assay precision was 4.5%. Assessments of IL10 and TTP were conducted using human ELISA kits from MyBioSource (San Diego, CA, USA). The mean CV% for intra-assay and inter-assay for IL10 were 4.44% and 6.27%, respectively, and for TTP were ≤8% and ≤12%, respectively. We were not able to read the data following standard instructions for IL10 and TTP ELISA tests, since our population had an exceptionally low concentration of these biomarkers. After consulting with the company's technical support, when evaluating IL10 using ELISA kits the first incubation time was increased to 2 h and the plate was put in a slow shaker during incubation. We also performed the test with standard curve assayed in serum diluent, and samples assayed and incubated at 4 °C overnight (20 h). No other changes were made to other procedures and protocols.

2.8. Statistical Analysis

Data were analyzed using the Statistical Package for Social Science (SPSS) version 25.0 (IBM SPSS). The statistical significance value was set at $\alpha < 0.05$ for all tests. Descriptive statistics and independent-sample t-tests were used to compare means of potential confounding variables between groups at baseline. Multilevel modeling (mixed model) was used for the analysis of inflammatory biomarkers. The effects of potential confounding factors for different variables were examined and if the effect was significant, they were added as covariates to the model. If a significant F-statistic was obtained, Bonferroni's post hoc test was used for pairwise comparisons.

3. Results

3.1. Baseline Characteristics

In total, 95 patients were screened, but only 42 met the inclusion criteria and were willing to participate in the study. Overall, 13 participants were excluded from the study due to a change in their clinic. Therefore, clinical, laboratory, and statistical analyses was performed on a total of 29 patients.

3.2. General Characteristics

Both groups had similar characteristics at baseline. The general characteristics of participants at baseline has been shown in Table 2.

Table 2. Baseline characteristics of participants by group.

Groups	Treatment (n = 15)	Control (n = 14)	p-Value
Men/women	8/7	11/3	0.08
Age (year) Means ± SD	52.9 ± 9.74	53.8 ± 12.8	0.84
Ethnicity African American/white	4/11	3/11	0.75
BMI [1] (kg/m^2) Means ± SD	33.5 ± 7.98	34.1 ± 6.04	0.84
Diabetes Duration (year) Means ± SD	14.40 ± 8.03	11.7 ± 6.17	0.32
Estimated wound age (m) Means ± SD	10.97 ± 15.09	10.58 ± 18.27	0.95
HbA1c [2] Means ± SD	7.95 ± 2.06	8.40 ± 2.16	0.57
Smoking (yes/no)	3/12	3/11	1.00

[1] BMI: Body Mass Index [2] HbA1C: Hemoglobin A1c.

The average age of the study population was 53.3 ± 11.1 years (mean ± SD). There were no statistically significant differences between participants regarding the duration of diabetes, estimated wound age, HbA1c, ethnicity, age, or body mass index (BMI) at baseline. Although the distribution of gender was different in the treatment and control groups, the effect of gender on each variable was examined and added to the model as a covariate if the effect was significant. The mean duration of diabetes was greater in the treatment group (14.4 ± 8 year) in comparison with the control group (11.7 ± 6 year); however, the difference was not statistically significant. There were no significant differences between the groups regarding indicators of socioeconomic status (SES) or other factors that could potentially affect the dietary intake of participants, including appetite problems or religious and cultural restrictions. Living, financial, and employment status, as well as a need for food assistance, were considered indicators of SES.

3.3. *Changes in Plasma Concentrations of Inflammatory and Anti-Inflammatory Biomarkers during the Study Period*

3.3.1. Changes in Plasma Concentrations of IL6 and CRP

There were no significant differences between the plasma concentrations of IL6 in the treatment and control groups at baseline (16.1 pg/mL vs. 15.8 pg/mL, respectively). The potential effects of confounding factors such as duration of diabetes, estimated wound age ($p = 0.004$), gender, HbA1c at baseline, age, smoking, and BMI were examined. Only estimated wound age had a statistically significant effect on the concentrations of IL6, therefore this was used as a covariate in the model. The mean concentration of IL6 significantly decreased in the plasma of the treatment group after adjustment of estimated wound age ($p = 0.001$). In the control group, the mean concentration of IL6 was 15 times higher than its concentration in the treatment group at the end of the study. Comparison of the mean concentrations of IL6 at different time-points of the study for the treatment and control groups are outlined in Figure 1.

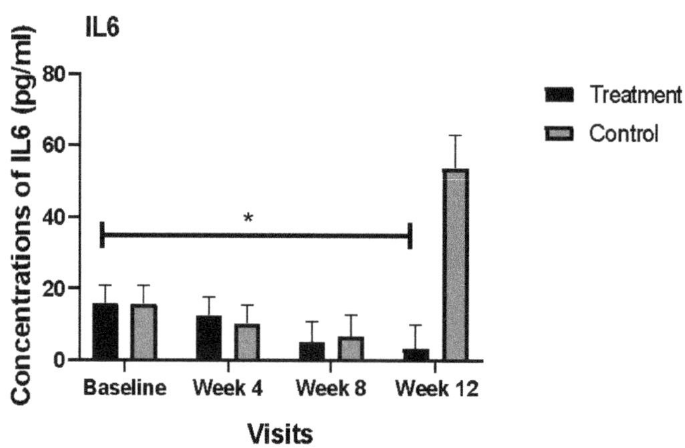

Figure 1. Comparison of mean plasma concentrations of IL6 (pg/mL) between the treatment and control group during the 12 weeks of the study. Bars represent the means ± SEM. * Denotes a significant time by group interaction ($p < 0.05$).

There was no statistically significant difference between the mean concentration of CRP among the two groups at baseline. We examined the effects of potential confounding factors, such as HbA1c at baseline, age, gender ($p = 0.02$), smoking, estimated wound age ($p < 0.001$), duration of diabetes ($p = 0.04$), and BMI. The significant confounding factors (gender, estimated wound age, and duration of diabetes) were kept in the model for further analyses. The interaction between group and time was not statistically significant after adjustment of confounding factors.

3.3.2. Changes in Plasma Concentrations of IL10 and TTP

There was no statistically significant difference between the mean concentration of IL10 among the two groups at baseline. The effects of potential confounding factors, such as HbA1c at baseline, age, gender, smoking, estimated wound age, duration of diabetes, and BMI, were evaluated; however, none of these factors had a significant effect on the plasma concentration of IL10, and, therefore, they were removed from the model. The interaction between group and time was not statistically significant for IL10.

Similar to IL10, the effects of potential confounding factors on the concentrations of the TTP were evaluated. Only HbA1c at baseline ($p = 0.03$) and BMI ($p = 0.01$) yielded significant on the concentrations of TTP. Additionally, the effect of gender on the concentration of TTP tended to be significant ($p = 0.06$); therefore, we included these factors as covariates in the model. The mean plasma concentration of TTP in the treatment group was increased numerically from 361 pg/mL at baseline to 1243 pg/mL at the end of the study. In the control group, the mean plasma concentration of TTP increased from 355 pg/mL to 479 pg/mL. Although the interaction between group and time was not statistically significant after adjustment of the confounding factors, the mean increase in the plasma concentrations of TTP in the treatment group was about seven times higher than the control group. Please see Figure 2.

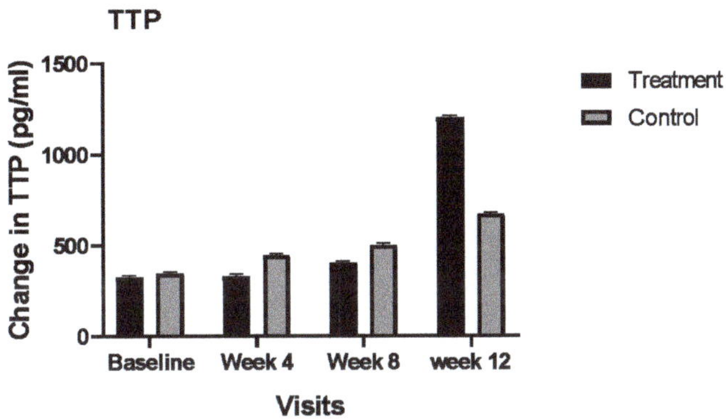

Figure 2. Comparison of mean plasma concentrations of TTP (pg/mL) between the treatment and control group during the 12 weeks of the study. Bars represent the means ± SEM.

4. Discussion

Our results showed that nutritional supplementation, used concurrently with nutrition education, can strongly decrease the plasma concentrations of IL6 in DFU patients. In contrast, the plasma concentrations of IL6 were increased drastically in our control group at the end of the study. Evidence has shown that an increase in IL6 could be an indicator of wound infection [13]; however, we did not collect data on wound infections to confirm this in our population.

The plasma concentration of TTP increased numerically in our intervention group almost seven times more than the control group. TTP is an RNA binding protein that enforces the degradation of mRNA encoding cytokines and chemokines [35], and, therefore, reduces systemic inflammation through the under-expression of inflammatory mediators, including IL6, TNF alpha, and IL18 [36–38]. Additionally, TTP negatively regulates NF-κB signaling at the transcriptional corepressor level, which represses inflammatory gene transcription [39]. Although the interaction between time and group was not statistically significant for plasma concentrations of TTP, the observed numerical increase in TTP, along with a strong significant decrease in the concentrations of IL6 in our treatment group, might suggest positive effects of our intervention on controlling inflammation status via regulation of TTP in patients with DFUs. The positive effects of the intervention on increasing (numerically) the plasma concentration of TTP might also suggest the potential of a positive effect of our intervention on improvement of plasma concentrations of TTP during long-term application. These results are supported by other studies where they showed that serum and urinary levels of IL6 and IL18 were significantly elevated while TTP was significantly depressed in patients with diabetes [36]. We are not aware of a similar study which evaluates the effects of dietary supplementation or quality of diet on the concentrations of TTP. More clinical trials with a larger population and longer duration of time are needed to validate our findings.

Our findings are consistent with the results of a study by Afzali et al., which showed that nutritional supplementation for 12 weeks reduced inflammation in patients with

DFUs [40]. Similarly, another study showed that nutritional supplementation could reduce rate of infection and antibiotic use in patients with DFUs [41]. We are not aware of any other study that examines the effects of dietary supplementation on inflammatory biomarkers in DFU patients. However, it has been reported that, in general, high-quality diets have beneficial effects on reducing concentrations of inflammatory biomarkers in various chronic conditions. An inverse association between intake of whole grains, fruits, nuts, and green leafy vegetables, and concentrations of IL6 and CRP, has been reported by several studies [42–45]. Therefore, providing nutrition education as part of the intervention in this study could have had synergistic effects on observed improvement of inflammation status in the intervention group. The strength of our study was to use both nutrition education and supplementation for improving inflammation status in patients with DFUs. To our knowledge, this is the first study to evaluate the effects of dietary intake of nutrients on concentrations of TTP as an inhibitory factor for inflammation. This anti-inflammatory protein could serve as the basis for novel approaches to chronic wound therapy.

Due to the relatively small sample size, we were not able to examine the effects of supplementation or education on inflammation status separately. Additionally, the potential effects of different blood-glucose-lowering medications, as well as the physical activity levels of the participants, were not evaluated. It has been reported that accumulation of advanced glycation end products (AGEs) contribute to DFUs [46,47]; however, we did not collect data on AGEs. Evidence has shown that inositol could decrease HbA1C in overweight patients with type 1 diabetes [48]. Moreover, it has been shown that inositol can reduce inflammation and oxidative stress related to diabetes in pregnant women with gestational diabetes [49]. Our supplement contained 200 mg of inositol per serving, which might have played a role in the observed improvement in inflammation status. Currently, there is no RDA for inositol. More studies need to be conducted to discover the optimum dosage of inositol for improving inflammation status in patients with DFUs. Analysis of dietary intake of our participants at baseline showed that our population had a significantly low intake of potent antioxidants such as vitamin E, vitamin A, zinc, copper, and manganese when compared with RDA recommendations; these data have been published elsewhere [50]. Presently, there are no recommendations for dietary intake of micronutrients/antioxidants for people with DFUs. Our aim in this study was to support participants in receiving an adequate amount of antioxidants based on the RDA recommendation; however, patients with DFUs might benefit from intakes above the RDA recommendation due to high production of ROS and the presence of an open wound. More clinical studies are needed to discover the optimum amount of each of these nutrients for improving inflammation status and wound healing in DFUs. In order to support the added benefits of other nutrients found in various foods, priority should be given to nutrition education to increase the quality of diet in this population. However, several studies showed that DFU patients could not receive an adequate amount of the essential nutrients from their food for several reasons [30,50–52]. When receiving enough nutrients from the diet is not possible, supplementing with optimum dosage of essential nutrients for improving conditions is recommended. Results of our study showed that educating patients with DFUs to consume high-quality diets, as well as providing them with supplements that support at least 50% of the RDA recommendation for antioxidants, have significant positive effects on improving inflammatory conditions. Currently, nutritional interventions or referral to dietitians are not part of standard care; however, as our results show, nutritional interventions are critical components of the wound healing process and should be an integral part of the treatment of DFUs.

5. Conclusions

Our findings showed that nutritional supplementation, along with nutrition education could significantly improve inflammation status in patients with DFUs. More clinical trials with larger sizes are needed to confirm our results.

Supplementary Materials: The following supporting information can be downloaded at: https://www.mdpi.com/article/10.3390/nu14122393/s1, Table S1: Nutrient composition of one serving (237 mL) of the nutritional supplement.

Author Contributions: Conceptualization, R.B. and M.S.; methodology, R.B., M.S. and C.L.; software, R.B.; formal analysis, R.B. and T.L.; investigation, R.B.; resources, R.B. and M.S.; data curation, R.B.; writing—original draft preparation, R.B.; writing—B.A., M.S., C.L. and N.A.; visualization, R.B.; supervision, M.S. and B.A.; project administration, R.B.; funding acquisition, R.B. and M.S. All authors have read and agreed to the published version of the manuscript.

Funding: This research received no external funding.

Institutional Review Board Statement: The study was conducted according to the guidelines of the Declaration of Helsinki and approved by the Institutional Review Board of FLORIDA STATE UNIVERSITY (2016.18608, 14 July 2016).

Informed Consent Statement: Informed consent was obtained from all subjects involved in the study.

Data Availability Statement: The datasets generated from this study are available from the corresponding author upon reasonable request.

Acknowledgments: We sincerely thank the nurses, physicians, staff, and patients at the Tallahassee Memorial Wound Healing Center. We recognize Joshua Kukus and Arthur Cooper for providing us with exceptional IT support.

Conflicts of Interest: The authors declare no conflict of interest.

References

1. Singh, G.; Chawla, S. Amputation in Diabetic Patients. *Med. J. Armed Forces India* **2006**, *62*, 36–39. [CrossRef]
2. Association for the Advancement of Wound Care (AAWC). How to Heal a Wound and Find a Wound Clinic. 2016. Available online: http://aawconline.org/ (accessed on 29 November 2016).
3. Calder, P.C.; Albers, R.; Antoine, J.-M.; Blum, S.; Bourdet-Sicard, R.; Ferns, G.A.; Folkerts, G.; Friedmann, P.S.; Frost, G.S.; Guarner, F.; et al. Inflammatory Disease Processes and Interactions with Nutrition. *Br. J. Nutr.* **2009**, *101*, S1–S45. [CrossRef] [PubMed]
4. Brem, H.; Tomic-Canic, M. Cellular and Molecular Basis of Wound Healing in Diabetes. *J. Clin. Investig.* **2007**, *117*, 1219–1222. [CrossRef] [PubMed]
5. Blakytny, R.; Jude, E.B. Altered Molecular Mechanisms of Diabetic Foot Ulcers. *Int. J. Low. Extrem. Wounds* **2009**, *8*, 95–104. [CrossRef]
6. Poljsak, B.; Šuput, D.; Milisav, I. Achieving the Balance between ROS and Antioxidants: When to Use the Synthetic Antioxidants. *Oxidative Med. Cell. Longev.* **2013**, *2013*, e956792. [CrossRef]
7. Svineng, G.; Ravuri, C.; Rikardsen, O.; Huseby, N.-E.; Winberg, J.-O. The Role of Reactive Oxygen Species in Integrin and Matrix Metalloproteinase Expression and Function. *Connect. Tissue Res.* **2008**, *49*, 197–202. [CrossRef]
8. Falanga, V. Wound Healing and Its Impairment in the Diabetic Foot. *Lancet* **2005**, *366*, 1736–1743. [CrossRef]
9. Patel, V.; Chivukula, I.V.; Roy, S.; Khanna, S.; He, G.; Ojha, N.; Mehrotra, A.; Dias, L.M.; Hunt, T.K.; Sen, C.K. Oxygen: From the Benefits of Inducing VEGF Expression to Managing the Risk of Hyperbaric Stress. *Antioxid. Redox Signal.* **2005**, *7*, 1377–1387. [CrossRef]
10. Morey, M.; O'Gaora, P.; Pandit, A.; Hélary, C. Hyperglycemia Acts in Synergy with Hypoxia to Maintain the Pro-Inflammatory Phenotype of Macrophages. *PLoS ONE* **2019**, *14*, e0220577. [CrossRef]
11. Teena, R.; Dhamodharan, U.; Ali, D.; Rajesh, K.; Ramkumar, K.M. Gene Expression Profiling of Multiple Histone Deacetylases (HDAC) and Its Correlation with NRF2-Mediated Redox Regulation in the Pathogenesis of Diabetic Foot Ulcers. *Biomolecules* **2020**, *10*, 1466. [CrossRef]
12. Buscemi, S.; Batsis, J.A.; Parrinello, G.; Massenti, F.M.; Rosafio, G.; Sciascia, V.; Costa, F.; Pollina Addario, S.; Mendola, S.; Barile, A.M.; et al. Nutritional Predictors of Mortality after Discharge in Elderly Patients on a Medical Ward. *Eur. J. Clin. Investig.* **2016**, *46*, 609–618. [CrossRef] [PubMed]
13. Korkmaz, P.; Koçak, H.; Onbaşı, K.; Biçici, P.; Özmen, A.; Uyar, C.; Özatağ, D.M. The Role of Serum Procalcitonin, Interleukin-6, and Fibrinogen Levels in Differential Diagnosis of Diabetic Foot Ulcer Infection. *J. Diabetes Res.* **2018**, *2018*, 7104352. [CrossRef] [PubMed]
14. Lee, E.G.; Luckett-Chastain, L.R.; Calhoun, K.N.; Frempah, B.; Bastian, A.; Gallucci, R.M. Interleukin 6 Function in the Skin and Isolated Keratinocytes Is Modulated by Hyperglycemia. *J. Immunol. Res.* **2019**, *2019*, e5087847. [CrossRef] [PubMed]
15. Weigelt, C.; Rose, B.; Poschen, U.; Ziegler, D.; Friese, G.; Kempf, K.; Koenig, W.; Martin, S.; Herder, C. Immune Mediators in Patients With Acute Diabetic Foot Syndrome. *Diabetes Care* **2009**, *32*, 1491–1496. [CrossRef] [PubMed]

16. Oyibo, S.O.; Jude, E.B.; Tarawneh, I.; Nguyen, H.C.; Harkless, L.B.; Boulton, A.J.M. A Comparison of Two Diabetic Foot Ulcer Classification Systems: The Wagner and the University of Texas Wound Classification Systems. *Diabetes Care* **2001**, *24*, 84–88. [CrossRef]
17. Galkowska, H.; Wojewodzka, U.; Olszewski, W.L. Chemokines, Cytokines, and Growth Factors in Keratinocytes and Dermal Endothelial Cells in the Margin of Chronic Diabetic Foot Ulcers. *Wound Repair Regen.* **2006**, *14*, 558–565. [CrossRef]
18. Kaur, P.; Choudhury, D. Modulation of Inflammatory Dynamics by Insulin to Promote Wound Recovery of Diabetic Ulcers. In *Wound Healing*; IntechOpen: London, UK, 2020. [CrossRef]
19. Azevedo, F.F.; Carvalho, C.R.O.; Saad, M.J.A.; Lima, M.H.M.; Da Silva Cunha Breder, J. Insulin Topical Modulates Inflammatory Phase and the Angiogenesis of the Burns Wound Healing in Diabetic-Induced Rats. *Diabetol. Metab. Syndr.* **2015**, *7*, A259. [CrossRef]
20. DeClue, C.E.; Shornick, L.P. The Cytokine Milieu of Diabetic Wounds. *Diabetes Manag.* **2015**, *5*, 525–537. [CrossRef]
21. Barbosa, E.; Faintuch, J.; Machado Moreira, E.A.; Gonçalves da Silva, V.R.; Lopes Pereima, M.J.; Martins Fagundes, R.L.; Filho, D.W. Supplementation of Vitamin E, Vitamin C, and Zinc Attenuates Oxidative Stress in Burned Children: A Randomized, Double-Blind, Placebo-Controlled Pilot Study. *J. Burn Care Res.* **2009**, *30*, 859–866. [CrossRef]
22. Salgado, R.M.; Cruz-Castañeda, O.; Elizondo-Vázquez, F.; Pat, L.; De la Garza, A.; Cano-Colín, S.; Baena-Ocampo, L.; Krötzsch, E. Maltodextrin/Ascorbic Acid Stimulates Wound Closure by Increasing Collagen Turnover and TGF-B1 Expression in Vitro and Changing the Stage of Inflammation from Chronic to Acute in Vivo. *J. Tissue Viability* **2017**, *26*, 131–137. [CrossRef]
23. Gross, R.L. The Effect of Ascorbate on Wound Healing. *Int. Ophthalmol. Clin.* **2000**, *40*, 51–57. [CrossRef] [PubMed]
24. Musalmah, M.; Fairuz, A.H.; Gapor, M.T.; Wan Ngah, W.Z. Effect of Vitamin E on Plasma Malondialdehyde, Antioxidant Enzyme Levels and the Rates of Wound Closures during Wound Healing in Normal and Diabetic Rats. *Asia Pac. J. Clin. Nutr.* **2002**, *11*, S448–S451. [CrossRef] [PubMed]
25. Sternberg, E.M. Neural Regulation of Innate Immunity: A Coordinated Nonspecific Host Response to Pathogens. *Nat. Rev. Immunol.* **2006**, *6*, 318–328. [CrossRef] [PubMed]
26. Holley, A.K.; Bakthavatchalu, V.; Velez-Roman, J.M.; St Clair, D.K. Manganese Superoxide Dismutase: Guardian of the Powerhouse. *Int. J. Mol. Sci.* **2011**, *12*, 7114–7162. [CrossRef]
27. Aschner, J.L.; Aschner, M. Nutritional Aspects of Manganese Homeostasis. *Mol. Asp. Med.* **2005**, *26*, 353–362. [CrossRef]
28. Das, A.; Chen, G.-F.; Kim, H.W.; Youn, S.-W.; Varadarajan, S.; McKinney, R.D.; Finney, L.; Ushio-Fukai, M.; Fukai, T. Abstract 16637: Antioxidant-1, a Novel Cu-Dependent Transcription Factor for NADPH Oxidase P47phox, Promotes Wound Healing by Regulating Inflammation and Angiogenesis. *Circulation* **2014**, *130*, A16637. [CrossRef]
29. Basiri, R.; Spicer, M.T.; Levenson, C.W.; Ormsbee, M.J.; Ledermann, T.; Arjmandi, B.H. Nutritional Supplementation Concurrent with Nutrition Education Accelerates the Wound Healing Process in Patients with Diabetic Foot Ulcers. Available online: https://pubmed.ncbi.nlm.nih.gov/32756299/ (accessed on 1 November 2020).
30. Maier, H.M.; Ilich-Ernst, J.; Arjmandi, B.; Kim, J.-S.; Spicer, M. Deficiencies in Nutritional Intake in Patients with Diabetic Foot Ulcers. *J. Nutr. Ther.* **2016**, *5*, 85–92. [CrossRef]
31. Nguyen, T.P.L.; Edwards, H.; Do, T.N.D.; Finlayson, K. Effectiveness of a Theory-Based Foot Care Education Program (3STEPFUN) in Improving Foot Self-Care Behaviours and Foot Risk Factors for Ulceration in People with Type 2 Diabetes. *Diabetes Res. Clin. Pract.* **2019**, *152*, 29–38. [CrossRef]
32. Defeudis, G.; Khazrai, Y.M.; Di Rosa, C.; Secchi, C.; Montedoro, A.; Maurizi, A.R.; Palermo, A.; Pozzilli, P.; Manfrini, S. Conversation Maps^{TM}, an Effective Tool for the Management of Males and Females with Type 2 Diabetes and Mildly Impaired Glycemic Control. *Hormones* **2018**, *17*, 113–117. [CrossRef]
33. Cohen, J. A Power Primer. *Psychol. Bull.* **1992**, *112*, 155–159. [CrossRef]
34. Plasma and Serum Preparation-US. Available online: https://www.thermofisher.com/us/en/home/references/protocols/cell-and-tissue-analysis/elisa-protocol/elisa-sample-preparation-protocols/plasma-and-serum-preparation.html (accessed on 18 September 2019).
35. Kratochvill, F.; Gratz, N.; Qualls, J.E.; Van De Velde, L.-A.; Chi, H.; Kovarik, P.; Murray, P.J. Tristetraprolin Limits Inflammatory Cytokine Production in Tumor-Associated Macrophages in an MRNA Decay Independent Manner. *Cancer Res.* **2015**, *75*, 3054–3064. [CrossRef] [PubMed]
36. Liu, F.; Guo, J.; Zhang, Q.; Liu, D.; Wen, L.; Yang, Y.; Yang, L.; Liu, Z. The Expression of Tristetraprolin and Its Relationship with Urinary Proteins in Patients with Diabetic Nephropathy. *PLoS ONE* **2015**, *10*, e0141471. [CrossRef] [PubMed]
37. Patial, S.; Curtis, A.D.; Lai, W.S.; Stumpo, D.J.; Hill, G.D.; Flake, G.P.; Mannie, M.D.; Blackshear, P.J. Enhanced Stability of Tristetraprolin MRNA Protects Mice against Immune-Mediated Inflammatory Pathologies. *Proc. Natl. Acad. Sci. USA* **2016**, *113*, 1865–1870. [CrossRef] [PubMed]
38. Ross, E.A.; Smallie, T.; Ding, Q.; O'Neil, J.D.; Cunliffe, H.E.; Tang, T.; Rosner, D.R.; Klevernic, I.; Morrice, N.A.; Monaco, C.; et al. Dominant Suppression of Inflammation via Targeted Mutation of the MRNA Destabilizing Protein Tristetraprolin. *J. Immunol.* **2015**, *195*, 265–276. [CrossRef]
39. Liang, J.; Lei, T.; Song, Y.; Yanes, N.; Qi, Y.; Fu, M. RNA-Destabilizing Factor Tristetraprolin Negatively Regulates NF-KB Signaling. *J. Biol. Chem.* **2009**, *284*, 29383–29390. [CrossRef]

40. Afzali, H.; Jafari Kashi, A.H.; Momen-Heravi, M.; Razzaghi, R.; Amirani, E.; Bahmani, F.; Gilasi, H.R.; Asemi, Z. The Effects of Magnesium and Vitamin E Co-Supplementation on Wound Healing and Metabolic Status in Patients with Diabetic Foot Ulcer: A Randomized, Double-Blind, Placebo-Controlled Trial. *Wound Repair Regen.* **2019**, *27*, 277–284. [CrossRef]
41. Tatti, P.; Barber, A.E. The Use of a Specialized Nutritional Supplement for Diabetic Foot Ulcers Reduces the Use of Antibiotics. *J. Endocrinol. Metab.* **2012**, *2*, 26–31. [CrossRef]
42. Nettleton, J.A.; Steffen, L.M.; Mayer-Davis, E.J.; Jenny, N.S.; Jiang, R.; Herrington, D.M.; Jacobs, D.R. Dietary Patterns Are Associated with Biochemical Markers of Inflammation and Endothelial Activation in the Multi-Ethnic Study of Atherosclerosis (MESA). *Am. J. Clin. Nutr.* **2006**, *83*, 1369–1379. [CrossRef]
43. Anderson, A.L.; Harris, T.B.; Tylavsky, F.A.; Perry, S.E.; Houston, D.K.; Lee, J.S.; Kanaya, A.M.; Sahyoun, N.R. Dietary Patterns, Insulin Sensitivity and Inflammation in Older Adults. *Eur. J. Clin. Nutr.* **2012**, *66*, 18–24. [CrossRef]
44. Julia, C.; Meunier, N.; Touvier, M.; Ahluwalia, N.; Sapin, V.; Papet, I.; Cano, N.; Hercberg, S.; Galan, P.; Kesse-Guyot, E. Dietary Patterns and Risk of Elevated C-Reactive Protein Concentrations 12 Years Later. *Br. J. Nutr.* **2013**, *110*, 747–754. [CrossRef]
45. Ahluwalia, N.; Andreeva, V.A.; Kesse-Guyot, E.; Hercberg, S. Dietary Patterns, Inflammation and the Metabolic Syndrome. *Diabetes Metab.* **2013**, *39*, 99–110. [CrossRef] [PubMed]
46. Maier, H.M. Nutritional Status and the Relationship of Dietary and Serum Advanced Glycation End-Products with Inflammation, Oxidative Stress and Healing of Diabetic Foot Ulcers. Ph.D. Thesis, The Florida State University, Tallahassee, FA, USA, 2016.
47. Vouillarmet, J.; Maucort-Boulch, D.; Michon, P.; Thivolet, C. Advanced Glycation End Products Assessed by Skin Autofluorescence: A New Marker of Diabetic Foot Ulceration. *Diabetes Technol.* **2013**, *15*, 601–605. [CrossRef] [PubMed]
48. Maurizi, A.R.; Menduni, M.; Del Toro, R.; Kyanvash, S.; Maggi, D.; Guglielmi, C.; Pantano, A.L.; Defeudis, G.; Fioriti, E.; Manfrini, S.; et al. A Pilot Study of D-Chiro-Inositol plus Folic Acid in Overweight Patients with Type 1 Diabetes. *Acta Diabetol.* **2017**, *54*, 361–365. [CrossRef] [PubMed]
49. Baldassarre, M.P.A.; Di Tomo, P.; Centorame, G.; Pandolfi, A.; Di Pietro, N.; Consoli, A.; Formoso, G. Myoinositol Reduces Inflammation and Oxidative Stress in Human Endothelial Cells Exposed In Vivo to Chronic Hyperglycemia. *Nutrients* **2021**, *13*, 2210. [CrossRef] [PubMed]
50. Basiri, R.; Spicer, M.; Munoz, J.; Arjmandi, B. Nutritional Intervention Improves the Dietary Intake of Essential Micronutrients in Patients with Diabetic Foot Ulcers. *Curr. Dev. Nutr.* **2020**, *4*, 8. [CrossRef]
51. Wheeler, M.L.; Dunbar, S.A.; Jaacks, L.M.; Karmally, W.; Mayer-Davis, E.J.; Wylie-Rosett, J.; Yancy, W.S. Macronutrients, Food Groups, and Eating Patterns in the Management of Diabetes: A Systematic Review of the Literature, 2010. *Diabetes Care* **2012**, *35*, 434–445. [CrossRef]
52. Collins, R.; Burrows, T.; Donnelly, H.; Tehan, P.E. Macronutrient and Micronutrient Intake of Individuals with Diabetic Foot Ulceration: A Short Report. *J. Hum. Nutr. Diet.* 2021, *epub ahead of print*. [CrossRef]

nutrients

MDPI

Review

Nutraceuticals/Drugs Promoting Mitophagy and Mitochondrial Biogenesis May Combat the Mitochondrial Dysfunction Driving Progression of Dry Age-Related Macular Degeneration

Lidianys María Lewis Luján [1], Mark F. McCarty [2], James J. Di Nicolantonio [3], Juan Carlos Gálvez Ruiz [4], Ema Carina Rosas-Burgos [1], Maribel Plascencia-Jatomea [1] and Simon Bernard Iloki Assanga [4,*]

[1] Department of Research and Postgraduate in Food, University of Sonora, Blvd. Luis Encinas y Rosales S/N, Col. Centro, Hermosillo 83000, Mexico; lidianys1@yahoo.es (L.M.L.L.); carina.rosas@unison.mx (E.C.R.-B.); maribel.plascencia@unison.mx (M.P.-J.)
[2] Catalytic Longevity Foundation, San Diego, CA 92109, USA; markfmccarty@gmail.com
[3] St. Luke's Mid America Heart Institute, Kansas City, MO 64111, USA; jjdinicol@gmail.com
[4] Department of Biological Chemical Sciences, Sonora University, Blvd. Luis Encinas y Rosales, Col. Centro, Hermosillo 83000, Mexico; juan.galvez@unison.mx
* Correspondence: ilokiassanga@gmail.com; Tel.: +52-(662)-1890-895

Abstract: In patients with age-related macular degeneration (AMD), the crucial retinal pigment epithelial (RPE) cells are characterized by mitochondria that are structurally and functionally defective. Moreover, deficient expression of the mRNA-editing enzyme Dicer is noted specifically in these cells. This Dicer deficit up-regulates expression of Alu RNA, which in turn damages mitochondria—inducing the loss of membrane potential, boosting oxidant generation, and causing mitochondrial DNA to translocate to the cytoplasmic region. The cytoplasmic mtDNA, in conjunction with induced oxidative stress, triggers a non-canonical pathway of NLRP3 inflammasome activation, leading to the production of interleukin-18 that acts in an autocrine manner to induce apoptotic death of RPE cells, thereby driving progression of dry AMD. It is proposed that measures which jointly up-regulate mitophagy and mitochondrial biogenesis (MB), by replacing damaged mitochondria with "healthy" new ones, may lessen the adverse impact of Alu RNA on RPE cells, enabling the prevention or control of dry AMD. An analysis of the molecular biology underlying mitophagy/MB and inflammasome activation suggests that nutraceuticals or drugs that can activate Sirt1, AMPK, Nrf2, and PPARα may be useful in this regard. These include ferulic acid, melatonin urolithin A and glucosamine (Sirt1), metformin and berberine (AMPK), lipoic acid and broccoli sprout extract (Nrf2), and fibrate drugs and astaxanthin (PPARα). Hence, nutraceutical regimens providing physiologically meaningful doses of several or all of the: ferulic acid, melatonin, glucosamine, berberine, lipoic acid, and astaxanthin, may have potential for control of dry AMD.

Keywords: nutraceuticals; age-related macular degeneration; mitochondrial biogenesis; Sirt1; AMPK; Nrf2; ferulic acid; melatonin; berberine; astaxanthin

Citation: Lewis Luján, L.M.; McCarty, M.F.; Di Nicolantonio, J.J.; Gálvez Ruiz, J.C.; Rosas-Burgos, E.C.; Plascencia-Jatomea, M.; Iloki Assanga, S.B. Nutraceuticals/Drugs Promoting Mitophagy and Mitochondrial Biogenesis May Combat the Mitochondrial Dysfunction Driving Progression of Dry Age-Related Macular Degeneration. *Nutrients* **2022**, *14*, 1985. https://doi.org/10.3390/nu14091985

Academic Editor: Sareen Gropper

Received: 27 March 2022
Accepted: 4 May 2022
Published: 9 May 2022

1. The Complex Molecular Biology Underlying the Pathogenesis of Dry AMD

Retinal pigment epithelium, which separates the neural retina from the underlying blood vessel-rich choroid, performs a number of tasks required for healthful ocular functions: phagocytizing and degrading the distal ends of photoreceptors while regenerating 11-cis retinal; regulating the flux of molecules from the choroid to the retina; and producing trophic factors which sustain the survival of retinal neurons while preventing over-proliferation of choroid blood vessels. The progressive loss of RPE cells over time in aging humans is responsible for the most common cause of irreversible visual impairment, so-called dry age-related macular degeneration (AMD)—also known as geographic atrophy.

Recent evidence points to the formation of NLRP3-dependent inflammasomes in retinal pigment epithelium (RPE) as a key factor driving RPE cell death in dry AMD. In particular, a series of elegant investigations, involving study of retinal pigment epithelium from deceased patients with dry AMD, studies with human and mouse RPE-derived cell lines, and studies in mice, have characterized a complex mechanism, consistent with the results of these studies, by which such inflammasomes can arise and induce the death of RPE cells [1–6]. For reasons yet to be clarified, expression of the microRNA-processing enzyme DICER1 is notably depressed in the RPE of patients with GA [1]. Moreover, knockdown of DICER1 leads to death of RPE in mice. While DICER1 plays a key role in generation of micro-RNAs, knockdown of other enzymes required for microRNA generation fails to induce death in RPE cells. However, DICER1 also functions to cleave the Alu RNA retrotransposon, and, as a consequence of DICER1 deficiency, this accumulates to excessive levels in the RPE [7]. In RPE cell lines, Alu RNA excess has been shown to lead to cell death which is dependent on the activation of NLRP3 inflammasomes. These generate interleukin-18, which acts in an autocrine fashion via MyD88 to induce cell death via a mechanism dependent on Fas ligand and caspase-8 [4]. These mechanisms seem likely to be pertinent to clinical AMD, as RPE from dry AMD patients displays increased levels of NRLP3 inflammasomes, IL-18, and caspases 1 and 8 [4].

The excess of Alu RNA induces structural and functional damage to RPE mitochondria that plays an obligate role in inflammasome activation. This damage entails the opening of mitochondrial permeability transition pores associated with increased mitochondrial generation of oxidants, a reduction in mitochondrial membrane potential, and escape of mitochondrial DNA (mtDNA) into the cytoplasm. Increased mitochondrial oxidant generation provides the oxidant stress required for NLRP3 inflammasome activation, as mitochondrially -targeted antioxidants—but not inhibitors of NADPH oxidase—block the impact of Alu RNA excess on inflammasome activation and cell death in RPE cells [2]. However, cytoplasmic mtDNA also plays a role in Alu RNA-mediated inflammasome activation. However, cytoplasmic mtDNA interacts with and activates cyclic GMP-AMP synthase (cGAS), which via interferon response factor 3 (IRF3) activation, promotes transcription of interferon-β [6]. Autocrine signaling by the latter drives non-canonical activation of NLRP3 inflammasomes via a mechanism dependent on caspase 4 expression/activation and on the presence of gasdermin [2].

Alu RNA also promotes inflammasome priming via NF-kappaB activation, another possible consequence of cGAS activation [3,8]. Activation of NF-kappaB might also reflect interaction of Alu RNA with double-stranded RNA receptors [9]. Stimulation of P2X7 receptors—leading to an efflux of cellular potassium also participates in Alu RNA-induced inflammasome activation. Extracellular ATP is the canonical mediator of P2X7 activation; however, whether ATP efflux mediates Alu RNA-induced P2X7 activation remains unclear. ATP release through gasdermin-mediated pore formation does not appear to play a role in this regard, even though the presence of gasdermin is required [6]. It has been suggested that Alu RNA may mediate P2X7 activation via an intracellular effect on the receptor, rather than ATP release [10]. Finally, activated NLRP3 generates interleukin-18 (IL-18), which via autocrine signaling leads to caspase-8-mediated apoptosis of RPE cells [4]. Figure 1 offers a simplified depiction of how DICER1 deficit and consequent Alu RNA excess lead to mitochondrial damage, inflammasome activation, and IL-18-mediated RPE cell death.

It is reasonable to presume that certain environmental factors linked to increased risk for GA, do so by amplifying some of the signaling mechanism outlined here. In particular, amyloid beta, complement component C5a, and the bisretinoid A2E have been found to promote NLRP3 inflammasome formation in ARPE-19 cells [11–13].

Accumulating evidence indicates that mitochondria are indeed structurally and functionally impaired in the RPE of AMD patients [14–17]. A loss of mitochondrial mass disrupted internal structure, diminished expression of electron transport chain proteins, and elevated oxidation of mtDNA have all been reported; the extent to which Alu RNA excess drives this phenomenon remains unclear [14,16]. These findings have prompted a number of investigators to suggest that measures which improve mitochondrial structure and function might prove useful for preventing and controlling dry AMD [14,17,18]. It is reasonable to suspect that the ability of Alu RNA to induce mitochondrial oxidant stress and promote extrusion of mtDNA is not just an acute effect observable with healthy mitochondria but may be a gradually evolving phenomenon in which mitochondrial function is progressively disrupted. If so, then insuring that RPE mitochondria are structurally and functionally sound by promoting mitophagy of unsound mitochondria, coupled with biogenesis of new mitochondria, might ameliorate the contribution of disrupted mitochondria to NLRP3 inflammasome activation and the consequent apoptotic death of RPE cells. And, measures which work in complementary ways to suppress NLRP3 inflammasome activation might also contribute to a rational strategy for combatting dry AMD. As explained below, nutraceuticals or drugs which boost the activity of Sirt1, AMPK, Nrf2, and PPARalpha could be expected to achieve these goals.

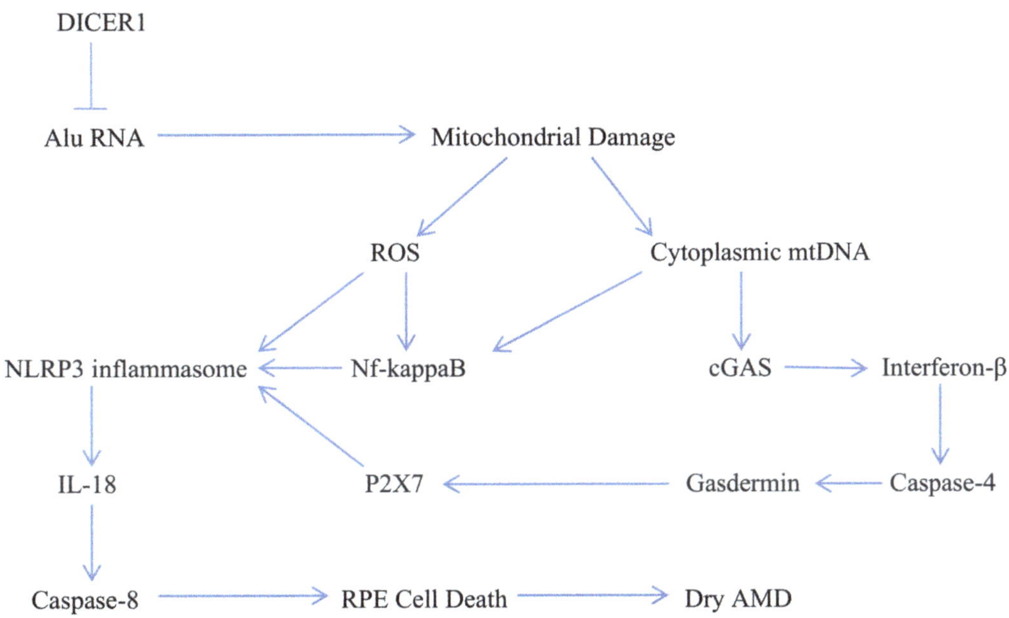

Figure 1. A simplified depiction of how DICER1 deficiency in RPE cells leads to Alu RNA excess, mitochondrial damage, NLRP3 inflammasome activation, autocrine IL-18 activity, and caspase-8-mediated cell death.

2. Regulation of Mitophagy, Mitochondrial Biogenesis, and Inflammasome Activation

The type III deacetylase Sirt1 and AMPK-activated kinase (AMPK) collaborate in the activation of both autophagy and mitophagy, while promoting mitochondrial biogenesis by boosting the activity of PPARg coactivator-1alpha (PGC-1a). They also can boost each other's activity. AMPK increases expression of nicotinamide ribosyltransferase (NAMPT), which promotes synthesis of Sirt1's obligate substrate NAD+. Sirt1, in turn, deacetylates and thereby stabilizes LKB1, which can phosphorylate and thereby activate AMPK [19–23]. With respect to autophagy, Sirt1 boosts the activity of number of proteins which participate in autophagosome formation; it also promotes the activity of Rab7, a G protein that enables fusion of autophagosomes and lysosomes [24–26]. Via deacetylation of the transcription factor FOXO1, Sirt1 enhances transcription of a number of genes which code for mediators of autophagy [26]. AMPK promotes autophagy by conferring an activating phosphorylation on ULK1, while inhibiting and suppressing the anti-autophagic activity of mTORC1 [27–29]. In regard to mitophagy, Sirt1 and AMPK collaborate in enhancing the expression of Parkin and Pink1, proteins which detect defective mitochondria with diminished membrane potential and mark them for incorporation into autophagic vacuoles; the mechanism of this effect is described below.

The coactivator activity of PGC-1a works in multiple ways to stimulate the synthesis of proteins that participate in mitochondrial biogenesis (MB), through its interactions with the transcription factors NRF-1, estrogen-related receptor-alpha (ERRα), and PPARα [30]. Aided by PGC-1α, NRF-1 promotes the transcription of genes coding for enzymes that mediate replication of mtDNA, such as Tfam, NRF-2, and TFB1 [31]. PGC-1α's interaction with ERRα boosts expression of Sirt3; the latter alters the transcriptional activity of FOXO3a, enabling it to boost transcription of mtDNA genes coding for certain proteins in the mitochondrial electron transport chain, as well as for the mitochondrial form of superoxide dismutase (SOD2) [32–35]. Sirt3 also enhances the activity of SOD2 by deacetylating it [36,37]. Hence, Sirt3 exerts a profound antioxidant effect within mitochondria. Importantly, Sirt3-deacetylated FOXO3a also promotes transcription of genes coding for Parkin and Pink1, the mediators of mitophagy [38–40]. Finally, via interaction with the transcription factor PPARα, PGC-1α boosts expression of the uncoupling protein UCP-2 (whose activity decreases mitochondrial superoxide generation) and a number of enzymes, including carnitine palmitoyltransferase-1, required for mitochondrial oxidation of fatty acids [41–43].

The contribution of Sir1 and AMPK to MB hinges on their ability to enhance both the expression and the activity of PGC-1α. Phosphorylation of PGC-1α by AMPK is a pre-requisite for its deacetylation by Sirt1; these effects importantly enhance PGC-1α's activity [44]. Sirt1 also promotes PGC-1α's expression at the transcriptional level by reducing activity of NF-kB, which can bind to the promoter of the PGC-1α gene and inhibit its transcription [45,46].

The transcriptional factor Nrf2 participates in MB by driving transcription of the transcription factor NRF-1, which, as we have seen, works with PGC-1a to promote expression of proteins vital for MB [47–49]. Sirt1 can enhance Nrf2's activity in this regard via deacetylation [50]. The reader is referred to Figure 2 for a depiction of the mechanisms whereby Sirt1, AMPK, Nrf2, and PPARα interact in the promotion of MB, while also enhancing expression of the mediators of mitophagy, Parkin and Pink1.

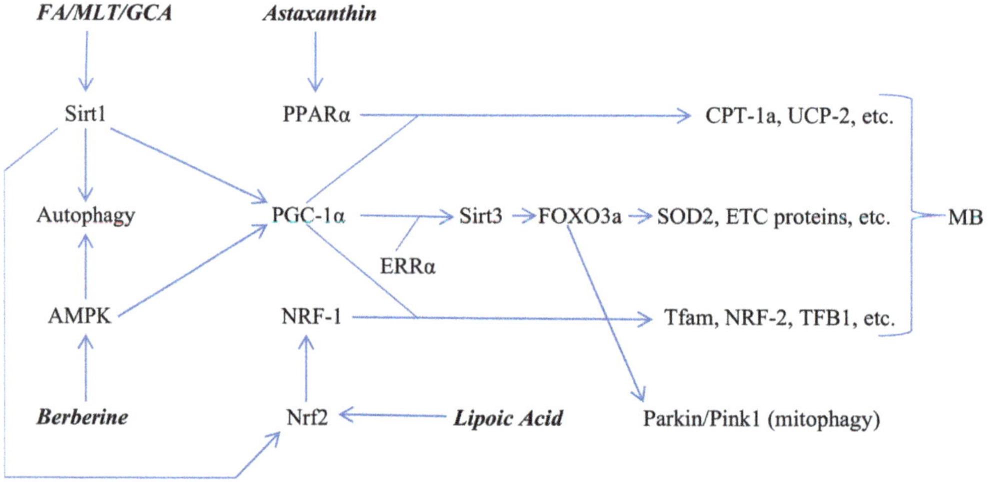

Figure 2. Roles for Sirt1, AMPK, Nrf2, and PPARα in promotion of mitochondrial biogenesis (MB) and autophagy/mitophagy. FA = ferulic acid; MLT = melatonin; GCA = glucosamine.

It is reasonable to suspect that activation of autophagy/mitophagy and MB will lessen activation of NLRP3 inflammasomes in RPE cells by suppressing mitochondrial oxidant generation and the cytoplasmic release of mtDNA. However, Nrf2 and AMPK can also inhibit inflammasome activation via their effects on expression and activity of thioredoxin (TRX) and thioredoxin interacting proteins (TXNIPs). The formation of NLRP3 inflammasomes hinges on the interaction between TXNIP and NLRP3; in cells with good redox control, this interaction is suppressed by complex formation between TXNIP and TRX [51,52]. In the context of oxidative stress, reversible oxidation of TRX prevents its interaction with TXNIP, freeing the latter to participate in inflammasome formation. Nrf2 activity, promoted by Sirt1, induces expression of both TRX and the enzyme which restores its reduced form, thioredoxin reductase, thereby decreasing the free pool of TXNIP [53,54]. The contribution of AMPK to suppression of inflammasome formation hinges on its ability to decrease the expression of TXNIP at the transcriptional level, possibly reflecting its inhibitory effect on activity of the transcription factor carbohydrate response element-binding protein (ChREBP) [55,56]. AMPK also has the potential to accelerate the proteasomal degradation of TXNIP [57].

The foregoing explains how Sir1, AMPK, Nrf2 and PPARα can interact to replace damaged mitochondria with healthy new ones, while amplifying mechanisms which protect mitochondria from oxidative stress and diminish their production of oxidants. Additionally, via the modulation of TRX/TXNIP, they work in an additional way to blunt NLRP3 inflammasome formation. Importantly, there are nutraceuticals and drugs capable of boosting the activity of each of these key enzymes.

3. Nutraceuticals and Drugs Can Activate Sirt1, AMPK, Nrf2 and PPARα

While resveratrol has gained a considerable reputation as an agent that can activate Sirt1, poor pharmacokinetics have to this point prevented it from being clinically useful for this purpose [58]. Moreover, its ability to directly activate Sirt1 has been challenged [59]. A more practical choice in this regard is ferulic acid. Ferulic acid is produced when gut bacteria metabolize dietary anthocyanins; the latter are not absorbed intact and ferulic acid seems likely to mediate most if not all of their protective properties [60]. Ferulic acid also occurs, mostly in conjugated forms, in a wide range of whole grains, fruits, vegetables and nuts. In animal and cell-culture studies, ferulic acid has demonstrated anti-inflammatory and antioxidant effects in a wide range of disease models; in Chinese medicine, sodium ferulate has been widely used in cardiovascular medicine [60,61]. The piperazine salt of ferulate is used in China for treatment of diabetic nephropathy [62]. In a recent controlled clinical trial, the administration of 1000 mg ferulic acid daily to hyperlipidemics was found to improve serum lipid profiles, decrease systemic markers of oxidative stress, and lower plasma C-reactive protein by one-third [63].

Only recently has it been recognized that its utilities in these regards may reflect its ability to up-regulate Sirt1 expression quite markedly at both the mRNA and protein level; this effect appears to be broad in scope, as it has been reported so far in chondrocytes, skeletal muscle fibers, testes, neural stem cells and hepatocytes [64–69]. How it achieves this effect still remains mysterious. In light of its efficient absorption, and its demonstrated benefits in cardiovascular medicine, ferulic acid, or salts thereof, may have considerable clinical potential for boosting Sirt1 activity. Of possible pertinence to AMD is the fact that rich dietary sources of anthocyanins—notably bilberry extract—have been traditionally used for retinal protection [70–72]. Whether they are genuinely beneficial for the prevention of dry AMD has not yet received adequate study.

Also useful in for Sirt1 activation is melatonin, which increases transcription of the gene encoding for Sirt1 via the activation of the BMAL1 transcription factor [73,74]. Melatonin also increases Nrf2 expression by the same mechanism, and hence may have considerable potential for dry AMD control [75,76]. Melatonin administration has been found to protect the retina in a mouse model of dry AMD (superior cervical ganglionectomy) and in a rat model of accelerated senescence associated with AMD [77,78]. Favorable anecdotal clinical experience with supplemental melatonin in AMD has also been reported, but formal clinical trials evaluating its efficacy in this regard are lacking [79].

Sirt1 is susceptible to O-GlcNAcylation at Ser 549, and this post-translational modification has been found to amplify its deacetylase activity [80]. Oral or parenteral administration of glucosamine can increase the cellular pool of UDP-N-acetylglucosamine;

this therefore up-regulates the O-GlcNAcylation of proteins, likely explaining the anti-inflammatory properties of supplemental glucosamine [81]. The ability of glucosamine to boost Sirt1 activity may hence explain recent epidemiology pointing to a lower risk for dry AMD in people who use this nutraceutical regularly; adjusted hazard rate for those using glucosamine for 3 or more years was found to be 0.493 (p = 0.003) [82]. Increased Sirt1 activity may also explain an increase in skeletal muscle MB and exercise performance in mice given oral glucosamine as an adjuvant to aerobic training [83].

Very recently, the bacterial ellagitannin metabolite urolithin A (which may mediate the protective benefits of pomegranate juice) has been shown to have potential for Sirt1 activation, as has the natural nicotinamide metabolite N1-methylnicotinamide, both of which unsurprisingly have anti-inflammatory properties [84–88]. Both urolithin A and MNA have recently become available in nutraceutical form, but have not yet been studied in regard to possible impact on AMD.

It is well know that the favorable impact of the drug metformin on diabetes control reflects its ability to activate AMPK [89,90]. Consistent with the thesis presented here, diabetics who use metformin have been reported to be at lower risk for AMD than diabetics who do not use this drug– albeit findings in this regard are not completely consistent [91–94]. The nutraceutical berberine, a key component of certain Chinese medicinal herbs, can also activate AMPK in a manner comparable to metformin, and is used widely in China for diabetes control [95–99].

A number of phytochemicals (known as phase 2 inducers) are capable of promoting the migration of Nrf2 to the nucleus—thereby promoting its transcriptional activity—by disrupting its binding to the cytoplasmic protein Keap1 [100,101]. The most clinically useful agents in this regard appear to be lipoic acid and sulforaphane—the latter can be supplied for clinical use in broccoli sprout extracts [102–104]. The ability of Sirt1 to deacetylate Nrf2 is complementary to the activity of phase 2 inducers in promoting the transcriptional activity of Nrf2 [50,105].

Drugs which can act as agonists for PPARα are used for the management of hyperlipidemia associated with metabolic syndrome; these include the fibrate drugs fenofibrate and gemfibrozil [106]. Curiously, it has been learned that the algae-derived phytonutrient astaxanthin—an outstanding scavenging antioxidant for biological membranes—can also function as a PPARα agonist, for which reason it can exert clinically useful hypolipidemic effects [107–111]. Astaxanthin may thus be useful in AMD both as an agonist for PPARα, and for protecting the mitochondrial electron transport chain from oxidative damage [112–114].

Overall, these considerations suggest that a nutraceutical regimens providing physiologically meaningful doses of several of these agents:ferulic acid, melatonin, glucosamine, berberine, lipoic acid and astaxanthin may have important potential for prevention and control of the dry form of AMD. Table 1 depicts dosage schedules of these agents that have proved to be clinically active for certain applications in past research. The drugs metformin and fenofibrate could function as alternatives to berberine and astaxanthin in this regard—albeit astaxanthin has scavenging antioxidant activity not possessed by fenofibrate. Potentially, these agents could be studied in RPE cell cultures in which Alu RNA expression has been artificially boosted.

Table 1. Suggested dose schedules for nutraceuticals with potential for controlling dry AMD.

Nutraceuticals	Dose Schedules
Ferulic Acid (or Sodium Ferulate)	250–500 mg twice daily [63]
Melatonin	3–20 mg at bedtime [115]
Glucosamine	1500–3000 mg once daily [81]
Berberine	500 mg, 2–3 times daily [99]
Lipoic Acid	600 mg 2 times daily [116]
Astaxanthin	12–20 mg daily [116]

It might also be noted that the protective impact of macular pigmentation—comprised of the xanthophyll carotenoids lutein, zeaxanthin, and meso-zeaxanthin, available as nutraceuticals—on AMD risk, suggests that photo-oxidative damage to retinal photoreceptors either promotes RPE deficit of Dicer1, or amplifies the pro-inflammatory signaling downstream from Dicer1 deficiency [117]. The key question remains: why is Dicer1 selectively depressed in RPE cells in AMD? Nutraceutical strategies for correcting the RPE deficit of DICER1 might cut to the root of AMD pathogenesis, but what these might be remains unclear.

Mention should be made of recent evidence that the adverse impact of Alu RNA accumulation on RPE survival requires cytoplasmic reverse transcription of this RNA into DNA within the cytoplasm [118]. This is mediated by the L1 reverse transcriptase, which can be inhibited by clinical concentrations of nucleotide reverse transcriptase inhibitor drugs used in management of HIV. An epidemiological analysis suggests that patients using such drugs regularly may be at lower risk for AMD [118]. Hence, a pharmaceutical strategy for managing dry AMD, potentially complementary to the measures suggested here, can be envisioned. However, rather perversely, these drugs have the potential to damage mitochondria [119].

4. Pertinence to Neovascular AMD

The neovascular, "wet" form of AMD is distinctly different from the dry form considered here. However, there is recent evidence that deficient genetic expression of Dicer1, whether general or specific to RPE cells, leads not only to geographic atrophy, but also to pathological neovascularization. Moreover, concurrent deficiency of proteins required for inflammasome signaling, or MyD88, required for IL-18 signaling, blunts this neovascularization [120]. This suggests that measures which oppose inflammasome activation in RPE cells—such as those suggested here—might also be expected to aid prevention of neovascular AMD.

Moreover, in light of a central role for endothelial NADPH oxidase activation in VEGF-mediated signaling that evoke retinal neoangiogenesis, the spirulina chromophore phycocyanobilin, a biliverdin derivative that can mimic the NADPH oxidase-inhibitory effects of its chemical relative bilirubin, may have potential for prevention and control of wet AMD [121–126]. Also, high intakes of glycine—which is clinically feasible, as this amino acid is highly soluble, pleasantly sweet, and inexpensive—has been demonstrated to have anti-angiogenic effects in rodent models of cancer and wound healing [127–130]. It has been suggested that indirect inhibition of endothelial NADPH oxidase activity may mediate this effect [131].

Author Contributions: Conceptualization, M.F.M., J.J.D.N., S.B.I.A. and L.M.L.L.; validation, S.B.I.A., M.F.M.; formal analysis, L.M.L.L., S.B.I.A. and J.C.G.R.; writing—original draft preparation, L.M.L.L., S.B.I.A., J.J.D.N. and M.F.M.; writing—review and editing, E.C.R.-B., M.P.-J. and L.M.L.L.; visualization, M.F.M. and S.B.I.A. All authors have read and agreed to the published version of the manuscript.

Funding: This research received no external funding.

Institutional Review Board Statement: Not applicable.

Informed Consent Statement: Not applicable.

Conflicts of Interest: The authors declare no conflict of interest.

References

1. Kaneko, H.; Dridi, S.; Tarallo, V.; Gelfand, B.D.; Fowler, B.J.; Cho, W.G.; Kleinman, M.E.; Ponicsan, S.L.; Hauswirth, W.W.; Chiodo, V.A.; et al. DICER1 deficit induces Alu RNA toxicity in age-related macular degeneration. *Nature* **2011**, *471*, 325–330. [CrossRef]
2. Tarallo, V.; Hirano, Y.; Gelfand, B.D.; Dridi, S.; Kerur, N.; Kim, Y.; Cho, W.G.; Kaneko, H.; Fowler, B.J.; Bogdanovich, S.; et al. DICER1 loss and Alu RNA induce age-related macular degeneration via the NLRP3 inflammasome and MyD88. *Cell* **2012**, *149*, 847–859. [CrossRef]

3. Kerur, N.; Hirano, Y.; Tarallo, V.; Fowler, B.J.; Bastos-Carvalho, A.; Yasuma, T.; Yasuma, R.; Kim, Y.; Hinton, D.R.; Kirschning, C.J.; et al. TLR-independent and P2X7-dependent signaling mediate Alu RNA-induced NLRP3 inflammasome activation in geographic atrophy. *Investig. Ophthalmol. Vis. Sci.* **2013**, *54*, 7395–7401. [CrossRef] [PubMed]
4. Kim, Y.; Tarallo, V.; Kerur, N.; Yasuma, T.; Gelfand, B.D.; Bastos-Carvalho, A.; Hirano, Y.; Yasuma, R.; Mizutani, T.; Fowler, B.J.; et al. DICER1/Alu RNA dysmetabolism induces Caspase-8-mediated cell death in age-related macular degeneration. *Proc. Natl. Acad. Sci. USA* **2014**, *111*, 16082–16087. [CrossRef] [PubMed]
5. Mao, X.; Fang, W.; Liu, Q. An emerging role of Alu RNA in geographic atrophy pathogenesis: The implication for novel therapeutic strategies. *Discov. Med.* **2016**, *22*, 337–349. [PubMed]
6. Kerur, N.; Fukuda, S.; Banerjee, D.; Kim, Y.; Fu, D.; Apicella, I.; Varshney, A.; Yasuma, R.; Fowler, B.J.; Baghdasaryan, E.; et al. cGAS drives noncanonical-inflammasome activation in age-related macular degeneration. *Nat. Med.* **2018**, *24*, 50–61. [CrossRef] [PubMed]
7. Ren, Y.-F.; Li, G.; Wu, J.; Xue, Y.-F.; Song, Y.-J.; Lv, L.; Zhang, X.-J.; Tang, K.-F. Dicer-dependent biogenesis of small RNAs derived from 7SL RNA. *PLoS ONE* **2012**, *7*, e40705. [CrossRef] [PubMed]
8. Fang, R.; Wang, C.; Jiang, Q.; Lv, M.; Gao, P.; Yu, X.; Mu, P.; Zhang, R.; Bi, S.; Feng, J.M.; et al. NEMO-IKKβ Are Essential for IRF3 and NF-κB Activation in the cGAS-STING Pathway. *J. Immunol.* **2017**, *199*, 3222–3233. [CrossRef] [PubMed]
9. Tossberg, J.T.; Heinrich, R.M.; Farley, V.M.; Crooke, P.S., III; Aune, T.M. Adenosine-to-Inosine RNA Editing of Alu Double-Stranded (ds)RNAs Is Markedly Decreased in Multiple Sclerosis and Unedited Alu dsRNAs Are Potent Activators of Proinflammatory Transcriptional Responses. *J. Immunol.* **2020**, *205*, 2606–2617. [CrossRef]
10. Di, V.F.; Giuliani, A.L.; Vultaggio-Poma, V.; Falzoni, S.; Sarti, A.C. Non-nucleotide Agonists Triggering P2 × 7 Receptor Activation and Pore Formation. *Front. Pharmacol.* **2018**, *9*, 39.
11. Wang, K.; Yao, Y.; Zhu, X.; Zhang, K.; Zhou, F.; Zhu, L. Amyloid beta induces NLRP3 inflammasome activation in retinal pigment epithelial cells via NADPH oxidase- and mitochondria-dependent ROS production. *J. Biochem. Mol. Toxicol* **2017**, *31*, e21887. [CrossRef]
12. Brandstetter, C.; Holz, F.G.; Krohne, T.U. Complement Component C5a Primes Retinal Pigment Epithelial Cells for Inflammasome Activation by Lipofuscin-mediated Photooxidative Damage. *J. Biol. Chem.* **2015**, *290*, 31189–31198. [CrossRef]
13. Anderson, O.A.; Finkelstein, A.; Shima, D.T. A2E induces IL-1ss production in retinal pigment epithelial cells via the NLRP3 inflammasome. *PLoS ONE* **2013**, *8*, e67263. [CrossRef]
14. Terluk, M.R.; Kapphahn, R.J.; Soukup, L.M.; Gong, H.; Gallardo, C.; Montezuma, S.R.; Ferrington, D.A. Investigating mitochondria as a target for treating age-related macular degeneration. *J. Neurosci.* **2015**, *35*, 7304–7311. [CrossRef]
15. Brown, E.E.; Lewin, A.S.; Ash, J.D. Mitochondria: Potential Targets for Protection in Age-Related Macular Degeneration. *Adv. Exp. Med. Biol.* **2018**, *1074*, 11–17.
16. Feher, J.; Kovacs, I.; Artico, M.; Cavallotti, C.; Papale, A.; Balacco, G.C. Mitochondrial alterations of retinal pigment epithelium in age-related macular degeneration. *Neurobiol. Aging* **2006**, *27*, 983–993. [CrossRef]
17. Ebeling, M.C.; Polanco, J.R.; Qu, J.; Tu, C.; Montezuma, S.R.; Ferrington, D.A. Improving retinal mitochondrial function as a treatment for age-related macular degeneration. *Redox Biol.* **2020**, *34*, 101552. [CrossRef]
18. Kaarniranta, K.; Uusitalo, H.; Blasiak, J.; Felszeghy, S.; Kannan, R.; Kauppinen, A.; Salminen, A.; Sinha, D.; Ferrington, D. Mechanisms of mitochondrial dysfunction and their impact on age-related macular degeneration. *Prog. Retin. Eye Res.* **2020**, *79*, 100858. [CrossRef]
19. Lan, F.; Cacicedo, J.M.; Ruderman, N.; Ido, Y. SIRT1 modulation of the acetylation status, cytosolic localization, and activity of LKB1. Possible role in AMP-activated protein kinase activation. *J. Biol. Chem.* **2008**, *283*, 27628–27635. [CrossRef]
20. Fulco, M.; Cen, Y.; Zhao, P.; Hoffman, E.P.; McBurney, M.W.; Sauve, A.A.; Sartorelli, V. Glucose restriction inhibits skeletal myoblast differentiation by activating SIRT1 through AMPK-mediated regulation of Nampt. *Dev Cell* **2008**, *14*, 661–673. [CrossRef]
21. Costford, S.R.; Bajpeyi, S.; Pasarica, M.; Albarado, D.C.; Thomas, S.C.; Xie, H.; Church, T.S.; Jubrias, S.A.; Conley, K.E.; Smith, S.R. Skeletal muscle NAMPT is induced by exercise in humans. *Am. J. Physiol. Endocrinol. Metab.* **2010**, *298*, E117–E126. [CrossRef]
22. Revollo, J.R.; Grimm, A.A.; Imai, S. The NAD biosynthesis pathway mediated by nicotinamide phosphoribosyltransferase regulates Sir2 activity in mammalian cells. *J. Biol. Chem.* **2004**, *279*, 50754–50763. [CrossRef]
23. Skokowa, J.; Lan, D.; Thakur, B.; Gupta, K.; Cario, G.; Brechlin, A.; Stanulla, M.; Welte, K. NAMPT is essential for the G-CSF-induced myeloid differentiation via a NAD(+)-sirtuin-1-dependent pathway. *Nat. Med.* **2009**, *15*, 151–158. [CrossRef]
24. Sacitharan, P.K.; Bou-Gharios, G.; Edwards, J.R. SIRT1 directly activates autophagy in human chondrocytes. *Cell Death Discov.* **2020**, *6*, 41. [CrossRef]
25. Jiang, Q.; Hao, R.; Wang, W.; Gao, H.; Wang, C. SIRT1/Atg5/autophagy are involved in the antiatherosclerosis effects of ursolic acid. *Mol. Cell Biochem.* **2016**, *420*, 171–184. [CrossRef]
26. Hariharan, N.; Maejima, Y.; Nakae, J.; Paik, J.; Depinho, R.A.; Sadoshima, J. Deacetylation of FoxO by Sirt1 Plays an Essential Role in Mediating Starvation-Induced Autophagy in Cardiac Myocytes. *Circ. Res.* **2010**, *107*, 1470–1482. [CrossRef]
27. Russell, R.C.; Yuan, H.X.; Guan, K.L. Autophagy regulation by nutrient signaling. *Cell Res.* **2014**, *24*, 42–57. [CrossRef]
28. Zhao, M.; Klionsky, D.J. AMPK-dependent phosphorylation of ULK1 induces autophagy. *Cell Metab.* **2011**, *13*, 119–120. [CrossRef]
29. Kim, J.; Guan, K.L. Regulation of the autophagy initiating kinase ULK1 by nutrients: Roles of mTORC1 and AMPK. *Cell Cycle* **2011**, *10*, 1337–1338. [CrossRef] [PubMed]

30. Scarpulla, R.C. Metabolic control of mitochondrial biogenesis through the PGC-1 family regulatory network. *Biochim. Biophys. Acta.* **2011**, *1813*, 1269–1278. [CrossRef]
31. Gleyzer, N.; Vercauteren, K.; Scarpulla, R.C. Control of mitochondrial transcription specificity factors (TFB1M and TFB2M) by nuclear respiratory factors (NRF-1 and NRF-2) and PGC-1 family coactivators. *Mol. Cell. Biol.* **2005**, *25*, 1354–1366. [CrossRef] [PubMed]
32. Kong, X.; Wang, R.; Xue, Y.; Liu, X.; Zhang, H.; Chen, Y.; Fang, F.; Chang, Y. Sirtuin 3, a new target of PGC-1alpha, plays an important role in the suppression of ROS and mitochondrial biogenesis. *PLoS ONE* **2010**, *5*, e11707. [CrossRef] [PubMed]
33. Giralt, A.; Hondares, E.; Villena, J.A.; Ribas, F.; DÃ az-DelfÃ n, J.; Giralt, M.; Iglesias, R.; Villarroya, F. Peroxisome proliferator-activated receptor-gamma coactivator-1alpha controls transcription of the Sirt3 gene, an essential component of the thermogenic brown adipocyte phenotype. *J. Biol. Chem.* **2011**, *286*, 16958–16966. [CrossRef] [PubMed]
34. Jacobs, K.M.; Pennington, J.D.; Bisht, K.S.; Aykin-Burns, N.; Kim, H.S.; Mishra, M.; Sun, L.; Nguyen, P.; Ahn, B.H.; Leclerc, J.; et al. SIRT3 interacts with the daf-16 homolog FOXO3a in the mitochondria, as well as increases FOXO3a dependent gene expression. *Int. J. Biol. Sci.* **2008**, *4*, 291–299. [CrossRef]
35. Sundaresan, N.R.; Gupta, M.; Kim, G.; Rajamohan, S.B.; Isbatan, A.; Gupta, M.P. Sirt3 blocks the cardiac hypertrophic response by augmenting Foxo3a-dependent antioxidant defense mechanisms in mice. *J. Clin. Investig.* **2009**, *119*, 2758–2771. [CrossRef]
36. Liu, X.; Zhang, L.; Wang, P.; Li, X.; Qiu, D.; Li, L.; Zhang, J.; Hou, X.; Han, L.; Ge, J.; et al. Sirt3-dependent deacetylation of SOD2 plays a protective role against oxidative stress in oocytes from diabetic mice. *Cell Cycle* **2017**, *16*, 1302–1308. [CrossRef]
37. Han, L.; Li, J.; Li, J.; Pan, C.; Xiao, Y.; Lan, X.; Wang, M. Activation of AMPK/Sirt3 pathway by phloretin reduces mitochondrial ROS in vascular endothelium by increasing the activity of MnSOD via deacetylation. *Food Funct.* **2020**, *11*, 3073–3083. [CrossRef]
38. Gupta, P.; Sharma, G.; Lahiri, A.; Barthwal, M.K. FOXO3a acetylation regulates PINK1, mitophagy, inflammasome activation inÂ murine palmitate-conditioned and diabetic macrophages. *J. Leukoc. Biol.* **2021**, *111*, 611–627. [CrossRef]
39. Ding, D.; Ao, X.; Li, M.; Miao, S.; Liu, Y.; Lin, Z.; Wang, M.; He, Y.; Wang, J. FOXO3a-dependent Parkin regulates the development of gastric cancer by targeting ATP-binding cassette transporter E1. *J. Cell Physiol.* **2021**, *236*, 2740–2755. [CrossRef]
40. Das, S.; Mitrovsky, G.; Vasanthi, H.R.; Das, D.K. Antiaging properties of a grape-derived antioxidant are regulated by mitochondrial balance of fusion and fission leading to mitophagy triggered by a signaling network of Sirt1-Sirt3-Foxo3-PINK1-PARKIN. *Oxid. Med. Cell. Longev.* **2014**, *2014*, 345105. [CrossRef]
41. Duncan, J.G.; Fong, J.L.; Medeiros, D.M.; Finck, B.N.; Kelly, D.P. Insulin-resistant heart exhibits a mitochondrial biogenic response driven by the peroxisome proliferator-activated receptor-alpha/PGC-1alpha gene regulatory pathway. *Circulation* **2007**, *115*, 909–917. [CrossRef]
42. Kersten, S. Integrated physiology and systems biology of PPARα. *Mol. Metab.* **2014**, *3*, 354–371. [CrossRef]
43. NikoliÄ, N.; Rhedin, M.; Rustan, A.C.; Storlien, L.; Thoresen, G.H.; Strömstedt, M. Overexpression of PGC-1α increases fatty acid oxidative capacity of human skeletal muscle cells. *Biochem. Res. Int.* **2012**, *2012*, 714074.
44. Canto, C.; Gerhart-Hines, Z.; Feige, J.N.; Lagouge, M.; Noriega, L.; Milne, J.C.; Elliott, P.J.; Puigserver, P.; Auwerx, J. AMPK regulates energy expenditure by modulating NAD+ metabolism and SIRT1 activity. *Nature* **2009**, *458*, 1056–1060. [CrossRef]
45. Remels, A.H.; Gosker, H.R.; Bakker, J.; Guttridge, D.C.; Schols, A.M.; Langen, R.C. Regulation of skeletal muscle oxidative phenotype by classical NF-κB signalling. *Biochim. Biophys. Acta.* **2013**, *1832*, 1313–1325. [CrossRef]
46. Álvarez-Guardia, D.; Palomer, X.; Coll, T.; Davidson, M.M.; Chan, T.O.; Feldman, A.M.; Laguna, J.C.; Vázquez-Carrera, M. The p65 subunit of NF-kappaB binds to PGC-1alpha, linking inflammation and metabolic disturbances in cardiac cells. *Cardiovasc. Res.* **2010**, *87*, 449–458. [CrossRef]
47. Selen, E.S.; Wolfgang, M.J. mTORC1 activation is not sufficient to suppress hepatic PPARα signaling or ketogenesis. *J. Biol. Chem.* **2021**, *297*, 100884. [CrossRef]
48. Botta, M.; Audano, M.; Sahebkar, A.; Sirtori, C.R.; Mitro, N.; Ruscica, M. PPAR Agonists and Metabolic Syndrome: An Established Role? *Int. J. Mol. Sci.* **2018**, *19*, 1197. [CrossRef]
49. Piantadosi, C.A.; Carraway, M.S.; Babiker, A.; Suliman, H.B. Heme oxygenase-1 regulates cardiac mitochondrial biogenesis via Nrf2-mediated transcriptional control of nuclear respiratory factor-1. *Circ. Res.* **2008**, *103*, 1232–1240. [CrossRef]
50. Ding, Y.-W.; Zhao, G.-J.; Li, X.-L.; Hong, G.-L.; Li, M.-F.; Qiu, Q.-M.; Wu, B.; Lu, Z.-Q. SIRT1 exerts protective effects against paraquat-induced injury in mouse type II alveolar epithelial cells by deacetylating NRF2 in vitro. *Int. J. Mol. Med.* **2016**, *37*, 1049–1058. [CrossRef]
51. Zhou, R.; Tardivel, A.; Thorens, B.; Choi, I.; Tschopp, J. Thioredoxin-interacting protein links oxidative stress to inflammasome activation. *Nat. Immunol.* **2010**, *11*, 136–140. [CrossRef]
52. Oslowski, C.M.; Hara, T.; O'Sullivan-Murphy, B.; Kanekura, K.; Lu, S.; Hara, M.; Ishigaki, S.; Zhu, L.J.; Hayashi, E.; Hui, S.T.; et al. Thioredoxin-interacting protein mediates ER stress-induced β cell death through initiation of the inflammasome. *Cell Metab.* **2012**, *16*, 265–273. [CrossRef]
53. Sakurai, A.; Nishimoto, M.; Himeno, S.; Imura, N.; Tsujimoto, M.; Kunimoto, M.; Hara, S. Transcriptional regulation of thioredoxin reductase 1 expression by cadmium in vascular endothelial cells: Role of NF-E2-related factor-2. *J. Cell Physiol.* **2005**, *203*, 529–537. [CrossRef] [PubMed]
54. Tanito, M.; Agbaga, M.P.; Anderson, R.E. Upregulation of thioredoxin system via Nrf2-antioxidant responsive element pathway in adaptive-retinal neuroprotection in vivo and in vitro. *Free Radic. Biol. Med.* **2007**, *42*, 1838–1850. [CrossRef]

55. Shaked, M.; Ketzinel-Gilad, M.; Cerasi, E.; Kaiser, N.; Leibowitz, G. AMP-activated protein kinase (AMPK) mediates nutrient regulation of thioredoxin-interacting protein (TXNIP) in pancreatic beta-cells. *PLoS ONE* **2011**, *6*, e28804. [CrossRef]
56. Chai, T.F.; Hong, S.Y.; He, H.; Zheng, L.; Hagen, T.; Luo, Y.; Yu, F.X. A potential mechanism of metformin-mediated regulation of glucose homeostasis: Inhibition of Thioredoxin-interacting protein (Txnip) gene expression. *Cell. Signal.* **2012**, *24*, 1700–1705. [CrossRef]
57. Wu, N.; Zheng, B.; Shaywitz, A.; Dagon, Y.; Tower, C.; Bellinger, G.; Shen, C.-H.; Wen, J.; Asara, J.; McGraw, T.E.; et al. AMPK-dependent degradation of TXNIP upon energy stress leads to enhanced glucose uptake via GLUT1. *Mol. Cell.* **2013**, *49*, 1167–1175. [CrossRef]
58. Chimento, A.; De Amicis, F.; Sirianni, R.; Sinicropi, M.S.; Puoci, F.; Casaburi, I.; Saturnino, C.; Pezzi, V. Progress to Improve Oral Bioavailability and Beneficial Effects of Resveratrol. *Int. J. Mol. Sci.* **2019**, *20*, 1381. [CrossRef]
59. Pacholec, M.; Bleasdale, J.E.; Chrunyk, B.; Cunningham, D.; Flynn, D.; Garofalo, R.S.; Griffith, D.; Griffor, M.; Loulakis, P.; Pabst, B.; et al. SRT1720, SRT2183, SRT1460, and resveratrol are not direct activators of SIRT1. *J. Biol. Chem.* **2010**, *285*, 8340–8351. [CrossRef]
60. McCarty, M.F.; Assanga, S.B.I. Ferulic acid may target MyD88-mediated pro-inflammatory signaling—Implications for the health protection afforded by whole grains, anthocyanins, and coffee. *Med. Hypotheses* **2018**, *118*, 114–120. [CrossRef]
61. Zhang, X.-X.; Zhao, D.-S.; Wang, J.; Zhou, H.; Wang, L.; Mao, J.-L.; He, J.-X. The treatment of cardiovascular diseases: A review of ferulic acid and its derivatives. *Pharmazie* **2021**, *76*, 55–60. [PubMed]
62. Yang, Y.Y.; Shi, L.X.; Li, J.H.; Yao, L.Y.; Xiang, D.X. Piperazine ferulate ameliorates the development of diabetic nephropathy by regulating endothelial nitric oxide synthase. *Mol. Med. Rep.* **2019**, *19*, 2245–2253. [CrossRef] [PubMed]
63. Bumrungpert, A.; Lilitchan, S.; Tuntipopipat, S.; Tirawanchai, N.; Komindr, S. Ferulic Acid Supplementation Improves Lipid Profiles, Oxidative Stress, and Inflammatory Status in Hyperlipidemic Subjects: A Randomized, Double-Blind, Placebo-Controlled Clinical Trial. *Nutrients* **2018**, *10*, 713. [CrossRef]
64. El-Mesallamy, H.O.; Gawish, R.A.; Sallam, A.M.; Fahmy, H.A.; Nada, A.S. Ferulic acid protects against radiation-induced testicular damage in male rats: Impact on SIRT1 and PARP1. *Environ. Sci. Pollut. Res. Int.* **2018**, *25*, 6218–6227. [CrossRef]
65. Moghadam, F.H.; Mesbah-Ardakani, M.; Nasr-Esfahani, M.H. Ferulic Acid exerts concentration-dependent anti-apoptotic and neuronal differentiation-inducing effects in PC12 and mouse neural stem cells. *Eur. J. Pharmacol.* **2018**, *841*, 104–112. [CrossRef]
66. Hou, T.; Zhang, L.; Yang, X. Ferulic acid, a natural polyphenol, protects against osteoporosis by activating SIRT1 and NF-κB in neonatal rats with glucocorticoid-induced osteoporosis. *Biomed. Pharm.* **2019**, *120*, 109205. [CrossRef]
67. Xu, T.; Song, Q.; Zhou, L.; Yang, W.; Wu, X.; Qian, Q.; Chai, H.; Han, Q.; Pan, H.; Dou, X.; et al. Ferulic acid alleviates lipotoxicity-induced hepatocellular death through the SIRT1-regulated autophagy pathway and independently of AMPK and Akt in AML-12 hepatocytes. *Nutr. Metab.* **2021**, *18*, 13. [CrossRef]
68. Wang, Y.; Chen, X.; Huang, Z.; Chen, D.; Yu, B.; Chen, H.; Yu, J.; Luo, Y.; Zheng, P.; He, J. Effects of dietary ferulic acid supplementation on growth performance and skeletal muscle fiber type conversion in weaned piglets. *J. Sci. Food. Agric.* **2021**, *101*, 5116–5123. [CrossRef]
69. Du, K.; Fang, X.; Li, Z. Ferulic acid suppresses interleukin-1β-induced degeneration of chondrocytes isolated from patients with osteoarthritis through the SIRT1/AMPK/PGC-1α signaling pathway. *Immun. Inflamm. Dis.* **2021**, *9*, 710–720. [CrossRef]
70. Gerding, H. Primary or Secondary Prophylaxis of AMD with Anthocyanins? *Klin. Monbl. Augenheilkd.* **2009**, *226*, 216–219. [CrossRef]
71. Pinelli, R.; Bertelli, M.; Scaffidi, E.; Polzella, M.; Fulceri, F.; Biagioni, F.; Fornai, F. Nutraceuticals for dry age-related macular degeneration: A case report based on novel pathogenic and morphological insights. *Arch. Ital. Biol.* **2020**, *158*, 24–34. [CrossRef]
72. Khoo, H.E.; Ng, H.S.; Yap, W.S.; Goh, H.J.H.; Yim, H.S. Nutrients for Prevention of Macular Degeneration and Eye-Related Diseases. *Antioxidants* **2019**, *8*, 85. [CrossRef]
73. Yu, L.; Sun, Y.; Cheng, L.; Jin, Z.; Yang, Y.; Zhai, M.; Pei, H.; Wang, X.; Zhang, H.; Meng, Q.; et al. Melatonin receptor-mediated protection against myocardial ischemia/reperfusion injury: Role of SIRT1. *J. Pineal. Res.* **2014**, *57*, 228–238. [CrossRef]
74. Zhou, B.; Zhang, Y.; Zhang, F.; Xia, Y.; Liu, J.; Huang, R.; Wang, Y.; Hu, Y.; Wu, J.; Dai, C.; et al. CLOCK/BMAL1 regulates circadian change of mouse hepatic insulin sensitivity by SIRT1. *Hepatology* **2014**, *59*, 2196–2206. [CrossRef]
75. Fang, J.; Yan, Y.; Teng, X.; Wen, X.; Li, N.; Peng, S.; Liu, W.; Donadeu, F.X.; Zhao, S.; Hua, J. Melatonin prevents senescence of canine adipose-derived mesenchymal stem cells through activating NRF2 and inhibiting ER stress. *Aging* **2018**, *10*, 2954–2972. [CrossRef]
76. Wang, Z.; Ma, C.; Meng, C.-J.; Zhu, G.-Q.; Sun, X.-B.; Huo, L.; Zhang, J.; Liu, H.-X.; He, W.-C.; Shen, X.-M.; et al. Melatonin activates the Nrf2-ARE pathway when it protects against early brain injury in a subarachnoid hemorrhage model. *J. Pineal. Res.* **2012**, *53*, 129–137. [CrossRef]
77. Diéguez, H.H.; Fleitas, M.F.G.; Aranda, M.L.; Calanni, J.S.; Sarmiento, M.I.K.; Chianelli, M.S.; Alaimo, A.; Sande, P.H.; Romeo, H.E.; Rosenstein, R.E.; et al. Melatonin protects the retina from experimental nonexudative age-related macular degeneration in mice. *J. Pineal. Res.* **2020**, *68*, e12643. [CrossRef]
78. Stefanova, N.A.; Zhdankina, A.A.; Fursova, A.Z.; Kolosova, N.G. Potential of melatonin for prevention of age-related macular degeneration: Experimental study. *Adv. Gerontol.* **2013**, *26*, 122–129. [CrossRef]
79. Yi, C.; Pan, X.; Yan, H.; Guo, M.; Pierpaoli, W. Effects of melatonin in age-related macular degeneration. *Ann. N. Y. Acad. Sci.* **2005**, *1057*, 384–392. [CrossRef]

80. Han, C.; Gu, Y.; Shan, H.; Mi, W.; Sun, J.; Shi, M.; Zhang, X.; Lu, X.; Han, F.; Gong, Q.; et al. O-GlcNAcylation of SIRT1 enhances its deacetylase activity and promotes cytoprotection under stress. *Nat. Commun.* **2017**, *8*, 1491. [CrossRef]
81. McCarty, M.F.; O'Keefe, J.H.; DiNicolantonio, J.J. Glucosamine for the Treatment of Osteoarthritis: The Time Has Come for Higher-Dose Trials. *J. Diet. Suppl.* **2019**, *16*, 179–192. [CrossRef] [PubMed]
82. Feng, K.M.; Chien, W.-C.; Chen, J.-T.; Chen, Y.-H.; Chung, C.-H.; Sun, C.-A.; Chen, C.-L. The impact of glucosamine on age-related macular degeneration in patients: A nationwide, population-based cohort study. *PLoS ONE* **2021**, *16*, e0251925. [CrossRef] [PubMed]
83. De La Rosa, A.; Olaso-Gonzalez, G.; Garcia-Dominguez, E.; Mastaloudis, A.; Hester, S.N.; Wood, S.M.; Gomez-Cabrera, M.C.; Viña, J. Glucosamine Supplementation Improves Physical Performance in Trained Mice. *Med. Sci. Sports Exerc.* **2022**, *54*, 466–474. [CrossRef] [PubMed]
84. Velagapudi, R.; Lepiarz, I.; El-Bakoush, A.; Katola, F.O.; Bhatia, H.; Fiebich, B.L.; Olajide, O.A. Induction of Autophagy and Activation of SIRT-1 Deacetylation Mechanisms Mediate Neuroprotection by the Pomegranate Metabolite Urolithin A in BV2 Microglia and Differentiated 3D Human Neural Progenitor Cells. *Mol. Nutr. Food Res.* **2019**, *63*, e1801237. [CrossRef]
85. Ghosh, N.; Das, A.; Biswas, N.; Gnyawali, S.; Singh, K.; Gorain, M.; Polcyn, C.; Khanna, S.; Roy, S.; Sen, C.K. Urolithin A augments angiogenic pathways in skeletal muscle by bolstering NAD(+) and SIRT1. *Sci. Rep.* **2020**, *10*, 20184. [CrossRef]
86. Hong, S.; Moreno-Navarrete, J.M.; Wei, X.; Kikukawa, Y.; Tzameli, I.; Prasad, D.; Lee, Y.; Asara, J.M.; Fernández-Real, J.M.; Maratos-Flier, E.; et al. Nicotinamide N-methyltransferase regulates hepatic nutrient metabolism through Sirt1 protein stabilization. *Nat. Med.* **2015**, *21*, 887–894. [CrossRef]
87. Chen, Y.; Zhang, J.; Li, P.; Liu, C.; Li, L. N1-methylnicotinamide ameliorates insulin resistance in skeletal muscle of type 2 diabetic mice by activating the SIRT1/PGC-1α signaling pathway. *Mol. Med. Rep.* **2021**, *23*, 270. [CrossRef]
88. Campagna, R.; Mateuszuk, Å.; Wojnar-Lason, K.; Kaczara, P.; TworzydÅ,, O.A.; Kij, A.; Bujok, R.; Mlynarski, J.; Wang, Y.; Sartini, D.; et al. Nicotinamide N-methyltransferase in endothelium protects against oxidant stress-induced endothelial injury. *Biochim. Biophys. Acta. Mol. Cell Res.* **2021**, *1868*, 119082. [CrossRef]
89. Zhou, G.; Myers, R.; Li, Y.; Chen, Y.; Shen, X.; Fenyk-Melody, J.; Wu, M.; Ventre, J.; Doebber, T.; Fujii, N.; et al. Role of AMP-activated protein kinase in mechanism of metformin action. *J. Clin. Investig.* **2001**, *108*, 1167–1174. [CrossRef]
90. Musi, N.; Hirshman, M.F.; Nygren, J.; Svanfeldt, M.; Bavenholm, P.; Rooyackers, O.; Zhou, G.; Williamson, J.M.; Ljunqvist, O.; Efendic, S.; et al. Metformin increases AMP-activated protein kinase activity in skeletal muscle of subjects with type 2 diabetes. *Diabetes* **2002**, *51*, 2074–2081. [CrossRef]
91. Brown, E.E.; Ball, J.D.; Chen, Z.; Khurshid, G.S.; Prosperi, M.; Ash, J.D. The Common Antidiabetic Drug Metformin Reduces Odds of Developing Age-Related Macular Degeneration. *Investig. Ophthalmol. Vis. Sci.* **2019**, *60*, 1470–1477. [CrossRef]
92. Chen, Y.-Y.; Shen, Y.-C.; Lai, Y.-J.; Wang, C.-Y.; Lin, K.-H.; Feng, S.-C.; Liang, C.-Y.; Wei, L.-C.; Chou, P. Association between Metformin and a Lower Risk of Age-Related Macular Degeneration in Patients with Type 2 Diabetes. *J. Ophthalmol.* **2019**, *2019*, 1649156. [CrossRef]
93. Blitzer, A.L.; Ham, S.A.; Colby, K.A.; Skondra, D. Association of Metformin Use with Age-Related Macular Degeneration: A Case-Control Study. *JAMA Ophthalmol.* **2021**, *139*, 302–309. [CrossRef]
94. Romdhoniyyah, D.F.; Harding, S.P.; Cheyne, C.P.; Beare, N.A.V. Metformin, a potential role in age-related macular degeneration: A systematic review and meta-analysis. *Ophthalmol. Ther.* **2021**, *10*, 245–260. [CrossRef]
95. Lee, Y.S.; Kim, W.S.; Kim, K.H.; Yoon, M.J.; Cho, H.J.; Shen, Y.; Ye, J.M.; Lee, C.H.; Oh, W.K.; Kim, C.T.; et al. Berberine, a natural plant product, activates AMP-activated protein kinase with beneficial metabolic effects in diabetic and insulin-resistant states. *Diabetes* **2006**, *55*, 2256–2264. [CrossRef]
96. Turner, N.; Li, J.Y.; Gosby, A.; To, S.W.; Cheng, Z.; Miyoshi, H.; Taketo, M.M.; Cooney, G.J.; Kraegen, E.W.; James, D.E.; et al. Berberine and its more biologically available derivative, dihydroberberine, inhibit mitochondrial respiratory complex I: A mechanism for the action of berberine to activate AMP-activated protein kinase and improve insulin action. *Diabetes* **2008**, *57*, 1414–1418. [CrossRef]
97. Hawley, S.A.; Ross, F.A.; Chevtzoff, C.; Green, K.A.; Evans, A.; Fogarty, S.; Towler, M.C.; Brown, L.J.; Ogunbayo, O.A.; Evans, A.M.; et al. Use of cells expressing gamma subunit variants to identify diverse mechanisms of AMPK activation. *Cell Metab.* **2010**, *11*, 554–565. [CrossRef]
98. Liang, Y.; Xu, X.; Yin, M.; Zhang, Y.; Huang, L.; Chen, R.; Ni, J. Effects of berberine on blood glucose in patients with type 2 diabetes mellitus: A systematic literature review and a meta-analysis. *Endocr. J.* **2019**, *66*, 51–63. [CrossRef]
99. Ye, Y.; Liu, X.; Wu, N.; Han, Y.; Wang, J.; Yu, Y.; Chen, Q. Efficacy and Safety of Berberine Alone for Several Metabolic Disorders: A Systematic Review and Meta-Analysis of Randomized Clinical Trials. *Front. Pharmacol.* **2021**, *12*, 653887. [CrossRef]
100. Dinkova-Kostova, A.T.; Holtzclaw, W.D.; Cole, R.N.; Itoh, K.; Wakabayashi, N.; Katoh, Y.; Yamamoto, M.; Talalay, P. Direct evidence that sulfhydryl groups of Keap1 are the sensors regulating induction of phase 2 enzymes that protect against carcinogens and oxidants. *Proc. Natl. Acad. Sci. USA* **2002**, *99*, 11908–11913. [CrossRef]
101. Wakabayashi, N.; Dinkova-Kostova, A.T.; Holtzclaw, W.D.; Kang, M.-I.; Kobayashi, A.; Yamamoto, M.; Kensler, T.W.; Talalay, P. Protection against electrophile and oxidant stress by induction of the phase 2 response: Fate of cysteines of the Keap1 sensor modified by inducers. *Proc. Natl. Acad. Sci. USA* **2004**, *101*, 2040–2045. [CrossRef]
102. Koriyama, Y.; Nakayama, Y.; Matsugo, S.; Kato, S. Protective effect of lipoic acid against oxidative stress is mediated by Keap1/Nrf2-dependent heme oxygenase-1 induction in the RGC-5 cellline. *Brain Res.* **2013**, *1499*, 145–157. [CrossRef]

103. Kyung, S.; Lim, J.W.; Kim, H. α-Lipoic Acid Inhibits IL-8 Expression by Activating Nrf2 Signaling in Helicobacter pylori-infected Gastric Epithelial Cells. *Nutrients* **2019**, *11*, 2524. [CrossRef]
104. Kensler, T.W.; Egner, P.A.; Agyeman, A.S.; Visvanathan, K.; Groopman, J.D.; Chen, J.-G.; Chen, T.-Y.; Fahey, J.W.; Talalay, P. Keap1-nrf2 signaling: A target for cancer prevention by sulforaphane. *Nat. Prod. Cancer Prev. Ther.* **2013**, *329*, 163–177.
105. Salminen, A.; Kauppinen, A.; Suuronen, T.; Kaarniranta, K. SIRT1 longevity factor suppresses NF-kappaB -driven immune responses: Regulation of aging via NF-kappaB acetylation? *Bioessays* **2008**, *30*, 939–942. [CrossRef]
106. Fruchart, J.C.; Duriez, P. Mode of action of fibrates in the regulation of triglyceride and HDL-cholesterol metabolism. *Drugs Today* **2006**, *42*, 39–64.
107. Jia, Y.; Kim, J.-Y.; Jun, H.-J.; Kim, S.-J.; Lee, J.-H.; Hoang, M.H.; Hwang, K.-Y.; Um, S.-J.; Chang, H.I.; Lee, S.-J. The natural carotenoid astaxanthin, a PPAR-α agonist and PPAR-γ antagonist, reduces hepatic lipid accumulation by rewiring the transcriptome in lipid-loaded hepatocytes. *Mol. Nutr. Food Res.* **2012**, *56*, 878–888. [CrossRef]
108. Jia, Y.; Wu, C.; Kim, J.; Kim, B.; Lee, S.J. Astaxanthin reduces hepatic lipid accumulations in high-fat-fed C57BL/6J mice via activation of peroxisome proliferator-activated receptor (PPAR) alpha and inhibition of PPAR gamma and Akt. *J. Nutr. Biochem.* **2016**, *28*, 9–18. [CrossRef] [PubMed]
109. Mashhadi, N.S.; Zakerkish, M.; Mohammadiasl, J.; Zarei, M.; Mohammadshahi, M.; Haghighizadeh, M.H. Astaxanthin improves glucose metabolism and reduces blood pressure in patients with type 2 diabetes mellitus. *Asia Pac. J. Clin. Nutr.* **2018**, *27*, 341–346. [PubMed]
110. Yoshida, H.; Yanai, H.; Ito, K.; Tomono, Y.; Koikeda, T.; Tsukahara, H.; Tada, N. Administration of natural astaxanthin increases serum HDL-cholesterol and adiponectin in subjects with mild hyperlipidemia. *Atherosclerosis* **2009**, *209*, 520–523. [CrossRef] [PubMed]
111. Choi, H.D.; Youn, Y.K.; Shin, W.G. Positive effects of astaxanthin on lipid profiles and oxidative stress in overweight subjects. *Plant Foods Hum. Nutr.* **2011**, *66*, 363–369. [CrossRef]
112. Krestinina, O.; Baburina, Y.; Krestinin, R.; Odinokova, I.; Fadeeva, I.; Sotnikova, L. Astaxanthin Prevents Mitochondrial Impairment Induced by Isoproterenol in Isolated Rat Heart Mitochondria. *Antioxidants* **2020**, *9*, 262. [CrossRef]
113. Kim, S.H.; Kim, H. Inhibitory Effect of Astaxanthin on Oxidative Stress-Induced Mitochondrial Dysfunction-A Mini-Review. *Nutrients* **2018**, *10*, 1137. [CrossRef]
114. Krestinina, O.; Baburina, Y.; Krestinin, R. Mitochondrion as a Target of Astaxanthin Therapy in Heart Failure. *Int. J. Mol. Sci.* **2021**, *22*, 7964. [CrossRef]
115. Lim, S.; Park, S.; Koyanagi, A.; Yang, J.W.; Jacob, L.; Yon, D.K.; Lee, S.W.; Kim, M.S.; Shin, J.I.; Smith, L. Effects of exogenous melatonin supplementation on health outcomes: An umbrella review of meta-analyses based on randomized controlled trials. *Pharmacol. Res.* **2022**, *176*, 106052. [CrossRef]
116. Jeffrey, S.; Samraj, P.I.; Raj, B.S. The Role of Alpha-lipoic Acid Supplementation in the Prevention of Diabetes Complications: A Comprehensive Review of Clinical Trials. *Curr. Diabetes Rev.* **2021**, *17*, e011821190404. [CrossRef]
117. Lem, D.W.; Davey, P.G.; Gierhart, D.L.; Rosen, R.B. A Systematic Review of Carotenoids in the Management of Age-Related Macular Degeneration. *Antioxidants* **2021**, *10*, 1255. [CrossRef]
118. Fukuda, S.; Varshney, A.; Fowler, B.J.; Wang, S.-B.; Narendran, S.; Ambati, K.; Yasuma, T.; Magagnoli, J.; Leung, H.; Hirahara, S.; et al. Cytoplasmic synthesis of endogenous Alu complementary DNA via reverse transcription and implications in age-related macular degeneration. *Proc. Natl. Acad. Sci. USA* **2021**, *118*, e2022751118. [CrossRef]
119. Koczor, C.A.; Jiao, Z.; Fields, E.; Russ, R.; Ludaway, T.; Lewis, W. AZT-induced mitochondrial toxicity: An epigenetic paradigm for dysregulation of gene expression through mitochondrial oxidative stress. *Physiol. Genom.* **2015**, *47*, 447–454. [CrossRef]
120. Wright, C.B.; Uehara, H.; Kim, Y.; Yasuma, T.; Yasuma, R.; Hirahara, S.; Makin, R.D.; Apicella, I.; Pereira, F.; Nagasaka, Y.; et al. Chronic Dicer1 deficiency promotes atrophic and neovascular outer retinal pathologies in mice. *Proc. Natl. Acad. Sci. USA* **2020**, *117*, 2579–2587. [CrossRef]
121. Chan, E.C.; van Wijngaarden, P.; Liu, G.S.; Jiang, F.; Peshavariya, H.; Dusting, G.J. Involvement of Nox2 NADPH oxidase in retinal neovascularization. *Investig. Ophthalmol. Vis. Sci.* **2013**, *54*, 7061–7067. [CrossRef] [PubMed]
122. Chan, E.C.; Liu, G.S.; Dusting, G.J. Redox mechanisms in pathological angiogenesis in the retina: Roles for NADPH oxidase. *Curr. Pharm. Des.* **2015**, *21*, 5988–5998. [CrossRef] [PubMed]
123. Tang, X.; Wang, J.J.; Wang, J.; Abboud, H.E.; Chen, Y.; Zhang, S.X. Endothelium-specific deletion of Nox4 delays retinal vascular development and mitigates pathological angiogenesis. *Angiogenesis* **2021**, *24*, 363–377. [CrossRef] [PubMed]
124. Kim, Y.M.; Kim, S.J.; Tatsunami, R.; Yamamura, H.; Fukai, T.; Ushio-Fukai, M. ROS-induced ROS release orchestrated by Nox4, Nox2, and mitochondria in VEGF signaling and angiogenesis. *Am. J. Physiol. Cell Physiol.* **2017**, *312*, C749–C764. [CrossRef]
125. McCarty, M.F. Clinical potential of Spirulina as a source of phycocyanobilin. *J. Med. Food* **2007**, *10*, 566–570. [CrossRef]
126. Zheng, J.; Inoguchi, T.; Sasaki, S.; Maeda, Y.; McCarty, M.F.; Fujii, M.; Ikeda, N.; Kobayashi, K.; Sonoda, N.; Takayanagi, R. Phycocyanin and phycocyanobilin from Spirulina platensis protect against diabetic nephropathy by inhibiting oxidative stress. *Am. J. Physiol. Regul. Integr. Comp. Physiol.* **2013**, *304*, R110–R120. [CrossRef]
127. Rose, M.L.; Madren, J.; Bunzendahl, H.; Thurman, R.G. Dietary glycine inhibits the growth of B16 melanoma tumors in mice. *Carcinogenesis* **1999**, *20*, 793–798. [CrossRef]
128. Yamashina, S.; Ikejima, K.; Rusyn, I.; Sato, N. Glycine as a potent anti-angiogenic nutrient for tumor growth. *J. Gastroenterol. Hepatol.* **2007**, *22* (Suppl. S1), S62–S64. [CrossRef]

129. Bruns, H.; Kazanavicius, D.; Schultze, D.; Saeedi, M.A.; Yamanaka, K.; Strupas, K.; Schemmer, P. Glycine inhibits angiogenesis in colorectal cancer: Role of endothelial cells. *Amino Acids* **2016**, *48*, 2549–2558. [CrossRef]
130. Amin, K.; Li, J.; Chao, W.R.; Dewhirst, M.W.; Haroon, Z.A. Dietary glycine inhibits angiogenesis during wound healing and tumor growth. *Cancer Biol. Ther.* **2003**, *2*, 173–178. [CrossRef]
131. McCarty, M.F.; Iloki-Assanga, S.; Lujan, L.M.L.; DiNicolantonio, J.J. Activated glycine receptors may decrease endosomal NADPH oxidase activity by opposing ClC-3-mediated efflux of chloride from endosomes. *Med. Hypotheses* **2019**, *123*, 125–129. [CrossRef]

nutrients

MDPI

Article

Extra Virgin Olive Oil Reduces Gut Permeability and Metabolic Endotoxemia in Diabetic Patients

Simona Bartimoccia [1], Vittoria Cammisotto [2], Cristina Nocella [2], Maria Del Ben [2], Alessandra D'Amico [3], Valentina Castellani [4], Francesco Baratta [2], Pasquale Pignatelli [2,5], Lorenzo Loffredo [2], Francesco Violi [5,*] and Roberto Carnevale [1,5]

1 Department of Medical-Surgical Sciences and Biotechnologies, Sapienza University of Rome, Corso della Repubblica 79, 40100 Latina, Italy; simona.bartimoccia@uniroma1.it (S.B.); roberto.carnevale@uniroma1.it (R.C.)
2 Department of Clinical Internal, Anaesthesiological and Cardiovascular Sciences, Sapienza University of Rome, Viale del Policlinico, 155, 00161 Rome, Italy; vittoria.cammisotto@uniroma1.it (V.C.); cristina.nocella@uniroma1.it (C.N.); maria.delben@uniroma1.it (M.D.B.); farmcesco.baratta@uniroma1.it (F.B.); pasquale.pignatelli@uniroma1.it (P.P.); lorenzo.loffredo@uniroma1.it (L.L.)
3 Department of Movement, Human and Health Sciences, University of Rome "Foro Italico", 00135 Rome, Italy; a.damico@studenti.uniroma4.it
4 Department of General Surgery and Surgical Speciality Paride Stefanini, Sapienza University of Rome, Viale del Policlinico, 155, 00161 Rome, Italy; valentina.castellani@uniroma1.it
5 Mediterranea Cardiocentro-Napoli, Via Orazio, 2, 80122 Naples, Italy
* Correspondence: francesco.violi@uniroma1.it; Tel.: +39-064461933; Fax: +39-0649970103

Abstract: Background: Extra virgin olive oil (EVOO) improves post-prandial glycemia, but the underlying mechanism has not been fully elucidated. We tested the hypothesis that EVOO improves post-prandial glycemia by reducing gut permeability-derived low-grade endotoxemia. Methods: Serum levels of lipopolysaccharides (LPS), zonulin, a marker of gut permeability, glucose, insulin and glucagon-like peptide 1 (GLP1) were measured in 20 patients with impaired fasting glucose (IFG) and 20 healthy subjects (HS) matched for sex and age. The same variables were measured in IFG patients (n = 20) and HS (n = 20) before and after a Mediterranean diet with 10 g EVOO added or not (n = 20) or in IFG patients (n = 20) before and after intake of 40 g chocolate with EVOO added or not. Results: Compared to HS, IFG had higher levels of LPS and zonulin. In HS, meal intake was associated with a significant increase of blood glucose, insulin, and GLP1 with no changes of blood LPS and zonulin. Two hours after a meal intake containing EVOO, IFG patients showed a less significant increase of blood glucose, a more marked increase of blood insulin and GLP1 and a significant reduction of LPS and zonulin compared to IFG patients not given EVOO. Correlation analysis showed that LPS directly correlated with blood glucose and zonulin and inversely with blood insulin. Similar findings were detected in IFG patients given a chocolate added or without EVOO. Conclusion: Addition of EVOO to a Mediterranean diet or chocolate improves gut permeability and low-grade endotoxemia.

Keywords: lipopolysaccharides (LPS); gut permeability; zonulin; Glucagon-like peptide1(GLP1); extra virgin olive oil (EVOO); oleuropein; Impaired fasting glucose (IFG)

Citation: Bartimoccia, S.; Cammisotto, V.; Nocella, C.; Del Ben, M.; D'Amico, A.; Castellani, V.; Baratta, F.; Pignatelli, P.; Loffredo, L.; Violi, F.; et al. Extra Virgin Olive Oil Reduces Gut Permeability and Metabolic Endotoxemia in Diabetic Patients. *Nutrients* **2022**, *14*, 2153. https://doi.org/10.3390/nu14102153

Academic Editor: Sareen Gropper

Received: 22 April 2022
Accepted: 19 May 2022
Published: 21 May 2022

1. Introduction

Circulating lipopolysaccharides (LPS), a membrane component of Gram-negative gut microbiota, may be detectable in human circulation in concentrations ranging from as low as 1 pg/mL to as high as 200 pg/mL [1]. In non-septic conditions, low-grade endotoxemia by LPS may be detectable in patients at risk of cardiovascular events such as patients with obesity or type 2 diabetes mellitus (T2DM) or in patients with clinically overt vascular disease [2,3]. In T2DM, the term "metabolic endotoxemia" has been coined to describe the association between low-grade endotoxemia and metabolic changes such as insulin resistance, hyperglycemia and lipid metabolism changes favoring obesity [4].

Metabolic endotoxemia has been suggested to occur as a consequence of enhanced gut permeability, which depends on transcellular or paracellular LPS translocation into systemic circulation [5]. Thus, modulation of gut permeability may have positive effects on glycemic control in T2DM. We and others have previously reported that in diabetic patients, intake of EVOO or its component oleuropein lowered post-prandial LPS and glycaemia, an effect of potentially clinical relevance, as post-prandial glycemia may increase the risk of cardiovascular disease [6–9]. However, we did not investigate the relationship, if any, between gut permeability and glycaemic control and if EVOO could improve the glycaemic profile by modulating gut permeability. EVOO is composed of 98–98.5% triglycerides [6] and 1.5–2% of minor components such as sterols [7], fatty alcohols [7], waxes [8], phenols, tocopherols, and carotenoids [9]; furthermore, EVOO is rich in polyphenols and vitamin E [10], which could exert a beneficial effect against cardiovascular events [11,12] and cancer [13], even if further randomized clinical studies are necessary to confirm these hypotheses. The antioxidant property of EVOO could turn useful in the context of gut permeability, which is, in fact, enhanced by oxidative stress-mediated zonulin activation and eventually down-regulation of adhesive proteins such as tight junctions [14]. Therefore, we hypothesized that acute administration of EVOO could improve metabolic endotoxemia, i.e., metabolic profile and endotoxemia by LSP, by affecting gut permeability as assessed by serum zonulin, an indirect marker of gut permeability [15]. To explore this issue, we compared the parallel changes of LPS, glycaemia, and serum zonulin in patients with impaired fasting glucose (IFG) and healthy subjects (HS).

2. Materials and Methods

2.1. Study Design and Participants

We performed a cross-sectional analysis of variables exploring gut permeability, low-grade endotoxemia by LPS and metabolic profile in HS (n = 20, 11 males, 9 females, aged 47.05 ± 6.41 years) and in patients with IFG (n = 20, 12 males, 8 females, aged 51.54 ± 8.02) (study 1). According to the American Diabetes Association guidelines [16], IFG was defined as a fasting blood sugar glucose concentration ≥100 and <126 mg/dL.

Additionally, we performed a secondary analysis of two previous reports in which EVOO was added to or not from a Mediterranean-type meal (studies 2 and 3) or to 40 g chocolate (study 4) [17–19].

Study 2: In this study, we analyzed 20 healthy subjects (HS; 11 males and 9 females) who received a typical Mediterranean lunch including or not 10 g of EVOO in a cross-over design; there was an interval of 30 days between the two phases of the study. Clinical and demographic characteristics of HS are reported in Table 1.

Table 1. Clinical characteristics of the study population.

	HS (n = 20)	IFG (n = 20)	p Value
Age (years)	47.05 ± 6.41	51.54 ± 8.02	0.057
Males n (%)	11 (55)	12 (60)	0.757
BMI (kg/m^2)	27.50 ± 3.91	28.79 ± 3.52	0.270
Systolic BP (mmHg)	121.40 ± 8.52	127 ± 11.29	0.085
Diastolic BP (mmHg)	76.85 ± 4.29	80.00 ± 6.28	0.072
Smokers n (%)	4 (20)	2 (10)	0.388

EVOO was provided by monoculture (Itri area, Latina) and its chemical characterization is reported in Supplementary Table S1.

Study 3: In this study, we performed an interventional study in 20 IFG patients (12 males and 8 females); clinical and demographic characteristics of IFG are reported in Table 1. We randomized 20 IFG to receive a typical Mediterranean lunch including or not 10 g of EVOO in a cross-over design; there was an interval of at least seven days between the two phases of the study. The Mediterranean-type lunch composition is reported in Supplementary Table S2. Briefly, the Mediterranean-type meal for both studies 2 and 3 consisted of pasta (g 100), chicken breast (150 g), salad (80 g), bread (80 g), and apple (200 g) for a total of 894 calories; blood analyses were performed before and 2 h after meal intake.

Study 4: We analyzed the acute effect of chocolate intake in a single blind, crossover study including IFG patients ($n = 20$), randomized to receive 40 g of chocolate spread mixed or not with EVOO. Blood analyses were carried out before and 2 h after the intake of 40 g chocolate (18.5% hazelnuts, 40% of cocoa, sugar, whole milk powder to reach 27% vitamin E over 100 g carbohydrates); the control chocolate had the same characteristics but without EVOO addition. Every blood determination was performed blind. None of the participants were receiving antioxidants supplements, statins, or dietary restrictions prior to the study. The study conformed to the ethical guidelines of the 1975 Declaration of Helsinki and was approved by the Ethical Committee of Sapienza University (Rif. N° 509/2016).

2.2. Serum Glucose and Insulin Assay

Glucose and insulin were analyzed in serum samples by a commercial ELISA Kit (Arbor Assay (Ann Arbor, MI, USA) and DRG International (Springfield, NJ, USA)). Glucose values were expressed as mg/dl and intra-assay and inter-assay coefficients of variation were 6% and 9%, respectively. Insulin values were expressed as μU/mL and intra-assay and inter-assay coefficients of variation were 2.2% and 4.5%, respectively.

2.3. Serum Glucagon Like Peptide-1 (GLP1) Assay

A commercial ELISA Kit (DRG International) was used for the quantitative determination of bioactive GLP1 (7e36) and (9e36) levels in serum. GLP-1 values are expressed as pmol/L and both intra- and inter-assay coefficients of variation were <10%.

2.4. Serum LPS Assay

LPS were measured in serum using a commercial ELISA kit (Cusabio, Wuhan, China). The standards and samples were plated for 2 h at room temperature into a micro-plate pre-coated with the antibody specific for LPS. After incubation, samples were read at 450 nm. Values were expressed as pg/mL; intra-assay and inter-assay coefficients of variation were <10%.

2.5. Serum Zonulin Assay

Zonulin concentration was detected in serum by a commercial ELISA kit (Elabscience Houston, TX, USA). Values were expressed as ng/mL; both intra-assay and inter-assay coefficients of variation were within 10%.

2.6. Statistical Methods

Categorical variables are reported as percentage and continuous variables as means ± SD unless otherwise indicated. Independence of categorical variables was tested by the chi-square test. Comparisons between groups were analyzed by Student's t-test and were replicated as appropriate with nonparametric tests (Kolmogorov–Smirnov (z) test in case of non-homogeneous variances as verified by Levene's test).

The cross-over study data were analysed for the assessment of treatment and period effects, by performing a split-plot ANOVA with one between-subject factor (treatment sequence) and two within-subject factors (pre- vs. post-treatment). The full model was considered, allowing for the assessment of all main effects and interactions. Pairwise comparisons were corrected by the Bonferroni test; results were expressed as means ± SD. Bivariate analysis was calculated by the Spearman rank correlation test. A value of $p < 0.05$ was considered statistically significant. All analyses were carried out with GraphPad Prism9.1.0 and IBM SPSS 25.0.

3. Results

Clinical characteristics of the population are reported in Table 1.

At baseline, patients with IFG had higher levels of blood glucose, insulin and GLP-1 compared to HS (Figure 1a–c); furthermore, patients with IFG had higher blood levels of LPS and zonulin compared to HS (Figure 1d,e).

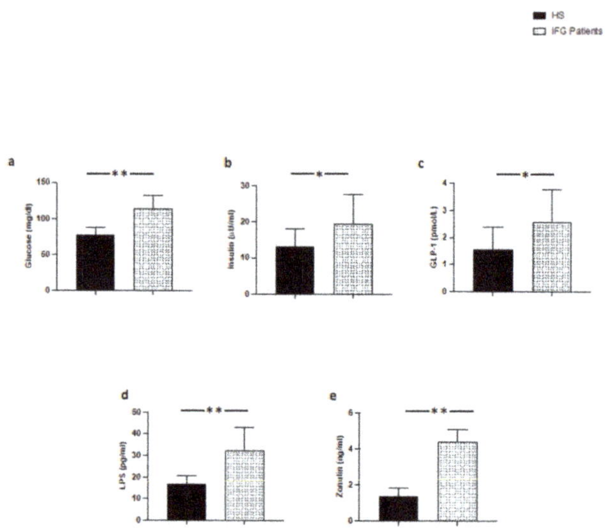

Figure 1. Serum levels of (**a**) glucose, (**b**) insulin, (**c**) GLP-1, (**d**) LPS, and (**e**) zonulin in patients with IFG (n = 20) and control subjects (n = 20); ** $p < 0.001$; * $p < 0.01$.

As previously reported [17] (Study 2), in HS meal intake was associated with a significant increase of blood glycaemia, insulin, and GLP-1 (Figure 2a–c). A trend of an increase of blood LPS and zonulin was detected but the difference was not significant (Figure 2d,e). LPS did not correlate with blood glucose, insulin, and zonulin; zonulin was not correlated with GLP-1 (Figure 2f–i).

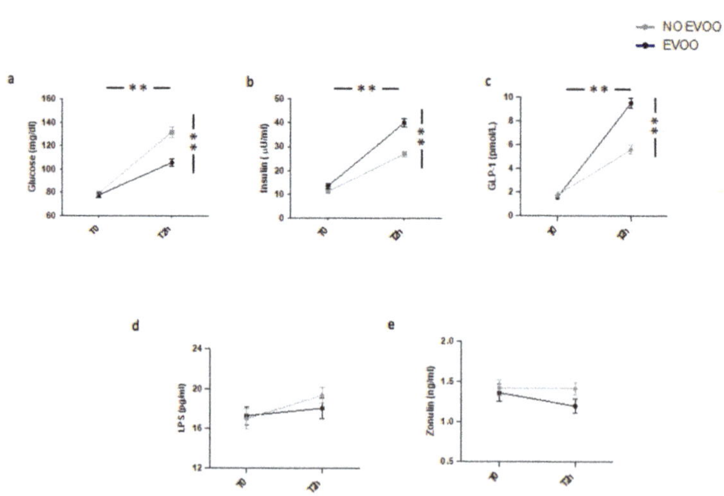

Figure 2. *Cont.*

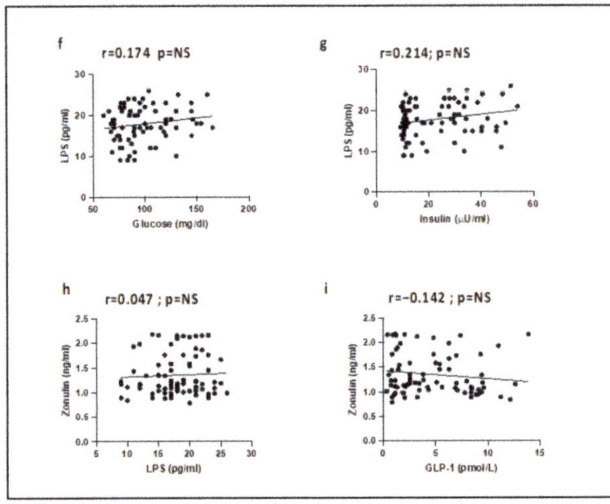

Figure 2. Serum levels of (**a**) glucose, (**b**) insulin, (**c**) GLP-1, (**d**) LPS, and (**e**) zonulin before (T0) and 2 h after a meal with (black line) or without (grey line) extra-virgin olive oil (EVOO) in HS ($n = 20$). Correlations of circulating LPS levels with (**f**) glucose, (**g**) insulin, and (**h**) zonulin in HS. Correlations of circulating Zonulin levels with (**i**) GLP-1 in HS. ** $p < 0.001$.

As previously reported [18] (Study 3), two hours after a meal intake containing EVOO, IFG patients showed a less significant increase of blood glucose and a more marked increase of blood insulin and GLP-1 (Figure 3a–c) compared to IFG not given EVOO. Additionally, compared to IFG patients not given EVOO, a significant reduction of blood LPS and zonulin was detected (Figure 3d,e). Correlation analysis showed that LPS directly correlated with blood glucose and zonulin and inversely with blood insulin; blood zonulin inversely correlated with GLP-1 (Figure 3f–i).

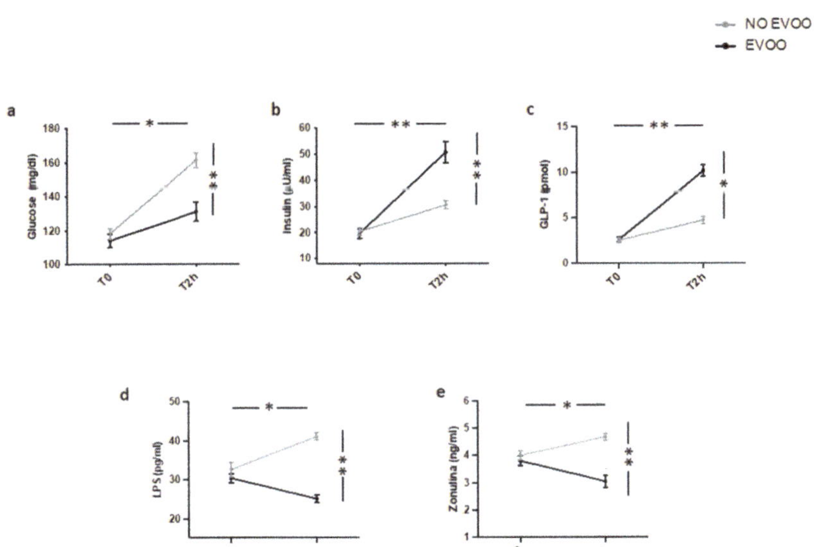

Figure 3. *Cont.*

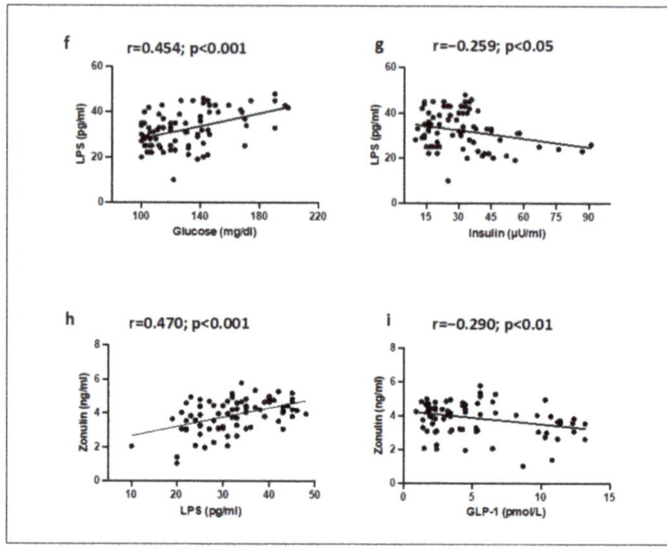

Figure 3. Serum levels of (**a**) glucose, (**b**) insulin, (**c**) GLP-1, (**d**) LPS, and (**e**) zonulin before (T0) and 2 h after (T2h) a meal with (black line) or without (grey line) extra-virgin olive oil (EVOO) in IFG patients (n = 20). Correlations of circulating LPS levels with (**f**) glucose, (**g**) insulin, and (**h**) zonulin in IFG. Correlations of circulating Zonulin levels with (i) GLP-1 in IFG. ** $p < 0.001$, * $p < 0.05$.

Similar modifications of metabolic profile, blood LPS, and zonulin as well as correlation analyses were observed in IFG patients (n = 20) taking chocolate with EVOO added compared to controls (Figure 4a–i).

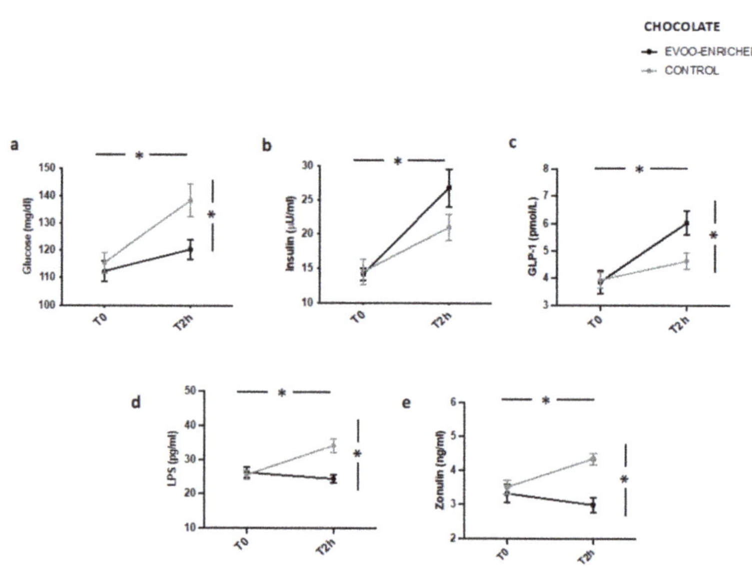

Figure 4. *Cont.*

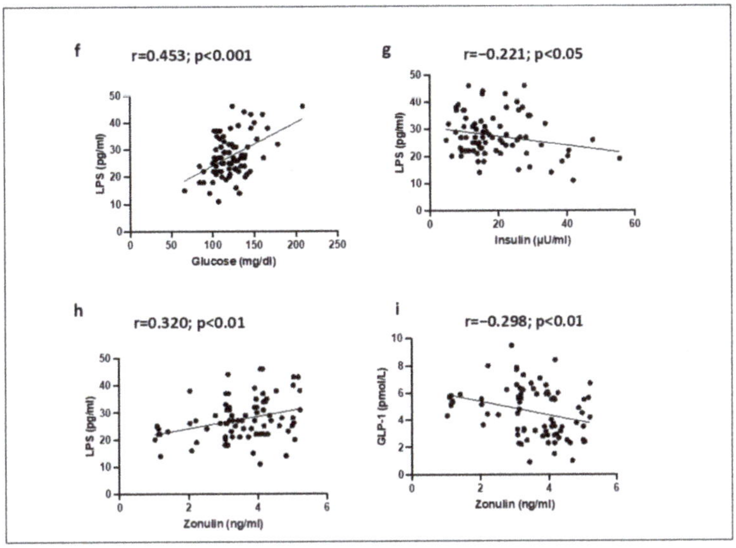

Figure 4. Serum levels of (**a**) glucose, (**b**) insulin, (**c**) GLP-1, (**d**) LPS, and (**e**) zonulin before (T0) and after 2 h (T2h) of oleuropein-enriched chocolate (black line) or control chocolate (grey line) in IFG patients ($n = 20$). Correlations of circulating LPS levels with (**f**) glucose, (**g**) insulin, and (**h**) zonulin in patients with IFG. Correlations of circulating zonulin levels with (**i**) GLP-1 in patients with * $p < 0.001$.

4. Discussion

The results of the present study demonstrate that, in IFG patients, EVOO counteracts LPS increases detected after meal or chocolate intake coincidentally with zonulin lowering, suggesting that this beneficial effect may be attributed to improvement of gut permeability.

Previous studies reported that patients with metabolic diseases such as those with T2DM or obesity display low-grade endotoxemia consequent to changes of proteins implicated in intestinal epithelial cell permeability, such as the tight junction (TJ) proteins ZO-1 or occludin [20]. In accordance with these reports, we found elevated levels of LPS in patients with IFG compared to controls coincidentally with zonulin elevation, which increases gut permeability by disassembling the TJ proteins [21]; this finding corroborates the hypothesis that T2DM is associated with metabolic endotoxemia consequent to changes of gut permeability [5,22]. Low-grade endotoxemia has a negative effect on metabolic profile as documented by an experimental study showing that LPS infusion increases glycaemia by negatively affecting insulin activity [5]; in accordance with this, we found that in IFG, but not in HS, LPS correlated directly with glycaemia and inversely with insulin. In two previous studies [21,23], we reported that a meal intake is associated with low-grade endotoxemia in IFG but not in HS but did not investigate the interplay between low-grade endotoxemia and the post-prandial metabolic profile or the relationship between post-prandial LPS and gut permeability. While the inverse relationship between low-grade endotoxemia and metabolic profile confirms previous data in T2DM [4,21], here we report that in patients with IFG intake of a Mediterranean-type meal or chocolate is associated with a coincident increase of LPS and zonulin, suggesting that changes of gut permeability may account for post-prandial low-grade endotoxemia. This finding would suggest, therefore, that in addition to LPS translocation to systemic circulation via chylomicron biosynthesis [24], post-prandial low-grade endotoxemia occurs via a change of gut permeability.

In two separate studies we also reported that EVOO lowers low-grade endotoxemia, but the underlying mechanism was not investigated [21,23]; furthermore, we did not analyze if the positive effect of EVOO on the metabolic profile could be related to a change of low-grade endotoxemia. The experiments conducted that adding EVOO to a Mediterranean-type meal or to chocolate showed a coincident decrease of LPS and zonulin, indicating that EVOO has a positive effect on gut permeability. In order to investigate the underlying mechanism, we focused on GLP-1, which is increased by EVOO administration [9,16,17], and is suggested to be implicated in the up-regulation of TJ proteins [23]; of note, we found a significant relationship between post-prandial increase of GLP-1 and the decrease of LPS and zonulin, suggesting GLP-1-mediated up-regulation of TJ proteins as a potential mechanism accounting for EVOO-related improvement of gut permeability. These data extend our previous report in this field, indicating that the positive effect of EVOO on metabolic profile may be related to changes of gut permeability and in turn to reduction of low-grade endotoxemia. In accordance with this, a coincident post-prandial LPS and glycaemia lowering along with an inverse relationship between LPS and insulin were detected.

The study has implications and limitations. We must acknowledge that serum zonulin is an indirect marker of gut permeability, and thereby further study is necessary to assess the relationship between EVOO intake and gut permeability. We cannot exclude that zonulin may also be implicated in LPS translocation to systemic circulation via mechanisms not related to TJ disassembling. Additionally, we cannot exclude that EVOO may lower circulating LPS by interfering with chylomicron biosynthesis. The study has been carried out in a single center and in Caucasian patients; further study is, therefore, necessary to support the present findings. The study has been conducted in acute conditions, and further study should be carried out to assess the effect of chronic EVOO on gut permeability and low-grade endotoxemia. Another limitation of the study is represented by the lack of further reference points (beyond 2 h from baseline values) that could have better described the post-prandial time course of the variables (glucose, insulin, GLP1, LPS, and zonulin).

5. Conclusions

The study reports that EVOO added to a Mediterranean-type diet or chocolate blunts the increase of glycaemia by interfering with markers of gut permeability and low-grade endotoxemia in IFG. Further study is necessary to assess if a similar effect can be detected by chronic administration of EVOO.

Supplementary Materials: The following supporting information can be downloaded at: https://www.mdpi.com/article/10.3390/nu14102153/s1, Table S1: Vitamin E and total polyphenols content in EVOO; Table S2: Meal composition.

Author Contributions: Conceptualization, F.V. and R.C.; study design, F.V. and R.C.; methodology, S.B., V.C. (Valentina Castellani), A.D., M.D.B., F.B.; validation, S.B., V.C. (Vittoria Cammisotto) and C.N.; investigation, S.B., V.C. (Vittoria Cammisotto) and C.N.; data curation, S.B.; formal analysis, L.L.; M.D.B. and F.B.; writing—original draft preparation, S.B.; R.C. and F.V.; writing—review and editing, P.P. and F.V.; supervision, P.P., R.C. and F.V. All authors have read and agreed to the published version of the manuscript.

Funding: This research received no external funding.

Institutional Review Board Statement: The study conformed to the ethical guidelines of the 1975 Declaration of Helsinki and was approved by the Ethical Committee of Sapienza University (Rif. N° 509/2016).

Informed Consent Statement: Informed consent was obtained from all subjects involved in the study.

Data Availability Statement: Not applicable.

Acknowledgments: Extra virgin olive oil (Itrano) was kindly provided by Antonio Zangrilli.

Conflicts of Interest: F.V. is the inventor of ChocOliv®, the mixture of EVOO with chocolate used in the present study to blunt post-prandial glycemia in IFG patients.

References

1. Saad, M.J.; Santos, A.; Prada, P.O. Linking Gut Microbiota and Inflammation to Obesity and Insulin Resistance. *Physiology* **2016**, *31*, 283–293. [CrossRef] [PubMed]
2. Gomes, J.M.G.; Costa, J.A.; Alfenas, R.C.G. Metabolic endotoxemia and diabetes mellitus: A systematic review. *Metabolism* **2017**, *68*, 133–144. [CrossRef] [PubMed]
3. Carnevale, R.; Sciarretta, S.; Valenti, V.; di Nonno, F.; Calvieri, C.; Nocella, C.; Frati, G.; Forte, M.; d'Amati, G.; Pignataro, M.G.; et al. Low-grade endotoxaemia enhances artery thrombus growth via Toll-like receptor 4: Implication for myocardial infarction. *Eur. Heart J.* **2020**, *41*, 3156–3165. [CrossRef] [PubMed]
4. Scheithauer, T.P.M.; Rampanelli, E.; Nieuwdorp, M.; Vallance, B.A.; Verchere, C.B.; van Raalte, D.H.; Herrema, H. Gut Microbiota as a Trigger for Metabolic Inflammation in Obesity and Type 2 Diabetes. *Front. Immunol.* **2020**, *11*, 571731. [CrossRef] [PubMed]
5. Cani, P.D.; Amar, J.; Iglesias, M.A.; Poggi, M.; Knauf, C.; Bastelica, D.; Neyrinck, A.M.; Fava, F.; Tuohy, K.M.; Chabo, C.; et al. Metabolic endotoxemia initiates obesity and insulin resistance. *Diabetes* **2007**, *56*, 1761–1772. [CrossRef]
6. Jimenez-Lopez, C.; Carpena, M.; Lourenco-Lopes, C.; Gallardo-Gomez, M.; Lorenzo, J.M.; Barba, F.J.; Prieto, M.A.; Simal-Gandara, J. Bioactive Compounds and Quality of Extra Virgin Olive Oil. *Foods* **2020**, *9*, 1014. [CrossRef]
7. Grob, K.; Lanfranchi, M.; Mariani, C. Evaluation of Olive Oils through the Fatty Alcohols, the Sterols and Their Esters by Coupled Lc-Gc. *J. Am. Oil Chem. Soc.* **1990**, *67*, 626–634. [CrossRef]
8. Hodaifa, G.; Martinez Nieto, L.; Lozano, J.L.; Sanchez, S. Changes of the wax contents in mixtures of olive oils as determined by gas chromatography with a flame ionization detector. *J. AOAC Int.* **2012**, *95*, 1720–1724. [CrossRef]
9. Mousavi, S.; Mariotti, R.; Stanzione, V.; Pandolfi, S.; Mastio, V.; Baldoni, L.; Cultrera, N.G.M. Evolution of Extra Virgin Olive Oil Quality under Different Storage Conditions. *Foods* **2021**, *10*, 1945. [CrossRef]
10. Carnevale, R.; Nocella, C.; Cammisotto, V.; Bartimoccia, S.; Monticolo, R.; D'Amico, A.; Stefanini, L.; Pagano, F.; Pastori, D.; Cangemi, R.; et al. Antioxidant activity from extra virgin olive oil via inhibition of hydrogen peroxide-mediated NADPH-oxidase 2 activation. *Nutrition* **2018**, *55*, 36–40. [CrossRef]
11. Violi, F.; Nocella, C.; Loffredo, L.; Carnevale, R.; Pignatelli, P. Interventional study with vitamin E in cardiovascular disease and meta-analysis. *Free Radic. Biol. Med.* **2022**, *178*, 26–41. [CrossRef] [PubMed]
12. Estruch, R.; Ros, E.; Salas-Salvado, J.; Covas, M.I.; Corella, D.; Aros, F.; Gomez-Gracia, E.; Ruiz-Gutierrez, V.; Fiol, M.; Lapetra, J.; et al. Primary prevention of cardiovascular disease with a Mediterranean diet. *N. Engl. J. Med.* **2013**, *368*, 1279–1290. [CrossRef] [PubMed]
13. Emma, M.R.; Augello, G.; Di Stefano, V.; Azzolina, A.; Giannitrapani, L.; Montalto, G.; Cervello, M.; Cusimano, A. Potential Uses of Olive Oil Secoiridoids for the Prevention and Treatment of Cancer: A Narrative Review of Preclinical Studies. *Int. J. Mol. Sci.* **2021**, *22*, 1234. [CrossRef] [PubMed]
14. Millman, J.F.; Okamoto, S.; Teruya, T.; Uema, T.; Ikematsu, S.; Shimabukuro, M.; Masuzaki, H. Extra-virgin olive oil and the gut-brain axis: Influence on gut microbiota, mucosal immunity, and cardiometabolic and cognitive health. *Nutr. Rev.* **2021**, *79*, 1362–1374. [CrossRef]
15. Fasano, A. Intestinal permeability and its regulation by zonulin: Diagnostic and therapeutic implications. *Clin. Gastroenterol. Hepatol.* **2012**, *10*, 1096–1100. [CrossRef] [PubMed]
16. American Diabetes Association. 2. Classification and Diagnosis of Diabetes: Standards of Medical Care in Diabetes-2018. *Diabetes Care* **2018**, *41*, S13–S27. [CrossRef]
17. Violi, F.; Loffredo, L.; Pignatelli, P.; Angelico, F.; Bartimoccia, S.; Nocella, C.; Cangemi, R.; Petruccioli, A.; Monticolo, R.; Pastori, D.; et al. Extra virgin olive oil use is associated with improved post-prandial blood glucose and LDL cholesterol in healthy subjects. *Nutr. Diabetes.* **2015**, *5*, e172. [CrossRef]
18. Carnevale, R.; Loffredo, L.; Del Ben, M.; Angelico, F.; Nocella, C.; Petruccioli, A.; Bartimoccia, S.; Monticolo, R.; Cava, E.; Violi, F. Extra virgin olive oil improves post-prandial glycemic and lipid profile in patients with impaired fasting glucose. *Clin. Nutr.* **2017**, *36*, 782–787. [CrossRef]
19. Del Ben, M.; Nocella, C.; Loffredo, L.; Bartimoccia, S.; Cammisotto, V.; Mancinella, M.; Angelico, F.; Valenti, V.; Cavarretta, E.; Carnevale, R.; et al. Oleuropein-enriched chocolate by extra virgin olive oil blunts hyperglycaemia in diabetic patients: Results from a one-time 2-hour post-prandial cross over study. *Clin. Nutr.* **2020**, *39*, 2187–2191. [CrossRef]
20. Ghosh, S.S.; Wang, J.; Yannie, P.J.; Ghosh, S. Intestinal Barrier Dysfunction, LPS Translocation, and Disease Development. *J. Endocr. Soc.* **2020**, *4*, bvz039. [CrossRef]
21. Carnevale, R.; Pastori, D.; Nocella, C.; Cammisotto, V.; Baratta, F.; Del Ben, M.; Angelico, F.; Sciarretta, S.; Bartimoccia, S.; Novo, M.; et al. Low-grade endotoxemia, gut permeability and platelet activation in patients with impaired fasting glucose. *Nutr. Metab. Cardiovasc. Dis.* **2017**, *27*, 890–895. [CrossRef] [PubMed]
22. Cani, P.D.; Bibiloni, R.; Knauf, C.; Waget, A.; Neyrinck, A.M.; Delzenne, N.M.; Burcelin, R. Changes in gut microbiota control metabolic endotoxemia-induced inflammation in high-fat diet-induced obesity and diabetes in mice. *Diabetes* **2008**, *57*, 1470–1481. [CrossRef] [PubMed]

23. Carnevale, R.; Pastori, D.; Nocella, C.; Cammisotto, V.; Bartimoccia, S.; Novo, M.; Del Ben, M.; Farcomeni, A.; Angelico, F.; Violi, F. Gut-derived lipopolysaccharides increase post-prandial oxidative stress via Nox2 activation in patients with impaired fasting glucose tolerance: Effect of extra-virgin olive oil. *Eur. J. Nutr.* **2019**, *58*, 843–851. [CrossRef] [PubMed]
24. Ghoshal, S.; Witta, J.; Zhong, J.; de Villiers, W.; Eckhardt, E. Chylomicrons promote intestinal absorption of lipopolysaccharides. *J. Lipid Res.* **2009**, *50*, 90–97. [CrossRef]

nutrients

MDPI

Article

Habitual Diet Pattern Associations with Gut Microbiome Diversity and Composition: Results from a Chinese Adult Cohort

Yuhan Zhang [1], Hongda Chen [1,*], Ming Lu [1], Jie Cai [2], Bin Lu [1], Chenyu Luo [1] and Min Dai [1,*]

1 Medical Research Center, Peking Union Medical College Hospital, Chinese Academy of Medical Sciences and Peking Union Medical College, Beijing 100730, China; zyhmsf426@student.pumc.edu.cn (Y.Z.); minglu@student.pumc.edu.cn (M.L.); lubin838744@student.pumc.edu.cn (B.L.); luochy23@163.com (C.L.)

2 Department of General Surgery, Peking Union Medical College Hospital, Chinese Academy of Medical Sciences and Peking Union Medical College, Beijing 100730, China; caijie1113@126.com

* Correspondence: chenhongda@pumch.cn (H.C.); daimin@pumch.cn (M.D.); Tel.: +86-10-6915-4660 (H.C.); +86-10-6915-4651 (M.D.)

Abstract: The influence of long-term diet on gut microbiota is an active area of investigation. The present work aimed to explore the associations between habitual diet patterns and gut microbiota in a large sample of asymptomatic Chinese adults. The gut microbiome was profiled through the sequencing of the 16S rRNA gene in stool samples from 702 Chinese adults aged 50–75 years who underwent colonoscopies and were diagnosed to be free of colorectal neoplasm. Long-term dietary consumption was assessed through a food-frequency questionnaire. The microbial associations with specific food groups and the posteriori dietary pattern were tested using the Kruskal–Wallis H test, permutational ANOVAs, and multivariate analyses with linear models. The Shannon indexes generally shared similar levels across different food intake frequency groups. Whole grain and vegetable intakes totally explained 1.46% of the microbiota compositional variance. Using the data-driven posteriori approach, a general dietary pattern characterized by lower intakes of refined grains was highlighted to be associated with higher abundances of the genus *Anaerostipes* and a species of it. We also observed 17 associations between various food group intakes and specific genera and species. For instance, the relative abundances of the genus *Weissella* and an uncultured species of it were negatively associated with red meat intake. The results of this study support the idea that the usual dietary consumption measured by certain food items or summary indexes is associated with gut microbial features. These results deepen the understanding of complex relationships of diet and gut microbiota, as well as their implications for gut microbiome studies of human chronic diseases.

Keywords: gut microbiota; habitual diet; 16S rRNA gene sequencing; Chinese; adults

Citation: Zhang, Y.; Chen, H.; Lu, M.; Cai, J.; Lu, B.; Luo, C.; Dai, M. Habitual Diet Pattern Associations with Gut Microbiome Diversity and Composition: Results from a Chinese Adult Cohort. *Nutrients* **2022**, *14*, 2639. https://doi.org/10.3390/nu14132639

Academic Editor: Michael J. Barratt

Received: 15 May 2022
Accepted: 22 June 2022
Published: 25 June 2022

1. Introduction

The gut microbiota play a vital role in the host homeostasis maintenance, ranging from the catabolism and biosynthesis of essential nutrients to immune regulations and nerve signals transmission [1–3]. Pathological alterations of the gut microbiota community have been shown to be involved in the development of a wide spectrum of health disorders [4–7]. Various human lifestyle and physiological variables exert differential impacts on the gut microbiota throughout the life span, with environmental factors outweighing the genetic ones [2,8]. Among these environmental variables, including living behaviors, food habits, and medication, diet has been a primary research focus recently due to its diversity and easily modifiable properties. Food intake is increasingly considered as an intervention target for disease treatment and health promotion, and it has further evolved into a hot research area called precision nutrition [9–11]. Additionally, inter-individual heterogeneity in gut microbiota mainly arising from differences in personal physiological and lifestyle variables (such as diet) may confound microbiota analyses, resulting in spurious associations in gut microbiome studies of human diseases [12,13].

Short-term dietary changes such as the introduction of specific nutrients, foodstuffs, or special diet patterns can rapidly and significantly influence gut microbial profiles [14]. The observed transient effect supports diet's causal role in gut microbiome alterations while necessitating the study of dietary habits' impact on the gut microbiota in the long run because previous studies have either focused on single nutritional factor at a time or only substances without deleterious effects on humans [15]. Additionally, specific changes induced by short-term interventions generally do not persist due to their limited duration, whereas long-term dietary habits may dominantly drive gut microbiota composition [14,15]. Large-scale observational studies have accordingly investigated the associations between usual diet and gut microbiota composition, unveiling the relationships between food intakes and the gut microbiota profiles, as well as some particular dietary patterns. Studies have suggested that plant-rich food intakes are associated with a more diverse and compositionally distinct microbiota, as well as elevated abundances of specific bacterial taxa with a greater potential to produce short chain fatty acids (SCFAs), including fruits, fiber-rich breads, and vegetarian or Mediterranean diets [16,17]. By comparison, the Western diet and high intakes of animal protein have been reported to be associated with lower microbiome diversity and the enrichment of harmful bacteria [16,18].

However, previous studies mainly focused on either some particular food groups [19–21], such as fiber, red meat and processed meat, or on Western-population-oriented predefined diet quality scores, such as the Healthy Eating Index [13,22,23] and the Mediterranean Diet Score [13,23,24]. Habitual dietary variables are multidimensional, with internal correlations. Summary dietary indices can simplify complexity by quantifying dietary variance in a single measure and possibly offer a potential means of diet control in microbiota studies. In addition, caution should be taken in extrapolating findings from European and American populations to other ethnic groups. To our knowledge, only two published studies have specifically looked into this topic among Chinese populations. Yu et al. observed that the long-term diet quality was positively associated with fecal microbiome diversity and an abundance of fiber-fermenting bacteria among people lived in urban communities in a single region (Shanghai, China) [25]. Lu et al. provided a nationwide gut microbiota baseline of the Chinese population and knowledge on important environmental covariates, though with a sole focus on the dominant staple food type (including rice and wheat) [26]. There remains great uncertainty with respect to the long-term dietary habits related gut microbiome profiles fluctuations among Chinese people, especially those over 50 years old who are prone to chronic diseases with the potential participation of the gut microbiota.

The aim of the present study was to explore the dietary associations of the posteriori long-term diet pattern and habitual food intakes with gut microbiota composition in a large sample of asymptomatic individuals aged 50–75 years from six cities of China.

2. Methods

2.1. Study Participants

This study was based on the TARGET-C study initiated in May, 2018. The rationale, design, and protocol have been published and extensively described elsewhere [27–29]. Briefly, the primary objective of the TARGET-C study was to compare the effectiveness of the colonoscopy-based fecal immunochemical test (FIT) and risk-adapted triage screening strategies for colorectal cancer in China. Epidemiological data and biological samples collected during this study were also used for interested investigations, such as the work presented here. After obtaining signed informed consent, the eligible participants were randomly assigned into three groups to undergo colonoscopy, FIT, and risk-adapted colorectal cancer screening (i.e., the colorectal cancer risk assessment followed by FITs for the low-risk group or colonoscopies for the high-risk group). Patients who had positive FIT results were also required to undergo a subsequent colonoscopy. All participants undergoing colonoscopy were required to collect stool samples within 24 h prior to bowel preparation for colonoscopy. This study was approved by the Ethics Committee of the National Cancer Center/Cancer Hospital, Chinese Academy of Medical Sciences, and Peking Union Medical College (18-013/1615).

For the present study, we included participants who had no abnormal findings at screening colonoscopy and had available stool samples for microbiota sequencing. Exclusion criteria comprised a history of cancer and any current administration of anticoagulants, analgesics, and anti-rheumatic drugs. In addition, patients exhibiting abnormal abdominal symptoms, such as abdominal pain, diarrhea, constipation, and hematochezia, within 1 month before the colonoscopy examinations were excluded. More details of the participants' enrollment can be found in Supplementary Figure S1.

2.2. Stool Sample Collection

Eligible participants for colonoscopy were instructed to collect two stool samples at home prior to bowel preparation for the scheduled colonoscopy within 24 h. One was collected using the FIT stool collection device for extended microbiome analysis. Existing evidence suggests that feces collected by these devices are stable at room temperature and can be used for gut microbiota studies [30]. The stool-filled containers in storage boxes were delivered to a central laboratory and immediately frozen at −20 °C until DNA extraction. In this study, we used these stool samples for 16S rRNA sequencing. The other collected stool samples were kept in stool container tubes, then packaged in insulated boxes equipped with ice packs, and brought to the clinical sites on the days of the colonoscopies. On receipt, the fecal samples were frozen at −80 °C and subsequently transported by a cold chain to the central biobanks for further research.

2.3. DNA Extraction and 16S rRNA Gene Sequencing

DNA was extracted using the QIAamp Fast DNA Stool Mini Kit (QIAGEN). The V4 region of the microbial 16S rRNA gene was amplified and sequenced on the Illumina MiSeq sequencing platform. To avoid end-read sequencing errors, all reads were truncated at the 150th base and a median Q score of >20. Noisy sequences, chimeric sequences, and singletons were removed, and then amplicon sequence variants (ASVs) were inferred from the clean sequencing reads using the DADA2 pipeline built into Qiime2 [31]. Taxonomy was assigned to each ASV using the classify-sklearn classification methods via the q2-feature-classifier plugin built from the Greengenes database (release 13.8). To quantify the taxonomic composition, all sequences were rarefied to an even sampling depth of 10,000. Only the taxa and taxa present in at least 1% of the samples with an average relative abundance greater than 0.01% were included in the downstream analyses. Diversity metrics were calculated using the R package vegan, including α-diversity index and distance-based β-diversity. The relative abundances of each taxon were used in the following analyses.

2.4. Dietary Data Collection

Information about food intake during the past 12 months was collected through a food-frequency questionnaire (FFQ). Dietary data covered 9 major food groups in China: red meat (pork, beef, lamb, etc.), white meat (fish, chicken, duck, goose, etc.), eggs, dairy products, cooked meat (e.g., sausage), refined grains (rice, wheat, etc.), whole grains (millet, corn, sorghum, etc.), fresh fruits, and fresh vegetables. All these foods were examined with 5 frequency levels of habitual consumption (monthly or never/rarely, once a week, more than 1 time per week, daily, or more than 1 time per day) during the past 12 months. For analysis purposes, we transformed the frequency to times per week (i.e., 0, 1, 4, 7, and 14, respectively).

2.5. Dietary Pattern Analysis

Posteriori dietary patterns were derived from the 9 food groups using factor analysis with a principal component method. We applied a factor analysis with the principal component method to identify the major common factors. Orthogonal varimax rotation was performed to attain mutually independent structure with great interpretability. The optimal number of factors was determined by the scree plot examination of the true dataset compared to random "parallel" matrices, factor interpretability, and the variance explained (5%) by each factor. Finally, we chose the three-factor solution, totally explaining 50% of the whole variance of food intake frequencies (see Supplementary Figure S2 and Table S1). Using the k-means clustering method, we finally clustered the participants into 3 groups according to the weighted factor scores from the factor analysis. For more details, see Supplementary Method 1.

2.6. Statistical Analysis

Covariates, including sociodemographic variables (sex and age), lifestyle factors (cigarette smoking, alcohol drinking, and physical activity) and BMI (in kg/m^2) were adjusted in the diet–microbiome association analysis. Distributions of ASV-based alpha-diversity (including Shannon, richness, chao1, Simpson, Pielou, ACE, and faith_pd index) by different food intake frequency groups were compared using the Kruskal–Wallis H test. Associations between dietary variables and the β-diversity dissimilarities were evaluated using a permutational multivariate ANOVA (PERMANOVA, 999 permutations) with adjustment for covariates, which was also used to measure the percentage of variation in microbial composition explained by the dietary variables. A p-value of <0.05 was considered to be significant. For a better visualization of the interindividual variation in gut microbiota composition, unconstrained principal coordinate analyses (PCoAs) of the Bray–Curtis distance were plotted and color-coded based on sex, age group, and BMI. Associations between dietary variables and gut microbiome profiles at the relative abundances of phyla, genera, and species level were tested using multivariate associations with linear models (MaAsLins). Detailed information regarding MaAsLins is provided in Supplementary Method 2. Models were multi-adjusted for the aforementioned covariates with a BH-adjusted p-value of <0.1 considered significant. All analyses were performed using R Version 4.0.5.

Although gut microbiota have been widely reported to geographically vary [32], it is hard to dissect the mixed effects of, for instance, lifestyle and long-term diets captured by the geographical variable. Thus, we conducted a sensitivity analysis among participants from the same province instead of regarding geography as a covariate to be adjusted, and we also considered the sample size. Additional sensitivity analyses were conducted by excluding (1) 223 participants who were assessed as at high risk of colorectal cancer or had positive FIT results and (2) 360 participants with BMI < 18.5 kg/m^2 or >24.0 kg/m^2.

3. Results

3.1. Study Sample Characteristics

A total of 702 participants were included in our final analysis, including 369 women and 333 men. Characteristics of the study population are presented in Table 1. The majority of the included individuals were aged between 50 and 70 years old, and they were evenly distributed by an age interval of 5 years, with only 5.56% aged over 70. The proportion of current smokers was 73.36%. Nearly two thirds of the population were non-drinkers. The BMI values were regrouped into three groups according to the Chinese definitions of "overweight" and "obesity", with more than a half having a BMI of less than 24 kg/m^2.

Table 1. Characteristics of the study population (N = 702).

	N	Percentage
Sex		
Female	369	52.56%
Male	333	47.44%
Age, years		
50–54	188	26.78%
55–59	153	21.79%
60–64	181	25.78%
65–69	141	20.09%
70–74	39	5.56%
Smoking status		
Current smoker	515	73.36%
Past smoker	45	6.41%
Nonsmoker	142	20.23%
Alcohol consumption		
No	463	65.95%
Seldom	102	14.53%
Regular	137	19.52%
BMI, kg/m^2		
<24.0	373	53.13%
24.0–27.9	282	40.17%
≥28.0	47	6.7%
Physical activity (MET, h/week)		
<33.60	175	24.93%
33.60–82.05	176	25.07%
82.05–147.80	175	24.93%
≥147.80	176	25.07%
Region		
Changsha, Hunan	190	27.07%
Hefei, Anhui	92	13.11%
Kunming, Yunnan	14	1.99%
Lanxi, Zhejiang	154	21.94%
Taizhou, Zhejiang	164	23.36%
Xuzhou, Jiangsu	88	12.54%

BMI, body mass index; MET, metabolic equivalents.

The usual dietary consumption of the participants is presented in Table 2. The amount of physical activity was evaluated using metabolic equivalent hours per day (MET-hours/day), which was regrouped into quantiles. The geographical distribution is also presented. The usual food intakes frequencies of the participants are presented in Table 2. Microbiota composition showed great interindividual variability at the phylum level (see Figure 1).

Table 2. Usual dietary consumption frequencies of the study population.

Food Group	>1 per Day	1 per Day	>1 per Week	1 per Week	<1 per Week
Red meat (pork, beef, lamb, etc.)	142 (20.23%)	272 (38.75%)	211 (30.06%)	64 (9.12%)	13 (1.85%)
White meat (fish and poultry)	57 (8.12%)	170 (24.22%)	280 (39.89%)	131 (18.66%)	64 (9.12%)
Eggs	51 (7.26%)	218 (31.05%)	263 (37.46%)	88 (12.54%)	82 (11.68%)
Dairy products (milk, yoghurt, etc.)	23 (3.28%)	123 (17.52%)	121 (17.24%)	66 (9.40%)	369 (52.56%)
Cooked and cured meats (e.g., sausages)	18 (2.56%)	22 (3.13%)	57 (8.12%)	53 (7.55%)	552 (78.63%)
Refined grains (rice, flour, etc.)	521 (74.22%)	103 (14.67%)	55 (7.83%)	11 (1.57%)	12 (1.71%)
Whole grains (millet, corn, sorghum, etc.)	62 (8.83%)	107 (15.24%)	234 (33.33%)	123 (17.52%)	176 (25.07%)
Fruits	98 (13.96%)	212 (30.20%)	167 (23.79%)	127 (18.09%)	98 (13.96%)
Vegetables	480 (68.38%)	159 (22.65%)	44 (6.27%)	14 (1.99%)	5 (0.71%)

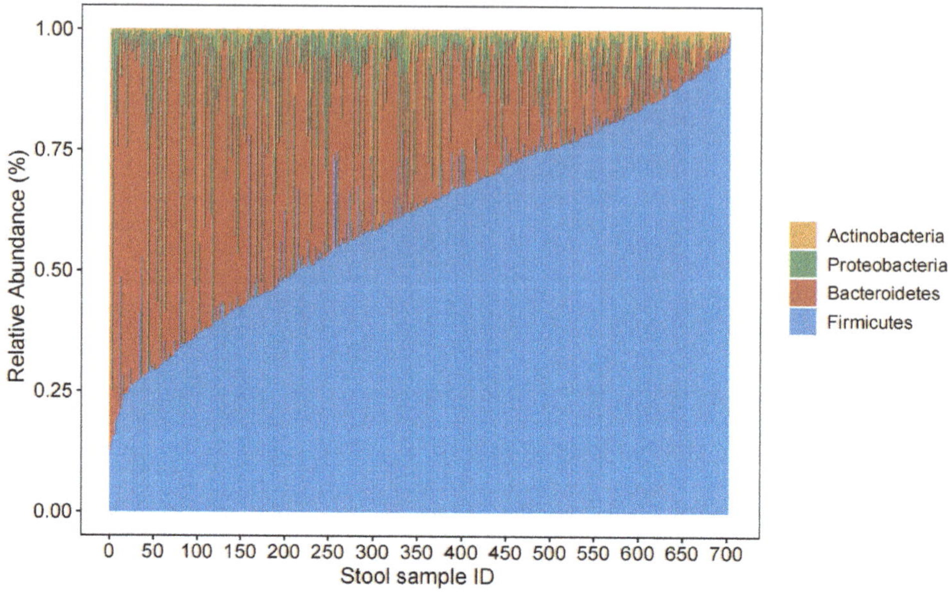

Figure 1. Relative abundances of the 4 most abundant phyla. Each thin vertical bar presents relative abundances determined in 1 individual stool sample, totaling 702.

3.2. Data-Driven Posteriori Dietary Patterns

Three dietary patterns were identified in the present Chinese population (Supplementary Tables S2 and S3). The first cluster, a traditional dietary pattern of the Yangtze River Delta, represented a typical traditional diet in South China characterized by high intakes of refined grains and vegetables but low intakes of cooked meat. A majority of participants from two sites of Zhejiang province, part of the Yangtze River Delta, followed this traditional Yangtze River Delta dietary pattern (indicated as Cluster A; see Supplementary Table S4). The second cluster was a modern dietary pattern that was characterized by specifically high intakes of eggs, dairy, fruits, vegetables and whole grains accompanied by medium intakes of red meat and white meat (indicated as Cluster B). The third cluster, labeled as the general dietary pattern, was characterized by the generally higher intake of each food group (4–6 times per week), except for the relatively lower consumption of cooked meat, compared to the other dietary patterns (indicated as Cluster C).

3.3. α-Diversity Indexes Distributed by Food Intake Frequencies

For the Shannon index, no significant differences were observed among different food intake frequencies for the nine food groups (Figure 2). Regarding red meat, white meat, cooked meat, dairy products, whole grains, and vegetables, the α-diversity index shared similar levels across different food intake frequency groups (Supplementary Figures S3–S5, S7, S9 and S11). The richness, chao1, ACE, faith_pd index were significantly distributed by egg intake frequencies (Supplementary Figures S6). For refined grain and fruit consumption, the faith_pd index presented different distributions among different food intake frequency groups (Supplementary Figures S8 and S10).

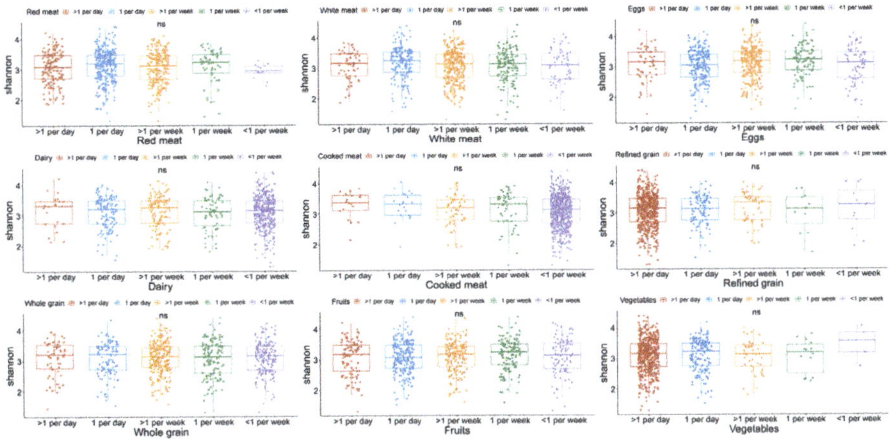

Figure 2. Boxplots for α-diversity Shannon index according to food intake frequencies in different food groups. ns: non-significant.

3.4. Associations between Dietary Variables and β-Diversity

Unconstrained PCoAs of the Bray–Curtis distance are shown in Figure 3. Compositional dissimilarities (β-diversity) of the gut microbiota between men and women and across different BMI groups were detected (Figure 3A,C). Although no clear clustering appeared among age groups, a grouping pattern along the gradient of age groups could be observed (Figure 3B). Arrows indicate the direction of gradient for covariates and were obtained via the envfit function (package "vegan"). Figure 3D presents the associations between dietary variables and β-diversity matrices found using PERMANOVAs. The Bray–Curtis distances of inter-individual dissimilarities were associated with whole grains and vegetables, explaining 1.46% of the total variation in the gut microbiota composition measured by the partial R^2 value with age, sex, BMI, smoking, alcohol consumption, and physical activity adjusted.

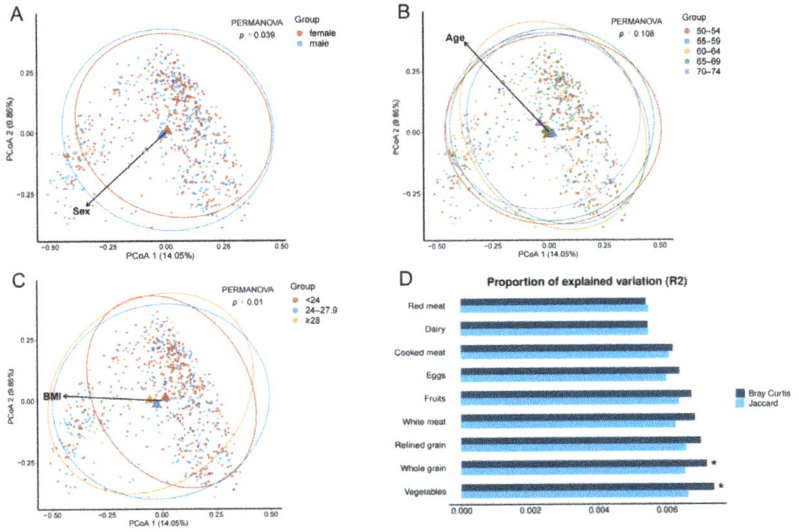

Figure 3. Variation in the gut microbiota composition represented by unconstrained PCoA based on the distance indexes. (**A–C**) present the grouping patterns of gut microbiota composition based on sex,

age, and BMI. (**D**) shows percentages of variation in gut microbiota composition explained by dietary variables using multi-adjusted permutational ANOVAs (999 permutations). PCoA, principal coordinate analysis. * *p*-value < 0.05.

3.5. Associations between Dietary Variables and Relative Abundances of Taxa

Taxa significantly associated with food groups and the posteriori dietary pattern are presented in Table 3. For instance, the genus *Weissella* and an unknown species of it were negatively associated with weekly red meat intake. Cooked meat was positively associated with an abundance of the genus *Coprobacter*. The relationships of *Weissella* and *Coprobacter* were kept consistent in the sensitivity analyses by restricting participants from a single province or removing individuals at a high risk of intestinal diseases, respectively (Supplementary Table S6).

Table 3. Associations between food intakes, posteriori dietary patterns, and gut microbial profiles using MaAsLins.

Food Group	Phylum	Class ǀ Order ǀ Family	Genus	Species	Value	Coef [1]	Coverage (%) [2]	Pval [3]	Qval [4]
Red meat	*Firmicutes*	*Bacilli* ǀ *Lactobacillales* ǀ *Leuconostocaceae*	*Weissella*	Uncultured organism	pd	−0.0379	28.35%	<0.0001	0.0300
Red meat	*Firmicutes*	*Bacilli* ǀ *Lactobacillales* ǀ *Leuconostocaceae*	*Weissella*		pd	−0.0379	29.91%	<0.0001	0.0308
Dairy	*Firmicutes*	*Clostridia* ǀ *Clostridiales* ǀ *Lachnospiraceae*	*Anaerostipes*	uncultured organism	pd	0.0146	67.95%	<0.0001	0.0261
Dairy	*Firmicutes*	*Clostridia* ǀ *Clostridiales* ǀ *Lachnospiraceae*	*Anaerostipes*		pd	0.0146	71.37%	<0.0001	0.0261
Cooked meat	*Bacteroidetes*	*Bacteroidia* ǀ *Bacteroidales* ǀ *Barnesiellaceae*	*Coprobacter*		pd	0.0118	11.97%	<0.0001	0.0044
Whole grains	*Firmicutes*	*Negativicutes* ǀ *Veillonellales* ǀ *Veillonellaceae*	*Megasphaera*	uncultured organism	mul_pd	0.0420	14.25%	<0.0001	0.0183
Refined grains	*Firmicutes*	*Bacilli* ǀ *Lactobacillales* ǀ *Lactobacillaceae*	*Lactobacillus*	uncultured organism	pw	0.0602	13.82%	0.0001	0.0763
Vegetables	*Firmicutes*	*Clostridia* ǀ *Clostridiales* ǀ *Ruminococcaceae*	*Eubacterium coprostanoligenes* group	uncultured organism	pd	−0.0767	23.50%	<0.0001	0.0123
Vegetables	*Firmicutes*	*Clostridia* ǀ *Clostridiales* ǀ *Ruminococcaceae*	*Eubacterium coprostanoligenes* group	uncultured organism	mul_pd	−0.0737	23.50%	<0.0001	0.0140
Vegetables	*Firmicutes*	*Clostridia* ǀ *Clostridiales* ǀ *Ruminococcaceae*	uncultured		pd	−0.0389	43.87%	<0.0001	0.0156
Vegetables	*Firmicutes*	*Clostridia* ǀ *Clostridiales* ǀ *Ruminococcaceae*	*Eubacterium coprostanoligenes* group	uncultured organism	mul_pw	−0.0746	23.50%	<0.0001	0.0173
Vegetables	*Firmicutes*	*Clostridia* ǀ *Clostridiales* ǀ *Ruminococcaceae*	uncultured		mul_pw	−0.0394	43.87%	<0.0001	0.0173
Vegetables	*Firmicutes*	*Clostridia* ǀ *Clostridiales* ǀ *Christensenellaceae*	*Christensenellaceae* R7 group	uncultured organism	pd	−0.0573	27.78%	<0.0001	0.0226
Vegetables	*Firmicutes*	*Clostridia* ǀ *Clostridiales* ǀ *Christensenellaceae*	*Christensenellaceae* R7 group	uncultured organism	mul_pd	−0.0561	27.78%	<0.0001	0.0256
Vegetables	*Firmicutes*	*Clostridia* ǀ *Clostridiales* ǀ *Ruminococcaceae*	*Eubacterium coprostanoligenes* group	uncultured organism	pw	−0.0754	23.50%	0.0001	0.0460
Vegetables	*Firmicutes*	*Clostridia* ǀ *Clostridiales* ǀ *Ruminococcaceae*	*Ruminococcaceae UCG 005*	uncultured organism	mul_pw	−0.0243	12.68%	0.0002	0.0588
Vegetables	*Firmicutes*	*Clostridia* ǀ *Clostridiales* ǀ *Ruminococcaceae*	uncultured		mul_pd	−0.0339	43.87%	0.0001	0.0588
Vegetables	*Firmicutes*	*Clostridia* ǀ *Clostridiales* ǀ *Ruminococcaceae*	uncultured		pw	−0.0388	43.87%	0.0002	0.0588
Vegetables	*Firmicutes*	*Clostridia* ǀ *Clostridiales* ǀ *Ruminococcaceae*	uncultured	uncultured organism	mul_pw	−0.0284	13.96%	0.0002	0.0595

Table 3. *Cont.*

Food Group	Phylum	Class ǀ Order ǀ Family	Genus	Species	Value	Coef [1]	Coverage (%) [2]	Pval [3]	Qval [4]
Vegetables	*Firmicutes*	*Clostridia* ǀ *Clostridiales* ǀ *Ruminococcaceae*	*Ruminococcaceae UCG 005*	uncultured organism	pd	−0.0230	12.68%	0.0002	0.0629
Vegetables	*Firmicutes*	*Clostridia* ǀ *Clostridiales* ǀ *Ruminococcaceae*	uncultured	uncultured organism	pd	−0.0267	13.96%	0.0003	0.0733
Vegetables	*Firmicutes*	*Clostridia* ǀ *Clostridiales* ǀ *Christensenellaceae*	*Christensenellaceae R7* group	uncultured organism	mul_pw	−0.0522	27.78%	0.0003	0.0743
Vegetables	*Firmicutes*	*Clostridia* ǀ *Clostridiales* ǀ *Christensenellaceae*	*Christensenellaceae R7* group	uncultured organism	pw	−0.0572	27.78%	0.0003	0.0743
Vegetables	*Firmicutes*	*Clostridia* ǀ *Clostridiales* ǀ *Ruminococcaceae*	*Ruminococcaceae UCG 005*	uncultured organism	mul_pd	−0.0222	12.68%	0.0003	0.0743
Vegetables	*Firmicutes*	*Bacilli* ǀ *Lactobacillales* ǀ *Leuconostocaceae*	*Leuconostoc*	uncultured organism	mul_pw	−0.0246	17.81%	0.0003	0.0745
Vegetables	*Firmicutes*	*Bacilli* ǀ *Lactobacillales* ǀ *Leuconostocaceae*	*Leuconostoc*		mul_pw	−0.0246	17.81%	0.0003	0.0745
Vegetables	*Firmicutes*	*Bacilli* ǀ *Lactobacillales* ǀ *Leuconostocaceae*	*Leuconostoc*	uncultured organism	pd	−0.0231	17.81%	0.0005	0.0867
Vegetables	*Firmicutes*	*Bacilli* ǀ *Lactobacillales* ǀ *Leuconostocaceae*	*Leuconostoc*		pd	−0.0231	17.81%	0.0005	0.0867
Vegetables	*Firmicutes*	*Clostridia* ǀ *Clostridiales* ǀ *Ruminococcaceae*	*Ruminococcaceae UCG 005*	uncultured organism	pw	−0.0246	12.68%	0.0005	0.0900
Vegetables	*Firmicutes*	*Bacilli* ǀ *Lactobacillales* ǀ *Leuconostocaceae*	*Leuconostoc*	uncultured organism	mul_pd	−0.0225	17.81%	0.0006	0.0958
Vegetables	*Firmicutes*	*Bacilli* ǀ *Lactobacillales* ǀ *Leuconostocaceae*	*Leuconostoc*		mul_pd	−0.0225	17.81%	0.0006	0.0958
Vegetables	*Firmicutes*	*Clostridia* ǀ *Clostridiales* ǀ *Ruminococcaceae*	*Eubacterium coprostanoligenes* group		pd	−0.0804	71.23%	0.0006	0.0958
Cluster	*Firmicutes*	*Clostridia* ǀ *Clostridiales* ǀ *Lachnospiraceae*	*Anaerostipes*	uncultured organism	C	0.0119	67.95%	0.0001	0.0749
Cluster	*Firmicutes*	*Clostridia* ǀ *Clostridiales* ǀ *Lachnospiraceae*	*Anaerostipes*		C	0.0115	71.37%	0.0001	0.0858

[1] For categorical features in MaAsLins analysis, the specific feature level for the coefficient and significance of association is reported. [2] Prevalence of bacterial taxa in the study sample is equal to the total of number of samples in which the feature is non-zero divided by the total number of samples used in the model. [3] *p*-value for MaAsLin adjusted for age, sex, BMI, smoking status, alcohol consumption, and physical activity; computed using the Maaslin2 package on R. [4] Corrected *p*-value by the Benjamini–Hochberg method (10% false discovery rate).

Dairy intake was positively associated with the genus *Anaerostipes* and an unknown species of it. Moreover, we found significant positive associations for whole grain intake with a species of the genus *Megasphaera* and refined grain intake with a species of the genus *Lactobacillus*, which were also observed in the sensitivity analyses (Supplementary Tables S5–S7). Vegetables were negatively inversely associated with the genus *Eubacterium coprostanoligenes* group and a species of it, a species of the genus *Christensenellaceae R7* group belonging to the family *Christensenellaceae*, and the genus *Leuconostoc*. For the whole picture of the habitual food intakes, individuals leading the general dietary style (Cluster C) had higher abundances of the genus *Anaerostipes* and a species of it compared to those who had the traditional Yangtze River Delta dietary pattern (Cluster A) characterized by higher intakes of refined grains and vegetables and lower intakes of dairy products.

4. Discussion

In this population-based study of 702 healthy Chinese adults free of colorectal neoplasm aged 50–75 years, we examined the associations between the habitual dietary pattern and the gut microbiome. Our data revealed that the α-diversity index generally shared similar levels across different food intake frequencies among nine major food groups, whereas whole grain and vegetable intakes drove the dissimilarities in gut microbial composition, as indicated by the distance-based β-diversity dissimilarities. Based on the data-driven posteriori dietary pattern analyses, our results also highlighted the relationship of the general dietary style with higher abundances of the genus *Anaerostipes* and a species of it, which was characterized by lower intakes of refined grains. Moreover, we observed a number of positive or inverse associations between usual food groups and abundances of certain taxa, concentrated in genera within the phylum *Firmicutes*.

Previously reported evidence supports our findings. Evidence from a randomized diet intervention trial aiming to examine the effect of carbohydrate type on gut microbial composition and function and metabolites showed that *Anaerostipes* had a higher abundance after a simple carbohydrate diet compared to a refined carbohydrate diet [33]. Due to the role of *Anaerostipes* as a butyrate producer, low abundance after the consumption of refined carbohydrate foods may contribute to the unfavorable effects of diets rich in refined carbohydrates. In addition, the authors of a recent study reported a myo-inositol pathway in *Anaerostipes spp.*, which was most abundantly present in mammalian tissues and fruits, suggesting a newly discovered benefit of intestinal *Anaerostipes spp.* for host health promotion [34]. In our study, participants consuming general diets had higher weekly fruit intakes than individuals with the traditional Yangtze River Delta dietary pattern.

For the specific food groups, our results showed that genus *Weissella* and an unknown species of it were negatively associated with weekly red meat intake. *Weissella* is a member of the lactic acid bacteria group, which has been well-studied and is best known for its potential in imparting beneficial human health effects [34]. Some strains of *Weissella* can prevent lipopolysaccharide-induced proinflammatory stress in murine macrophages and human colonic epithelial cells [35]. Dairy has presented a positive association with the abundance of *Anaerostipes*, which warrants further investigation, whereas mice model studies have suggested that *Anaerostipes caccae* may be involved in the protective process against the allergic response to cow's milk [36]. The association between vegetable intake and *Christensenellaceae* disappeared after excluding individuals with abnormal BMI levels, predominantly overweight and obese people. This phenomenon could be explained by previously reported evidence that suggests that the relative abundance of *Christensenellaceae* in the human gut is inversely related to host BMI in different populations, making its relationship with BMI the most robust and reproducible link between the microbial ecology of the human gut and metabolic disease [37].

The *Eubacterium coprostanoligenes* group is characterized as one of the hub genera in the fecal micro-ecosystem of high-fat diets, and studies have shown that the *Eubacterium coprostanoligenes* mediates the effect of high-fat diets on dyslipidemia through sphingosine [38]. The requirement of lecithin for the growth of *Eubacterium coprostanoligenes* [39], which is primarily rich in animal foods, may partly explain the negative association between the relative abundances of *Eubacterium coprostanoligenes* and vegetable intakes found here.

In the present study, we used aggregated items to collect information on broad dietary habits of participants for the sake of convenient dietary data collection. This led to high variability in terms of specific food types and nutrient composition, as well as the population-specific findings. For example, people residing in Europe consume different types of vegetables than Chinese people. Moreover, the complexity of food composition including macronutrients, micronutrients, and food additives made it difficult to elucidate the intricate diet–microbiota relationship. The significant findings in our study need to be cautiously interpreted, and some associations could be explained from a biological mechanistic standpoint. Thus, additional efforts and deeper insights regarding the underlying mechanisms are required before considering translating such knowledge to personalized diet intervention strategies. Nevertheless, we have confirmed that future studies should consider dietary variables as covariates in analyses of disease-microbiome associations to disentangle the effects of diet on the gut microbiome from disease-related associations. To simplify the complexity of multidimensional diet data with internal correlations, researchers can the dietary index as a summary measure when quantifying dietary variance in microbiota studies instead of individual dietary features [13], including priori or posteriori dietary indices [40].

The presented study is the so-far largest multi-center study of the association between the gut microbiota and the habitual diet with unitary and general measurements in the Chinese population. However, some points should be considered in interpreting our findings.

Firstly, we only assessed nine commonly consumed food groups in the Chinese diet using aggregated items, so food groups that could be further classified were broadly considered, e.g., milk and yogurt were considered as general dairy products. Although the major dietary patterns in the studied Chinese adults were well captured, only the frequencies (not the quantities of the major food groups) were collected, which made it less feasible to completely quantify food intakes. In addition, given the potentially rapid and transient effects of food on the human gut microbiome [14], the bacterial profiles characterized in a single fecal sample will likely reflect the effect of food consumption patterns in the period immediately prior to sample collection and not necessarily a participant's long-term steady state. Additionally, long-term dietary habits were coarsely assessed using a one-time FFQ that collected data on food intake patterns over the prior 12 months. Thus, large-scale observational studies using accurate frameworks to capture long-term dietary exposures and stable gut microbiota composition and to reduce random within-person variation are needed for exploration of associations between habitual food intakes and gut microbiome. Subsequent time points with both dietary and microbiota data would be of utmost interest to investigate the stability of the studied relations over time. Though the participants of our study were from multiple regions, we performed a sensitivity analysis with individuals from a single province instead of regarding region as a covariate since the dietary information partly captured the geographical characteristics of the studied population (Supplementary Table S8). However, we cannot rule out residual confounding effects due to imperfectly measured covariates and unmeasured confounders, despite multivariable adjustments and sensitivities analyses. Furthermore, the TARGET-C study was initially established to evaluate the effectiveness of different colorectal cancer screening strategies. Participants enrolled in this study were apparently healthy upon recruitment according to stringent inclusion criteria but no systematic physical examination, thus providing a less pure foundation to investigate the diet–microbiota relationship. Extensive studies in a completely disease-free context are needed. Finally, the annotation resolution of the 16S rRNA amplicon sequencing was limited, so future efforts focusing on a broader picture of microbiome variability and the potential functional capability of the gut microbiome through shotgun metagenomics may provide deeper insight into the diet–gut microbiome relationship.

5. Conclusions

In summary, in a large sample of the Chinese population free of colorectal neoplasm, we found that the long-term dietary pattern characterized by lower intakes of refined grains was associated with higher abundances of the genus *Anaerostipes* and a species of it. The dietary pattern can act as a summary measure that captures gut microbiota variance attributable to habitual diet in microbiome studies. Future studies are needed to investigate whether and to what extent the gut microbiota may mediate or modify the effects of habitual diets on human physiological and pathological processes.

Supplementary Materials: The following supporting information can be downloaded at: https://www.mdpi.com/article/10.3390/nu14132639/s1, Figure S1. Workflow diagram for the subject enrollment and exclusion, Figure S2. Parallel analysis scree plot to determine the number of factors for factor analysis, Figure S3. α-diversity indexes (richness, chao 1, simpson index, pielou index, ACE and faith-pd index in different intake frequency groups of red meat, Figure S4. α-diversity indexes (richness, chao 1, simpson index, pielou index, ACE and faith-pd index in different intake frequency groups of white meat, Figure S5. α-diversity indexes (richness, chao 1, simpson index, pielou index, ACE and faith-pd index in different intake frequency groups of cooked meat, Figure S6. α-diversity indexes (richness, chao 1, simpson index, pielou index, ACE and faith-pd index in different intake frequency groups of eggs, Figure S7. α-diversity indexes (richness, chao 1, simpson index, pielou index, ACE and faith-pd index in different intake frequency groups of dairy products, Figure S8. α-diversity indexes (richness, chao 1, simpson index, pielou index, ACE and faith-pd index in different intake frequency groups of refined grain, Figure S9. α-diversity indexes (richness, chao 1, simpson index, pielou index, ACE and faith-pd index in different intake frequency groups of whole grain,

Figure S10. α-diversity indexes (richness, chao 1, simpson index, pielou index, ACE and faith-pd index in different intake frequency groups of fruits, Figure S11. α-diversity indexes (richness, chao 1, simpson index, pielou index, ACE and faith-pd index in different intake frequency groups of vegetables; Table S1. Factor loading matrix of major factors by principal component analysis with varimax rotation, Table S2. Classification of subjects by cluster analysis using factor score, Table S3. Dietary patterns identified by K-means clustering, Table S4. Region distributions of 702 participants according to the established dietary patterns, Table S5. Sensitivity analysis of associations between food intakes and gut microbial profiles using MaAsLins with population at a high risk of intestinal diseases removed, $N = 479$, Table S6. Sensitivity analysis of associations between food intakes and gut microbial profiles using MaAsLins among population from a single province, $N = 318$, Table S7. Sensitivity analysis of associations between food intakes and gut microbial profiles using MaAsLins among population with normal BMI values, $N = 342$, Table S8. Contingency correlation coefficient between the region variable and dietary variables.

Author Contributions: Guarantors of the article: Y.Z., H.C. and M.D. Specific author contributions: M.D. and H.C. conceptualized and designed the study. Y.Z., M.L., B.L. and H.C. participated in acquisition of data and quality control; Y.Z., J.C., C.L. and H.C. participated in data analysis and interpretation. Y.Z. drafted the manuscript. M.D. and H.C. revised the manuscript for important intellectual content. All authors critically revised the manuscript and approved the final version. All authors have read and agreed to the published version of the manuscript.

Funding: This work was supported by the Natural Science Foundation of Beijing Municipality (7202169), the National Natural Science Foundation of China (82173606) and the Beijing Nova Program of Science and Technology (Z191100001119065).

Institutional Review Board Statement: The study was conducted in accordance with the Declaration of Helsinki, and approved by the Ethics Committee of the National Cancer Center/Cancer Hospital, Chinese Academy of Medical Sciences, and Peking Union Medical College (18-013/1615).

Informed Consent Statement: Informed consent was obtained from all subjects involved in the study.

Data Availability Statement: Access to individual-level data, including the microbial DNA sequences encoding the 16S rRNA V4 region and associated demographic and lifestyle metadata, can be obtained upon reasonable request to the corresponding author (daimin@pumch.cn).

Conflicts of Interest: The authors declare no conflict of interest.

References

1. Schroeder, B.O.; Bäckhed, F. Signals from the gut microbiota to distant organs in physiology and disease. *Nat. Med.* **2016**, *22*, 1079–1089. [CrossRef]
2. Kundu, P.; Blacher, E.; Elinav, E.; Pettersson, S. Our Gut Microbiome: The Evolving Inner Self. *Cell* **2017**, *171*, 1481–1493. [CrossRef]
3. Muller, P.A.; Matheis, F.; Schneeberger, M.; Kerner, Z.; Jové, V.; Mucida, D. Microbiota-modulated CART(+) enteric neurons autonomously regulate blood glucose. *Science* **2020**, *370*, 314–321. [CrossRef]
4. Janney, A.; Powrie, F.; Mann, E.H. Host-microbiota maladaptation in colorectal cancer. *Nature* **2020**, *585*, 509–517. [CrossRef]
5. Ruff, W.E.; Greiling, T.M.; Kriegel, M.A. Host-microbiota interactions in immune-mediated diseases. *Nat. Rev. Microbiol.* **2020**, *18*, 521–538. [CrossRef]
6. Fan, Y.; Pedersen, O. Gut microbiota in human metabolic health and disease. *Nat. Rev. Microbiol.* **2021**, *19*, 55–71. [CrossRef]
7. Morais, L.H.; Schreiber, H.L.t.; Mazmanian, S.K. The gut microbiota-brain axis in behaviour and brain disorders. *Nat. Rev. Microbiol.* **2021**, *19*, 241–255. [CrossRef]
8. Rothschild, D.; Weissbrod, O.; Barkan, E.; Kurilshikov, A.; Korem, T.; Zeevi, D.; Costea, P.I.; Godneva, A.; Kalka, I.N.; Bar, N.; et al. Environment dominates over host genetics in shaping human gut microbiota. *Nature* **2018**, *555*, 210–215. [CrossRef]
9. Aron-Wisnewsky, J.; Clément, K. The gut microbiome, diet, and links to cardiometabolic and chronic disorders. *Nat. Rev. Nephrol.* **2016**, *12*, 169–181. [CrossRef]
10. Wolter, M.; Grant, E.T.; Boudaud, M.; Steimle, A.; Pereira, G.V.; Martens, E.C.; Desai, M.S. Leveraging diet to engineer the gut microbiome. *Nat. Rev. Gastroenterol. Hepatol.* **2021**, *18*, 885–902. [CrossRef]
11. Kolodziejczyk, A.A.; Zheng, D.; Elinav, E. Diet-microbiota interactions and personalized nutrition. *Nat. Rev. Microbiol.* **2019**, *17*, 742–753. [CrossRef] [PubMed]
12. Vujkovic-Cvijin, I.; Sklar, J.; Jiang, L.; Natarajan, L.; Knight, R.; Belkaid, Y. Host variables confound gut microbiota studies of human disease. *Nature* **2020**, *587*, 448–454. [CrossRef] [PubMed]
13. Bowyer, R.C.E.; Jackson, M.A.; Pallister, T.; Skinner, J.; Spector, T.D.; Welch, A.A.; Steves, C.J. Use of dietary indices to control for diet in human gut microbiota studies. *Microbiome* **2018**, *6*, 77. [CrossRef]

14. David, L.A.; Maurice, C.F.; Carmody, R.N.; Gootenberg, D.B.; Button, J.E.; Wolfe, B.E.; Ling, A.V.; Devlin, A.S.; Varma, Y.; Fischbach, M.A.; et al. Diet rapidly and reproducibly alters the human gut microbiome. *Nature* **2014**, *505*, 559–563. [CrossRef]
15. Fragiadakis, G.K.; Wastyk, H.C.; Robinson, J.L.; Sonnenburg, E.D.; Sonnenburg, J.L.; Gardner, C.D. Long-term dietary intervention reveals resilience of the gut microbiota despite changes in diet and weight. *Am. J. Clin. Nutr.* **2020**, *111*, 1127–1136. [CrossRef] [PubMed]
16. Zmora, N.; Suez, J.; Elinav, E. You are what you eat: Diet, health and the gut microbiota. *Nat. Rev. Gastroenterol. Hepatol.* **2019**, *16*, 35–56. [CrossRef]
17. Koponen, K.K.; Salosensaari, A.; Ruuskanen, M.O.; Havulinna, A.S.; Männistö, S.; Jousilahti, P.; Palmu, J.; Salido, R.; Sanders, K.; Brennan, C.; et al. Associations of healthy food choices with gut microbiota profiles. *Am. J. Clin. Nutr.* **2021**, *114*, 605–616. [CrossRef]
18. Zhernakova, A.; Kurilshikov, A.; Bonder, M.J.; Tigchelaar, E.F.; Schirmer, M.; Vatanen, T.; Mujagic, Z.; Vila, A.V.; Falony, G.; Vieira-Silva, S.; et al. Population-based metagenomics analysis reveals markers for gut microbiome composition and diversity. *Science* **2016**, *352*, 565–569. [CrossRef]
19. Partula, V.; Mondot, S.; Torres, M.J.; Kesse-Guyot, E.; Deschasaux, M.; Assmann, K.; Latino-Martel, P.; Buscail, C.; Julia, C.; Galan, P.; et al. Associations between usual diet and gut microbiota composition: Results from the Milieu Intérieur cross-sectional study. *Am. J. Clin. Nutr.* **2019**, *109*, 1472–1483. [CrossRef]
20. Falony, G.; Joossens, M.; Vieira-Silva, S.; Wang, J.; Darzi, Y.; Faust, K.; Kurilshikov, A.; Bonder, M.J.; Valles-Colomer, M.; Vandeputte, D.; et al. Population-level analysis of gut microbiome variation. *Science* **2016**, *352*, 560–564. [CrossRef]
21. Taylor, B.C.; Lejzerowicz, F.; Poirel, M.; Shaffer, J.P.; Jiang, L.; Aksenov, A.; Litwin, N.; Humphrey, G.; Martino, C.; Miller-Montgomery, S.; et al. Consumption of Fermented Foods Is Associated with Systematic Differences in the Gut Microbiome and Metabolome. *mSystems* **2020**, *5*, e00901-19. [CrossRef] [PubMed]
22. Kaplan, R.C.; Wang, Z.; Usyk, M.; Sotres-Alvarez, D.; Daviglus, M.L.; Schneiderman, N.; Talavera, G.A.; Gellman, M.D.; Thyagarajan, B.; Moon, J.Y.; et al. Gut microbiome composition in the Hispanic Community Health Study/Study of Latinos is shaped by geographic relocation, environmental factors, and obesity. *Genome Biol.* **2019**, *20*, 219. [CrossRef] [PubMed]
23. Maskarinec, G.; Hullar, M.A.J.; Monroe, K.R.; Shepherd, J.A.; Hunt, J.; Randolph, T.W.; Wilkens, L.R.; Boushey, C.J.; Le Marchand, L.; Lim, U.; et al. Fecal Microbial Diversity and Structure Are Associated with Diet Quality in the Multiethnic Cohort Adiposity Phenotype Study. *J. Nutr.* **2019**, *149*, 1575–1584. [CrossRef] [PubMed]
24. Gallè, F.; Valeriani, F.; Cattaruzza, M.S.; Gianfranceschi, G.; Liguori, R.; Antinozzi, M.; Mederer, B.; Liguori, G.; Romano Spica, V. Mediterranean Diet, Physical Activity and Gut Microbiome Composition: A Cross-Sectional Study among Healthy Young Italian Adults. *Nutrients* **2020**, *12*, 2164. [CrossRef]
25. Yu, D.; Nguyen, S.M.; Yang, Y.; Xu, W.; Cai, H.; Wu, J.; Cai, Q.; Long, J.; Zheng, W.; Shu, X.O. Long-term diet quality is associated with gut microbiome diversity and composition among urban Chinese adults. *Am. J. Clin. Nutr.* **2021**, *113*, 684–694. [CrossRef]
26. Lu, J.; Zhang, L.; Zhai, Q.; Zhao, J.; Zhang, H.; Lee, Y.K.; Lu, W.; Li, M.; Chen, W. Chinese gut microbiota and its associations with staple food type, ethnicity, and urbanization. *NPJ Biofilm. Microbiomes* **2021**, *7*, 71. [CrossRef]
27. Chen, H.; Li, N.; Shi, J.; Ren, J.; Liu, C.; Zhang, Y.; Jiang, Z.; Zhang, Z.; Dai, M. Comparative evaluation of novel screening strategies for colorectal cancer screening in China (TARGET-C): A study protocol for a multicentre randomised controlled trial. *BMJ Open* **2019**, *9*, e025935. [CrossRef]
28. Chen, H.; Lu, M.; Liu, C.; Zou, S.; Du, L.; Liao, X.; Dong, D.; Wei, D.; Gao, Y.; Zhu, C.; et al. Comparative Evaluation of Participation and Diagnostic Yield of Colonoscopy vs Fecal Immunochemical Test vs Risk-Adapted Screening in Colorectal Cancer Screening: Interim Analysis of a Multicenter Randomized Controlled Trial (TARGET-C). *Am. J. Gastroenterol.* **2020**, *115*, 1264–1274. [CrossRef]
29. Zhang, Y.; Lu, M.; Lu, B.; Liu, C.; Ma, Y.; Liu, L.; Miao, X.; Qin, J.; Chen, H.; Dai, M. Leveraging Fecal Microbial Markers to Improve the Diagnostic Accuracy of the Fecal Immunochemical Test for Advanced Colorectal Adenoma. *Clin. Transl. Gastroenterol.* **2021**, *12*, e00389. [CrossRef]
30. Vogtmann, E.; Chen, J.; Kibriya, M.G.; Chen, Y.; Islam, T.; Eunes, M.; Ahmed, A.; Naher, J.; Rahman, A.; Amir, A.; et al. Comparison of Fecal Collection Methods for Microbiota Studies in Bangladesh. *Appl. Environ. Microbiol.* **2017**, *83*, e00361-17. [CrossRef]
31. Callahan, B.J.; McMurdie, P.J.; Rosen, M.J.; Han, A.W.; Johnson, A.J.; Holmes, S.P. DADA2: High-resolution sample inference from Illumina amplicon data. *Nat. Methods* **2016**, *13*, 581–583. [CrossRef] [PubMed]
32. Gupta, V.K.; Paul, S.; Dutta, C. Geography, Ethnicity or Subsistence-Specific Variations in Human Microbiome Composition and Diversity. *Front. Microbiol.* **2017**, *8*, 1162. [CrossRef] [PubMed]
33. Faits, T.; Walker, M.E.; Rodriguez-Morato, J.; Meng, H.; Gervis, J.E.; Galluccio, J.M.; Lichtenstein, A.H.; Johnson, W.E.; Matthan, N.R. Exploring changes in the human gut microbiota and microbial-derived metabolites in response to diets enriched in simple, refined, or unrefined carbohydrate-containing foods: A post hoc analysis of a randomized clinical trial. *Am. J. Clin. Nutr.* **2020**, *112*, 1631–1641. [CrossRef] [PubMed]
34. Bui, T.P.N.; Mannerås-Holm, L.; Puschmann, R.; Wu, H.; Troise, A.D.; Nijsse, B.; Boeren, S.; Bäckhed, F.; Fiedler, D.; deVos, W.M. Conversion of dietary inositol into propionate and acetate by commensal Anaerostipes associates with host health. *Nat. Commun.* **2021**, *12*, 4798. [CrossRef] [PubMed]

35. Singh, S.; Bhatia, R.; Singh, A.; Singh, P.; Kaur, R.; Khare, P.; Purama, R.K.; Boparai, R.K.; Rishi, P.; Ambalam, P.; et al. Probiotic attributes and prevention of LPS-induced pro-inflammatory stress in RAW264.7 macrophages and human intestinal epithelial cell line (Caco-2) by newly isolated Weissella cibaria strains. *Food Funct.* **2018**, *9*, 1254–1264. [CrossRef]
36. Feehley, T.; Plunkett, C.H.; Bao, R.; Choi Hong, S.M.; Culleen, E.; Belda-Ferre, P.; Campbell, E.; Aitoro, R.; Nocerino, R.; Paparo, L.; et al. Healthy infants harbor intestinal bacteria that protect against food allergy. *Nat. Med.* **2019**, *25*, 448–453. [CrossRef]
37. Waters, J.L.; Ley, R.E. The human gut bacteria *Christensenellaceae* are widespread, heritable, and associated with health. *BMC Biol.* **2019**, *17*, 83. [CrossRef]
38. Wei, W.; Jiang, W.; Tian, Z.; Wu, H.; Ning, H.; Yan, G.; Zhang, Z.; Li, Z.; Dong, F.; Sun, Y.; et al. Fecal *g. Streptococcus* and *g. Eubacterium_coprostanoligenes_group* combined with sphingosine to modulate the serum dyslipidemia in high-fat diet mice. *Clin. Nutr.* **2021**, *40*, 4234–4245. [CrossRef]
39. Freier, T.A.; Beitz, D.C.; Li, L.; Hartman, P.A. Characterization of *Eubacterium coprostanoligenes* sp. nov., a cholesterol-reducing anaerobe. *Int. J. Syst. Bacteriol.* **1994**, *44*, 137–142. [CrossRef]
40. Cotillard, A.; Cartier-Meheust, A.; Litwin, N.S.; Chaumont, S.; Saccareau, M.; Lejzerowicz, F.; Tap, J.; Koutnikova, H.; Lopez, D.G.; McDonald, D.; et al. A posteriori dietary patterns better explain variations of the gut microbiome than individual markers in the American Gut Project. *Am. J. Clin. Nutr.* **2021**, *115*, 432–443. [CrossRef]

nutrients

Review

Microbiota Modulation in Patients with Metabolic Syndrome

Ricardo Araujo [1,2,*], Marta Borges-Canha [3,4] and Pedro Pimentel-Nunes [4,5]

1 Nephrology & Infectious Diseases R&D Group, i3S—Instituto de Investigação e Inovação em Saúde, Universidade do Porto, 4200-135 Porto, Portugal
2 INEB—Instituto de Engenharia Biomédica, Universidade do Porto, 4200-135 Porto, Portugal
3 Department of Surgery and Physiology, Faculty of Medicine, University of Porto, 4200-319 Porto, Portugal
4 Department of Endocrinology, Diabetes and Metabolism, Centro Hospitalar Universitário de São João, 4200-319 Porto, Portugal
5 RISE@CI-IPOP (Health Research Network, IPO Porto), Porto Comprehensive Cancer Center (Porto CCC), 4200-072 Porto, Portugal
* Correspondence: ricjparaujo@yahoo.com; Tel.: +351-2260-74900

Abstract: Metabolic syndrome (MS) comprises a vast range of metabolic dysfunctions, which can be associated to cardiovascular disease risk factors. MS is reaching pandemic levels worldwide and it currently affects around 25% in the adult population of developed countries. The definition states for the diagnosis of MS may be clear, but it is also relevant to interpret the patient data and realize whether similar criteria were used by different clinicians. The different criteria explain, at least in part, the controversies on the theme. Several studies are presently focusing on the microbiota changes according to the components of MS. It is widely accepted that the gut microbiota is a regulator of metabolic homeostasis, being the gut microbiome in MS described as dysbiotic and certain taxonomic groups associated to metabolic changes. Probiotics, and more recently synbiotics, arise as promising therapeutic alternatives that can mitigate some metabolic disturbances, namely by correcting the microbiome and bringing homeostasis to the gut. The most recent studies were revised and the promising results and perspectives revealed in this review.

Keywords: inflammation; gut metabolites; gut microbiome; metabolic syndrome; obesity; probiotics; synbiotics

Citation: Araujo, R.; Borges-Canha, M.; Pimentel-Nunes, P. Microbiota Modulation in Patients with Metabolic Syndrome. *Nutrients* **2022**, *14*, 4490. https://doi.org/10.3390/nu14214490

Academic Editor: Sareen Gropper

Received: 6 October 2022
Accepted: 18 October 2022
Published: 25 October 2022

1. Introduction

Metabolic syndrome (MS), also known as syndrome X or insulin resistance syndrome, comprises a constellation of metabolic dysfunctions, which represent cardiovascular (CV) disease risk factors [1]. Its definition may be controversial according to various entities [1]. One of the most accepted and used definition is the one recommended by the National Cholesterol Education Program (NCEP), 2005 [2]. The definition states that the diagnosis may be made in the presence of any three or more of the following: (1) fasting blood glucose greater than 100 mg/dL or drug treatment for elevated blood glucose; (2) high-density lipoprotein (HDL) cholesterol <140 mg/dL in men or <50 mg/dL in women, or drug treatment for low HDL cholesterol; (3) blood triglycerides > 150 mg/dL or drug treatment for elevated triglycerides; (4) waist circumference > 102 cm in men or >88 cm in women; (5) blood pressure > 130/85 mmHg or drug treatment for hypertension [2]. When interpreting data, it is important to realize whether this or other criteria were used by the authors. The different criteria used in the existing literature explain, at least in part, the controversies on this theme.

The complex and not entirely clear pathophysiology of MS is largely acknowledged [1]. Abdominal adiposity and insulin resistance are thought to be central elements for its development [3]. Data shows complex interactions between internal factors, as genetic backgrounds, as well as external factors, such as physical activity and diet [4,5]. Nonetheless, genetic background is believed to be only a minor component for MS development, given the epidemic grow of such metabolic disturbance, which is unlikely related to genetics [6]. On the other hand, epigenetic changes namely in the spermatozoa, oocytes or in utero may have an important role [1]. Nutrition (both intrauterine and postnatal) and growth have also shown strong associations with MS in the adulthood [1]. Inflammation may also be an important contributing factor to the metabolic dysfunction [7]. This led to the concept of immunometabolism, linking inflammation, and metabolic defects [7,8]. For instance, MS is now known for being a milieu of a chronic pro-inflammatory state namely presenting with elevated inflammatory cytokines (such as tumour necrosis factor-α and interleukin-6) and acute-phase reactants (such as C-reactive protein and fibrinogen) [5]. Data shows that inflammatory cytokines associated to MS stimulate insulin resistance in adipose tissue and muscle [5].

MS is reaching pandemic levels worldwide and it currently affects around 25% in the adult population of developed countries [1]. The rising prevalence of MS parallels obesity and type 2 diabetes prevalence's, which are often coincidental [1]. Identifying these patients is crucial to achieve their optimal CV risk management. MS components are independent risk factors for CV disease and the combination of them may be synergic [9]. Given the uncertainty on its pathophysiology and aetiology, as well as the great variability among different individuals, the best treatment approach is not known [3]. It is consensual that prevention rather than treating should be the targeted, and that no single medication can eradicate it [1]. Currently, lifestyle changes (namely concerning diet and exercise) are basilar in the treatment of patients with MS [10]. Different recommendations are available and most include the goal of 7-10% weight loss, regular moderate intensity physical activity (according to the patient's clinical status) and adopting a diet with low intake of saturated fat, transfat, and cholesterol [5]. Individual pharmacological therapy may address central adiposity, insulin resistance, dyslipidaemia, hypertension, and hypercoagulable state [3]. Additionally, in the setting of severe obesity, bariatric surgery is a greatly effective treatment of multiple risk factors [3].

In this review we will explore the relationship between MS and the gut microbiome and the potential of microbial modulators (probiotics or synbiotics) to interfere with the disease and improve patients' health. In addition, a systematic review on the randomized control trials conducted using probiotics or synbiotics in patients with MS will be shown.

2. Metabolic Syndrome and Microbiota

The human gut is known for its wide microbiota composition, which usually lives in a symbiotic relationship with the host. These microorganisms use the undigested nutrients reaching the colon as substrates to live, and some of the microbes are important to final product degradation and for vitamin formation, among other crucial functions related to host's immunity [11–13].

It is widely accepted that the gut microbiota is a regulator of metabolic homeostasis [14–17]. Particularly, multiple latest studies aimed to characterize the role of the microbiota in the pathogenesis of MS, given that these two are thought to be highly correlated. Although the specific microorganism profile in patients with MS is not yet known, it seems likely that these patients have a different microbiota composition (dysbiosis), when compared to patients without MS (Figure 1). This different milieu, including different bacterial metabolites, may regulate inflammation and immunity, as well as the metabolic homeostasis [18]. The recognition of the microbiome impact on metabolism is recent and yet to be elucidated. Possible explanations for this regulation, which likely act together, may embrace the regulation by the microbiome of epithelial lipid uptake, hepatic gluconeogenesis, circadian host biology, and insulin signalling, among other possible mechanisms [17].

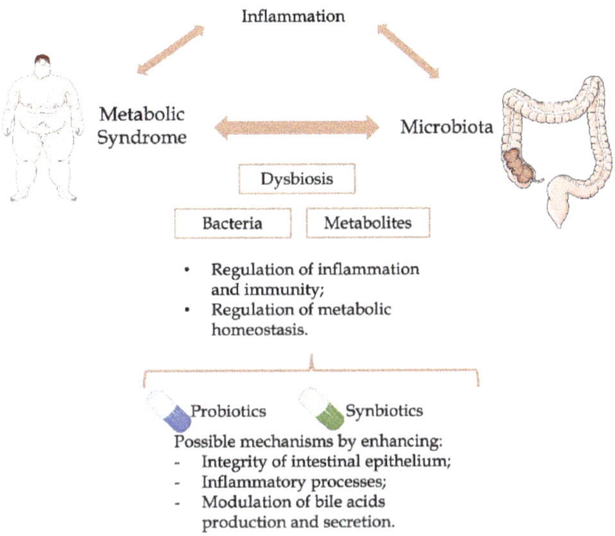

Figure 1. Mechanisms and modulation of the relationship between metabolic syndrome, human microbiome, and inflammation.

Concerning the microbiome profile, the HELIUS study, a multi-ethnic population study, reported higher proportion of Enterobacteriaceae and lower of Peptostreptococcaceae in patients with MS [19]. Also, enrichment of Enterobacteriaceae, as well as in *Turicibacter* sp., *Clostridium coccoides, Clostridium leptum*, and decrease of *Butyricicoccus* sp., *Akkermansia muciniphila*, and *Faecalibacterium prausnitzii* was reported in Romanian patients with MS [20]. Similarly, Qin and colleagues [21] reported microbiota changes in patients with MS namely decreased abundance of *Alistipes onderdonkii, Clostridium asparagiforme, Clostridium citroniae, Clostridium scindens, Roseburia intestinalis*, and *Bacteroides thetaiotaomicron*. Walker and colleagues [22] performed a population cross-sectional analysis in which from the 8 operational taxonomic units (OTUs) associated with diabetes, 3 OTUs (identified as belonging to Ruminococcaceae, Clostridiales, and Lachnospiraceae) were also significantly associated with MS and CV disease risk. These results advocate that microbiota may mediate mechanisms that contribute to cardiometabolic phenotypes through common mechanisms.

There are also studies focusing on the microbiota changes according to the components of MS. For example, Atzeni and colleagues [23] aimed to determine different faecal microbiota signatures associated with insulin resistance in a population with MS and concluded that differences in insulin resistance associated to a singular microbiota profile. These authors reported a negative association between insulin resistance and *Desulfovibrio, Odoribacter*, and Oscillospiraceae UCG-002, through mechanism of amino acid degradation, gluconeogenesis, immunomodulation and acetate, and a positive association between insulin resistance and *Feacalibaterium* and *Butyricicoccus* linked with the production of butyrate [23].

Yan and colleagues [24] studied 41 patients to identify gut microbiota changes in patients with visceral obesity. These authors found strong correlations between 16 species and visceral adiposity, being the strongest one with *Escherichia coli*. Additionally, the degradation of short-chain fatty acids (SCFAs) may be related to visceral adipose accumulation. The authors underline the hypothesis of an intrinsic connection between the gut microbiota and visceral adiposity, as well as the related metabolic disorders.

The METISM cohort is a Finland population cohort composed by unrelated man primarily designed to determine the prevalence and genetic determinants of metabolic and CV diseases. Org and colleagues [25] aimed to investigate the associations between gut microbiota and its plasma metabolites, with MS features. These authors identified a panoply of associations between gut microbiota composition and circulation metabolites, and MS features. For instance, these authors report an association between the microbiota metabolite trimethylamine N-oxide (TMAO, in the fasting plasma), associated with coronary artery disease and stroke, and the abundance of Peptococcaceae and *Prevotella*, and a negative association between TMAO and the abundance *F. prausnitzii*. These results underline that gut microbiota may modulate several cardio-metabolically traits [25].

Concerning microbiota metabolites, Xiaomin and colleagues [18] summarized current knowledge on the role of gut microbiota-derived tryptophan metabolites in the development of several diseases, including MS. Tryptophan is an essential amino acid, obtained from dietary proteins, and its metabolites, such as such as indole-3-lactate, indole-3-acrylate, indole-3-propionate, indole-3-aldehyde, indoleacetic acid, indole-3-acetaldehyde, and kynurenine (Kyn), can be produced by multiple taxa resident in the gut microbiota, and may have a role in MS pathogenesis. The metabolites can promote the differentiation and function of anti-inflammatory cells (such as anti-inflammatory macrophages and Treg cells) and are involved in maintaining the gut mucosal homeostasis [18]. Namely, blood levels of specific tryptophan metabolites are lower in patients with type 2 diabetes, when compared to the lean controls [17,26]. Also, a study using high fat fed rodents showed that increased acetate production, which occurs when microbiota is exposed to calorically dense nutrients, and particularly in the setting of chronic exposure to calorically dense food, promotes obesity and its related consequences of hyperlipidaemia, fatty liver disease, and insulin resistance [27].

On the other hand, Qin and colleagues [21] described that microbiota profile changes in patients with MS were associated with increased inflammation, through the inhibition of SCFAs production. A significantly lower microbiota diversity was observed in patients with MS. Namely, the relative abundance of Clostridiales (*Chlorobium phaeobacteroides, Clostridium asparagiforme, Clostridium bartlettii, Clostridium leptum, Clostridium scindens*, and *Collinsella aerofaciens*), five species from the order Bacteroidales (*Bacteroides fragilis, Roseburia intestinalis, Bacteroides nordii, Bacteroides thetaiotaomicron*, and *Bacteroides xylanisolven*), species from the genus *Alistipes* (*Alistipes onderdonkii, Alistipes hadrus, Alistipes colihominis*, and unclassified), and three species belonging to the family Ruminococcaceae (bacterium D16, *Ruminococcus lactaris*, and *Ruminococcus obeum*) were enriched in controls, when compared to MS patients. In addition, 28 bacterial species were negatively correlated with waist circumstance, being the strongest correlation with *Alistipes onderdonkii*. In line with these findings is the study from Vriezze and colleagues [28], in which microbiota transfer from lean donors to individuals with obesity and MS led to an increase in the abundance of butyrate-producing microbes and to an increase in insulin sensitivity six weeks after the procedure.

Given the data presented above, an association between microbiota and MS seems very likely and plausible. Despite the gap in knowledge regarding the specific microbiota profile in patients with MS, multiple data on modulation of microbiota in these patients is quickly arising.

3. Administration of Probiotic Supplements

3.1. Effects and Mechanisms of Action

Multiple factors associated to patients with MS, such as age and genetic background, cannot be changed, while other factors, such as weight and body mass index (BMI), triglycerides and high-density lipoprotein, or hypertension, can be somehow modifiable in order to improve the metabolic status of patients with MS [29]. Probiotics are alternatives which have been shown to be able to help to mitigate some of the described risk factors by enhancing the integrity of intestinal epithelium, adjusting inflammatory processes and endotoxin levels, modulating the bile acids production and secretion, and/or releasing antimicrobial peptides, among other mechanisms [30,31]. Therefore, it is important to know the mechanisms of action usually associated to the administration of probiotic supplements to the diet of patients with MS to understand and clarify its impact on metabolic health (Figure 1).

Improvements of the gut epithelial barrier, specifically among tight-junction proteins, can reduce bacterial translocation, inflammation, and metabolic endotoxaemia at the gut in patients with MS and these patients have been described with gut epithelium impairment [32–34]. Such gut impairment can be stimulated with poor diets and lack of certain nutrients. In the absence of fibers in the diet, the mucus barrier can work as source of nutrients for mucin-degrading bacteria, therefore affecting the epithelial thickness [35]. A firm inner structure associated to balanced microbiota, confers protection to the host [36]. *Lactobacillus reuteri* may compensate for impaired of aryl hydrocarbon receptors (related to some hormonal and immune responses) by increasing the availability of intestinal metabolites and improving metabolic homeostasis, being such results related to the restoration of the intestinal barrier function in animal models [37]. The Mediterranean diet, rich in polyunsaturated fats, polyphenols, carotenoids, and vitamins, was shown to be effective in reducing the risk of MS through the reinforcement of the gut barrier and the reduction of endotoxaemia in patients with in non-alcoholic fatty liver disease [38].

The most popular probiotics are members of lactobacilli and bifidobacteria groups, which are capable of interfering with dysbiotic gut biodiversity [39]. A higher Bacteroidetes/Firmicutes ratio is important in the gut and multiple probiotics have been showing the ability to modulate and normalize such ratio in murine models, as well as the abundance of Proteobacteria [40,41]. Specific gut bacteria, such as *Bilophila wadsworthia*, can also worsen the host metabolism in patients with high fat diets, being directly and indirectly related to inflammation mechanisms [42]. The probiotic *Lactobacillus rhamnosus* CNCM I-3690 was capable of reducing *B. wadsworthia*-induced immune and metabolic impairment by limiting its proliferation in the gut, reducing inflammation, and reinforcing intestinal barrier. The administration of multiple probiotics can also increase anti-inflammatory bacteria, such as *Prevotella*, in murine models of hepatocellular carcinoma along with their metabolites (i.e., propionate), shifting the bacteria community to Bacteroidetes, *Prevotella* and *Oscillibacter*, in addition to promoting IL-10 signalling and inhibiting pro-inflammatory helper T cell secretion from the gut to the liver [43].

By increasing proinflammatory molecules, such as lipopolysaccharides (LPS), it can be speculated that endotoxaemia can be promoted and metabolic disorders induced, therefore increasing the body fat mass and other metabolic parameters in obese patients. These effects can be reduced by probiotics through the preservation of gut permeability interfering with endotoxin levels [44]. Probiotic supplementation in rats may increase fatty acid oxidation, correct energy metabolism, plasma glucose and insulin resistance, inhibit cholesterol synthesis, prevent bile salt recycling, and modulate proinflammatory cytokines, therefore improving functional integrity of liver through the reduction of lipid reabsorption at the intestine [45]. Plasma bile acids, such as glycocholic acid, glycoursodeoxycholic acid, taurohyodeoxycholic acid, and tauroursodeoxycholic acid, were reduced in overweight adults taking synbiotics, supporting the effects of dietary supplements on certain metabolic pathways [46]. SCFAs, such as acetate, propionate and butyrate, can be released during the degradation of dietary fibers and are responsible for activities on the intestinal epithelial barrier, the immune system and the gut microbiota, sometimes working as bacterial inhibitors and quorum-sensing signaling molecules to regulate bacterial cell density and biofilm formation [36]. Nevertheless, it is important to decipher the potential beneficial anti-obesogenic, hypocholesterolemic, antihypertensive, and antiinflammatory properties of SCFAs and other metabolites produced and released by bacteria [47].

There are multiple probiotic strains described in the literature as presenting interesting and potential impact on MS. For example, *L. rhamnosus* BFE5264 resulted in a significant reduction of the serum cholesterol level that was accompanied by changes in intestinal microbiota and the production of SCFA in animal models [41]. *Bacillus licheniformis* Zhengchangsheng® significantly decreased body weight gain and fat accumulation, serum lipid profiles, and proinflammatory cytokine levels, and improved glucose and lipid metabolism in obese mice [48]. *Lactobacillus gerneri* BNR17 was shown to inhibit the secretion of adiponectin and serum leptin and reduce mesenteric adipose tissue mass and adipocyte size in obese mice [49]. *Lactobacillus pentosus* GSSK2 and *Lactobacillus plantarum* GS26A exhibited improved glucose tolerance, liver biomarkers, alleviated oxidative stress, and restored the histoarchitechture of adipose tissue, colon, and liver, compared with high fat diet animals [45]. *L. reuteri* ATCC treated mice gained significantly less body weight than the control mice [50] and another strain of *L. reuteri* increased the expression of Cpt1a (gene involved in fatty acid oxidation pathway) in obese mice, although the lipogenic genes in the liver of mice were not altered by the probiotics [50]. *L. rhamnosus* NCIMB 8010 and *Pediococcus acidilactici* NCIMB 8018 improved the viability of human hepatocellular carcinoma cell line HepG2, protected against apoptosis under normal and insulin resistance conditions and attenuated oxidative stress by improving mitochondrial metabolism and dynamics [51]. *Bifidobacterium* supplementation ameliorated visceral fat accumulation and insulin sensitivity of the metabolic syndrome in rats under high fat diet [52]. Among the next-generation probiotics, *A. muciniphila* and *F. prausnitzii* are also promising candidates, being their abundance found reduced in different intestinal disorders [53] and increased in patients with MS [54].

3.2. Probiotics in MS

The search for Clinical Trials and Randomized Controlled Trials was conducted on PUBMED/MEDLINE, considering eligible articles published in English, French, Spanish, or Portuguese between January 1990 and September 2022. The terms used were "metabolic syndrome" and "probiotics" or "synbiotics". Figure 2 shows the diagram for the selection of sources included in this systematic review.

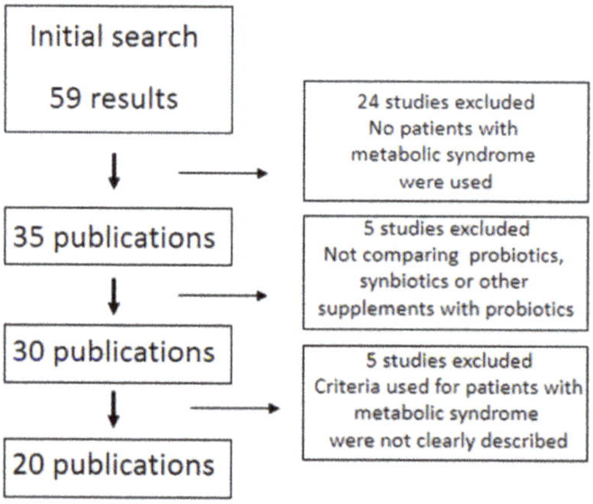

Figure 2. Diagram with the search results and criteria for selection of sources.

The prophylactic potential of isolated probiotics in patients with MS has been tested in randomized clinical trials, but the results are still scarce. The results can be promising for particular probiotics, but the initial trials were not enthusiastic. *Lactobacillus salivarius* Ls-33 was tested on a series of biomarkers related to inflammation in adolescents with obesity and MS and no differences were observed after 12 weeks of treatment regarding anthropometric evaluation, blood pressure (systolic and diastolic), fasting glucose and insulin, homeostasis model assessment of insulin resistance, C-peptide, cholesterol, high-density lipoprotein cholesterol, low-density lipoprotein cholesterol, triglyceride, free fatty acids, C-reactive protein, interleukin-6, tumour necrosis factor-α, or faecal calprotectin [55]. In addition, *Lactobacillus casei* Shirota was tested by multiple studies regarding its effects on gut permeability, microbiome biodiversity and metabolite production, presence of endotoxin and neutrophil function in MS. Gut permeability can be significantly increased in

MS as described above, but the treatment with *L. casei* Shirota did not show different results between patient and control groups [56]. Bacteroidetes/Firmicutes ratio was significantly higher in healthy controls compared to patients with MS, but the gut microbiome was not influenced by the probiotic. In addition, the proteins zonulin and calprotectin, usually higher in patients with MS, was not modified by the probiotic [32]; TMAO was not affected by *L. casei* Shirota either [57]. The insulin sensitivity index significantly improved after 3 months of probiotic supplementation, but the values were not different from the controls, as well as the values for β-cell and endothelial functions, or the inflammation markers [56,58].

More recently, other probiotics showed more success in clinical trials. The individual strain *L. reuteri* V3401 was tested by Tenorio-Jiménez and colleagues [59,60] and, although the decrease of Bacteroidetes/Firmicutes ratio was not corrected in obese patients, a rise of Verrucomicrobia was observed in patients receiving the probiotic. In addition, interleukin-6 and soluble vascular cell adhesion molecule 1 diminished following the treatment with the probiotic. Nevertheless, no significant correlation was observed between Verrucomicrobia abundance, and any inflammatory biomarker and subsequent studies are needed to complement the observations. Microbes4U© is a pilot study performed in patients with prediabetes and MS conducted to evaluate the tolerance, safety, and feasibility of the Gram-negative bacterium *A. muciniphila*, ingested either alive or pasteurized for 12 weeks, as a next-generation probiotic [61]. Beneficial impacts were shown on anthropometric measurements, as well as on the lipid profile, glycaemic parameters, such as insulin resistance, hepatic profile, and endotoxaemia, possibly due to interference with amino acids metabolism especially of alanine and arginine.

Multispecies probiotics may be more effective than single strain on metabolic disorders. Kassaian and colleagues [62] tested the effects of multiple probiotics (freeze-dried *Lactobacillus acidophilus, Bifidobacterium bifidum, Bifidobacterium lactis*, and *Bifidobacterium longum* with maltodextrin as filler) and synbiotics (the previous probiotics plus inulin as prebiotic) in individuals with prediabetes and MS. A clear reduction of hyperglycaemia in the groups treated with probiotic and synbiotic, as well as a reduction in hypertension in the group treated with probiotic, were reported.

3.3. Synbiotics in MS

The potential benefit of prebiotics can be conjugated with probiotics to potentiate its effects and support its adaptation and growth in challenging gut environments. Multiple sets of synbiotics have been tested in patients with MS, and the results have been clearly positive as described above by the study of Kassaian and colleagues [62]. Additional studies have been published and the results are in accordance.

Synbiotic capsules containing *L. casei, L. rhamnosus, L. acidophilus, Lactobacillus bulgaricus, B. longum, Bifidobacterium breve*, and *Streptococcus thermophiles*, plus the prebiotic short chain fructo-oligosaccharide were tested on patients with MS [63]. The synbiotic treatment significantly reduced fasting blood glucose in the MS group versus placebo, but no differences were observed in other metabolic factors, including insulin level, homeostatic model assessment for insulin resistance, homoeostatic model assessment-β, and insulin/glucagon ratio. In another study, 38 patients with MS were supplemented with either synbiotic capsules containing seven strains of friendly bacteria (*L. casei, L. rhamnosus, L. acidophilus, L. bulgaricus B. longum, B. breve*, and *S. thermophilus*) plus fructo-oligosaccharide or placebo and increased the efficacy of diet therapy and the management of insulin resistance, although no significant differences were observed in low-density lipoprotein (LDL) levels, waist circumference, BMI, metabolism, and energy intake between the groups [64].

More relevant differences were reported by Rabiei and colleagues [65] by testing seven probiotic strains (*L. casei, L. rhamnosus, L. acidophilus, L. bulgaricus, B. longum, B. breve*, and *S. thermophilus*), plus fructo-oligosaccharide as prebiotic in patients with MS. The synbiotic treatment improved the status of BMI, fasting blood sugar, insulin resistance, homeostatic model assessment for insulin resistance, glucagon-like peptide-1, and peptide YY in patients, and interestingly, the trend of weight loss in the synbiotic group was significant until the end of the study. Cicero and colleagues [66] also tested a synbiotic formula comprising of *L. plantarum* PBS067, *L. acidophilus* PBS066, and *L. reuteri* PBS072 with active prebiotics in elderly patients with MS (aged 65–80 years). Patients receiving synbiotics improved waist circumference and fasting plasma insulin, arterial pressure, total cholesterol, high-density lipoprotein cholesterol, non-high-density lipoprotein cholesterol, triglycerides, low-density lipoprotein cholesterol, high-sensitivity C-reactive protein, and tumour necrosis factor-α serum levels. Compared to placebo, the patients receiving synbiotic treatment improved visceral adiposity index and triglycerides either. The EQ-5D Visual Analogue Scale (VAS) questionnaire confirmed an increase of quality of life in patients treated with synbiotics.

3.4. Other Foods with Probiotics in MS

The probiotics can be added to other foods and supplements and its effects have also been described in multiple studies and trials. The beneficial effects of functional yogurt NY-YP901 supplemented with mixture of *S. thermophilus, L. acidophilus, Bifidobacterium infantis,* and extra-ingredients containing *B. breve* CBG-C2, *Enterococcus faecalis* FK-23, fibersol-2 and other compounds, was tested in patients with MS [67]. In the group consuming NY-YP901, improvements were observed in body weight, BMI, and low-density lipoprotein-cholesterol after 8 weeks. A fortified yogurt containing the starter cultures of *S. thermophiles* and *L. bulgaricus* enriched with *B. lactis* Bb-12 was tested in overweight and obese patients with MS under a caloric-restricted diet [68]. The fortified yogurt reduced the body fat mass, body fat percentage, waist circumference, homoeostasis model of assessment-insulin resistance, triglyceride concentration versus patients consuming low fat yogurt, and led to a significant increase in total 25-hydroxyvitamin D, high density lipoprotein-cholesterol and quantitative insulin sensitivity check index. A probiotic yogurt containing *L. acidophilus* La5 and *B. lactis* Bb12 was compared with a regular yogurt for 2 months in patients with MS and significant reduction in the blood glucose and vascular cell adhesion molecule-1 was observed [69]. The probiotic yogurt induced changes in plasminogen activator inhibitor-1, insulin, homoeostasis model of assessment-insulin resistance, and quantitative insulin sensitivity check index compared to baseline, as well as improved fasting blood glucose and some serum markers associated to the endothelial function.

The influence of fermented milk with *L. plantarum* was tested in postmenopausal women with MS and showed positive results regarding CV risk factors by decreasing total cholesterol levels and fasting glucose levels [70]. In another study, the daily ingestion of fermented milk with *B. lactis* HN019 was tested in patients with MS and showed significant reduction in BMI, total cholesterol, low-density lipoprotein, tumour necrosis factor-α, and interleukin-6 pro-inflammatory cytokines when compared to baseline and control group values [71].

Probiotic kefir, comprising *Lactococcus lactis subsp. lactis, Lactococcus lactis subsp. cremoris, Lactococcus lactis subsp. diacetylactis, Leuconostoc mesenteroides subsp. cremoris, Lactobacillus kefyr, Kluyveromyces marxianus*, and *Saccharomyces unisporus*, was tested on patients with MS [72]. A significant increase in serum apolipoprotein A1 concentrations was provided by kefir compared to milk consumption. The regular kefir consumption did not provide superior effects compared with milk consumption on anthropometrical measurements, glycaemic control, inflammatory parameters, or blood pressure. Another study showed a decrease in fasting blood glucose without a change in glycated haemoglobin concentration after kefir (with more than 30 species of bacteria and more than 12 species of yeast and fungi) was administrated to patients with MS [73].

4. Discussion

In humans, data is being concordant towards beneficial effects of probiotics on patients with MS especially concerning weight loss, despite the effect is not transversal to all patients as described above. There are three important points to take into account when studying probiotics and its impacts on health. First, the individualized response to the consumption of probiotics may be dependent on microbiome variations and the ability of the probiotic strain(s) to interact and modify the host gut microbiome [74]. The gut microbiome and its variability is one of the first variables that need to be monitored in clinical trials in order to correctly compare patients. Patients should be carefully grouped, not only based on similar clinical features, but also taking into account the variability of the human microbiome as the response to modulatory treatments can be discrepant. Second, the variability of metabolic responses found among bacterial strains can be vast. For example, the strains *L. rhamnosus* LGG and *L. rhamnosus* BFE5264 belong to the same species, yet these strains may impact the gut microbiome of murine models for MS very differently and result in distinct cholesterol reduction levels [41]. These results strongly emphasise the importance of strain-specificity and metabolic networks potentially available in each strain. Third, the features of one probiotic formulation should not be generalized to multiple probiotics. The colony forming counts, type of strains, ratio of strains or the manufacturing processes of one probiotic product should be carefully considered and studied individually [75].

Although the mechanism of some probiotics has been clearly described and its impacts studied, it may be possible to combine probiotics strains via the complementary of mechanisms of action, therefore putting them to work together to achieve healthy goals. The metabolic deterioration of liver can be associated with excessive accumulation of free fatty acids, exhaustive oxidative stress, cellular apoptosis and inflammation, impairment of some insulin pathways and lipotoxicity [76], and probiotics may act on these multiple points as described above. Alternative mechanisms of action have been described for other probiotics in animal models and considerable advances may be soon seen in this topic. For example, *L. plantarum* PCS 26 might act as a liver X receptor agonist and help to improve lipid profiles in hypercholesterolemic patients with complex diseases, such as MS [77]. More recently, synbiotic supplementation showed recovering of nitric oxide function associated to hypertension in rats under high fat diets and correction of systolic blood pressure [78] and this represents a new and additional mechanism of action to be targeted. *L. plantarum* strains may also be capable to stimulate hepatic and renal nuclear factor-erythroid 2-related factor 2 (Nrf2) expression in hyperlipidemic mice and alleviate MS [79].

Engineered strains represent a dynamic and interesting new option for probiotics with specific activities and targets. An engineered *L. reuteri* secreting interleukin -22 was developed based on the probiotic *L. reuteri* ATCC PTA 6475 and could ameliorate non-alcoholic fatty liver disease [80]. Treatment with *L. reuteri* expressing interleukin-22 yielded subtle changes in the expression of reg3 genes in the small intestine and interleukin-22 levels in the plasma in some animal models. Ongoing research projects aimed to identify specific bacterial targets in the gut microbiome and then create phage cocktails designed to eliminate particular bacterial strains are also underway [36] and may represent a valid alternative for clinical cases associated to the proliferation of specific bacteria.

In this review it was described the effect of some probiotics and synbiotics currently available that were tested on patients with MS. Current results are very promising. In addition, it was observed that multiple strains (synbiotics) may be presenting better results on patients with MS due to the multitude of mechanisms of action that be working together in such cases. The number of trials available is still limited and the number of tested patients in each trial (some dozens) is also reduced. The ethnicity and nutritional habits tend to be similar as most of the studies were conducted in occidental countries, therefore, some differences may be observed when other populations are tested.

5. Conclusions

Although the specific microorganism profile in patients with MS is not yet properly known, these patients seem to have a different microbiota composition, when compared to patients without MS. Despite the gap in knowledge regarding the specific microbiota profile in patients with MS, multiple data on modulation of microbiota in these patients is quickly arising. It has been clearly described differences in the gut microbiome of patients with MS compared with healthy individuals, and such differences can be mitigated in some patients by the administration of probiotics or synbiotics. The number of published studies is still limited, and additional results can be expected soon as multiple randomized studies are currently being conducted. Nevertheless, there are multiple factors capable to affect the microbiota of patients with MS that should be considered simultaneously. Therefore, it is extremely difficult to associate particular microbial and metabolic changes to single factors. As more studies are published and both the diversity and stability of gut microbiome is revealed in patients with MS, a clear picture of the intricate relationship between microbiome and disease can become clear and additional therapeutic options can be explored.

Author Contributions: R.A., M.B.-C. and P.P.-N. designed the outline and performed the writing. All authors have read and agreed to the published version of the manuscript.

Funding: R.A. was supported by Individual Call to Scientific Employment Stimulus—Second Edition (grant number CEECIND/01070/2018).

Institutional Review Board Statement: Not applicable.

Informed Consent Statement: Not applicable.

Conflicts of Interest: The authors declare no conflict of interest.

References

1. Saklayen, M.G. The Global Epidemic of the Metabolic Syndrome. *Curr. Hypertens. Rep.* **2018**, *20*, 12. [CrossRef] [PubMed]
2. Lorenzo, C.; Williams, K.; Hunt, K.J.; Haffner, S.M. The National Cholesterol Education Program—Adult Treatment Panel III, International Diabetes Federation, and World Health Organization Definitions of the Metabolic Syndrome as Predictors of Incident Cardiovascular Disease and Diabetes. *Diabetes Care* **2007**, *30*, 8–13. [CrossRef] [PubMed]
3. Cornier, M.-A.; Dabelea, D.; Hernandez, T.L.; Lindstrom, R.C.; Steig, A.J.; Stob, N.R.; Van Pelt, R.E.; Wang, H.; Eckel, R.H. The Metabolic Syndrome. *Endocr. Rev.* **2008**, *29*, 777–822. [CrossRef] [PubMed]
4. McCracken, E.; Monaghan, M.; Sreenivasan, S. Pathophysiology of the metabolic syndrome. *Clin. Dermatol.* **2018**, *36*, 14–20. [CrossRef] [PubMed]
5. Grundy, S.M.; Cleeman, J.I.; Daniels, S.R.; Donato, K.A.; Eckel, R.H.; Franklin, B.A.; Gordon, D.J.; Krauss, R.M.; Savage, P.J.; Smith, S.C., Jr.; et al. Diagnosis and management of the metabolic syndrome: An American Heart Association/National Heart, Lung, and Blood Institute scientific statement. *Circulation* **2005**, *112*, 2735–2752. [CrossRef]
6. Locke, A.E.; Kahali, B.; Berndt, S.I.; Justice, A.E.; Pers, T.H.; Day, F.R.; Powell, C.; Vedantam, S.; Buchkovich, M.L.; Yang, J.; et al. Genetic studies of body mass index yield new insights for obesity biology. *Nature* **2015**, *518*, 197–206. [CrossRef]
7. Hotamisligil, G.S. Inflammation, metaflammation and immunometabolic disorders. *Nature* **2017**, *542*, 177–185. [CrossRef]
8. Lee, Y.S.; Olefsky, J. Chronic tissue inflammation and metabolic disease. *Genes Dev.* **2021**, *35*, 307–328. [CrossRef]
9. Ilkun, O.; Boudina, S. Cardiac Dysfunction and Oxidative Stress in the Metabolic Syndrome: An Update on Antioxidant Therapies. *Curr. Pharm. Des.* **2013**, *19*, 4806–4817. [CrossRef]
10. Grundy, S.M.; Stone, N.; Bailey, A.L.; Beam, C.; Birtcher, K.K.; Blumenthal, R.S.; Braun, L.T.; De Ferranti, S.; Faiella-Tommasino, J.; Forman, D.E.; et al. AHA/ACC/AACVPR/AAPA/ABC/ACPM/ADA/AGS/APhA/ASPC/NLA/PCNA Guideline on the Management of Blood Cholesterol: A Report of the American College of Cardiology/American Heart Association Task Force on Clinical Practice Guidelines. *Circulation* **2019**, *139*, e1082–e1143. [CrossRef]
11. Gill, S.R.; Pop, M.; Deboy, R.T.; Eckburg, P.B.; Turnbaugh, P.J.; Samuel, B.S.; Gordon, J.I.; Relman, D.A.; Fraser-Liggett, C.M.; Nelson, K.E. Metagenomic Analysis of the Human Distal Gut Microbiome. *Science* **2006**, *312*, 1355–1359. [CrossRef] [PubMed]
12. Turnbaugh, P.J.; Ley, R.E.; Hamady, M.; Fraser-Liggett, C.M.; Knight, R.; Gordon, J.I. The Human Microbiome Project. *Nature* **2007**, *449*, 804–810. [CrossRef] [PubMed]
13. Jasim, S.A.; Opulencia, M.J.C.; Ramírez-Coronel, A.A.; Abdelbasset, W.K.; Abed, M.H.; Markov, A.; Al-Awsi, G.R.L.; Shamsiev, J.A.; Hammid, A.T.; Shalaby, M.N.; et al. The emerging role of microbiota-derived short-chain fatty acids in immunometabolism. *Int. Immunopharmacol.* **2022**, *110*, 108983. [CrossRef] [PubMed]
14. Bäckhed, F.; Ding, H.; Wang, T.; Hooper, L.V.; Koh, G.Y.; Nagy, A.; Semenkovich, C.F.; Gordon, J.I. The Gut Microbiota as an Environmental Factor That Regulates Fat Storage. *Proc. Natl. Acad. Sci. USA* **2004**, *101*, 15718–15723. [CrossRef]
15. Bäckhed, F.; Manchester, J.K.; Semenkovich, C.F.; Gordon, J.I. Mechanisms underlying the resistance to diet-induced obesity in germ-free mice. *Proc. Natl. Acad. Sci. USA* **2007**, *104*, 979–984. [CrossRef]
16. Khan, M.T.; Nieuwdorp, M.; Bäckhed, F. Microbial Modulation of Insulin Sensitivity. *Cell Metab.* **2014**, *20*, 753–760. [CrossRef] [PubMed]
17. Cox, T.O.; Lundgren, P.; Nath, K.; Thaiss, C.A. Metabolic control by the microbiome. *Genome Med.* **2022**, *14*, 80. [CrossRef]
18. Su, X.; Gao, Y.; Yang, R. Gut Microbiota-Derived Tryptophan Metabolites Maintain Gut and Systemic Homeostasis. *Cells* **2022**, *11*, 2296. [CrossRef]

19. Deschasaux, M.; Bouter, K.; Prodan, A.; Levin, E.; Groen, A.; Herrema, H.; Tremaroli, V.; Snijder, M.; Nicolaou, M.; Zwinderman, A.; et al. Differences in gut microbiota composition in metabolic syndrome and type 2 diabetes subjects in a multi-ethnic population: The HELIUS study. *Proc. Nutr. Soc.* **2020**, *79*, E183. [CrossRef]
20. Pircalabioru, G.G.; Ilie, I.; Oprea, L.; Picu, A.; Petcu, L.M.; Burlibasa, L.; Chifiriuc, M.-C.; Musat, M. Microbiome, Mycobiome and Related Metabolites Alterations in Patients with Metabolic Syndrome—A Pilot Study. *Metabolites* **2022**, *12*, 218. [CrossRef]
21. Qin, Q.; Yan, S.; Yang, Y.; Chen, J.; Li, T.; Gao, X.; Yan, H.; Wang, Y.; Wang, J.; Wang, S.; et al. A Metagenome-Wide Association Study of the Gut Microbiome and Metabolic Syndrome. *Front. Microbiol.* **2021**, *12*, 682721. [CrossRef] [PubMed]
22. Walker, R.L.; Vlamakis, H.; Lee, J.W.J.; Besse, L.A.; Xanthakis, V.; Vasan, R.S.; Shaw, S.Y.; Xavier, R.J. Population study of the gut microbiome: Associations with diet, lifestyle, and cardiometabolic disease. *Genome Med.* **2021**, *13*, 188. [CrossRef] [PubMed]
23. Atzeni, A.; Bastiaanssen, T.F.S.; Cryan, J.F.; Tinahones, F.J.; Vioque, J.; Corella, D.; Fito, M.; Vidal, J.; Moreno-Indias, I.; Gomez-Perez, A.M.; et al. Taxonomic and Functional Fecal Microbiota Signatures Associated With Insulin Resistance in Non-Diabetic Subjects With Overweight/Obesity Within the Frame of the PREDIMED-Plus Study. *Front. Endocrinol.* **2022**, *13*, 804455. [CrossRef] [PubMed]
24. Yan, H.; Qin, Q.; Chen, J.; Yan, S.; Li, T.; Gao, X.; Yang, Y.; Li, A.; Ding, S. Gut Microbiome Alterations in Patients With Visceral Obesity Based on Quantitative Computed Tomography. *Front. Cell. Infect. Microbiol.* **2021**, *11*, 823262. [CrossRef]
25. Org, E.; Blum, Y.; Kasela, S.; Mehrabian, M.; Kuusisto, J.; Kangas, A.J.; Soininen, P.; Wang, Z.; Ala-Korpela, M.; Hazen, S.L.; et al. Relationships between gut microbiota, plasma metabolites, and metabolic syndrome traits in the METSIM cohort. *Genome Biol.* **2017**, *18*, 70. [CrossRef]
26. De Mello, V.D.; Paananen, J.; Lindström, J.; Lankinen, M.A.; Shi, L.; Kuusisto, J.; Pihlajamäki, J.; Auriola, S.; Lehtonen, M.; Rolandsson, O.; et al. Indolepropionic acid and novel lipid metabolites are associated with a lower risk of type 2 diabetes in the Finnish Diabetes Prevention Study. *Sci. Rep.* **2017**, *7*, 46337. [CrossRef]
27. Perry, R.J.; Peng, L.; Barry, N.A.; Cline, G.W.; Zhang, D.; Cardone, R.L.; Petersen, K.F.; Kibbey, R.G.; Goodman, A.L.; Shulman, G.I. Acetate mediates a microbiome–brain–β-cell axis to promote metabolic syndrome. *Nature* **2016**, *534*, 213–217. [CrossRef]
28. Vrieze, A.; Van Nood, E.; Holleman, F.; Salojärvi, J.; Kootte, R.S.; Bartelsman, J.F.; Dallinga-Thie, G.M.; Ackermans, M.T.; Serlie, M.J.; Oozeer, R.; et al. Transfer of Intestinal Microbiota From Lean Donors Increases Insulin Sensitivity in Individuals With Metabolic Syndrome. *Gastroenterology* **2012**, *143*, 913–916.e7. [CrossRef]
29. Czarnecka, K.; Czarnecka, P.; Tronina, O.; Bączkowska, T.; Durlik, M. Multidirectional facets of obesity management in the metabolic syndrome population after liver transplantation. *Immun. Inflamm. Dis.* **2022**, *10*, 3–21. [CrossRef]
30. Chang, C.-J.; Lin, T.-L.; Tsai, Y.-L.; Wu, T.-R.; Lai, W.-F.; Lu, C.-C.; Lai, H.-C. Next generation probiotics in disease amelioration. *J. Food Drug Anal.* **2019**, *27*, 615–622. [CrossRef]
31. Lee, S.J.; Bose, S.; Seo, J.-G.; Chung, W.-S.; Lim, C.-Y.; Kim, H. The effects of co-administration of probiotics with herbal medicine on obesity, metabolic endotoxemia and dysbiosis: A randomized double-blind controlled clinical trial. *Clin. Nutr.* **2014**, *33*, 973–981. [CrossRef] [PubMed]
32. Stadlbauer, V.; Leber, B.; Lemesch, S.; Trajanoski, S.; Bashir, M.; Horvath, A.; Tawdrous, M.; Stojakovic, T.; Fauler, G.; Fickert, P.; et al. Lactobacillus casei Shirota Supplementation Does Not Restore Gut Microbiota Composition and Gut Barrier in Metabolic Syndrome: A Randomized Pilot Study. *PLoS ONE* **2015**, *10*, e0141399. [CrossRef] [PubMed]
33. Festi, D.; Schiumerini, R.; Eusebi, L.H.; Marasco, G.; Taddia, M.; Colecchia, A. Gut microbiota and metabolic syndrome. *World J. Gastroenterol.* **2014**, *20*, 16079–16094. [CrossRef] [PubMed]
34. Amar, J.; Chabo, C.; Waget, A.; Klopp, P.; Vachoux, C.; Bermúdez-Humarán, L.G.; Smirnova, N.; Bergé, M.; Sulpice, T.; Lahtinen, S.; et al. Intestinal mucosal adherence and translocation of commensal bacteria at the early onset of type 2 diabetes: Molecular mechanisms and probiotic treatment. *EMBO Mol. Med.* **2011**, *3*, 559–572. [CrossRef] [PubMed]
35. Desai, M.S.; Seekatz, A.M.; Koropatkin, N.M.; Kamada, N.; Hickey, C.A.; Wolter, M.; Pudlo, N.A.; Kitamoto, S.; Terrapon, N.; Muller, A.; et al. A Dietary Fiber-Deprived Gut Microbiota Degrades the Colonic Mucus Barrier and Enhances Pathogen Susceptibility. *Cell* **2016**, *167*, 1339–1353. [CrossRef] [PubMed]
36. Albillos, A.; de Gottardi, A.; Rescigno, M. The gut-liver axis in liver disease: Pathophysiological basis for therapy. *J. Hepatol.* **2020**, *72*, 558–577. [CrossRef]
37. Natividad, J.M.; Agus, A.; Planchais, J.; Lamas, B.; Jarry, A.C.; Martin, R.; Michel, M.-L.; Chong-Nguyen, C.; Roussel, R.; Straube, M.; et al. Impaired Aryl Hydrocarbon Receptor Ligand Production by the Gut Microbiota Is a Key Factor in Metabolic Syndrome. *Cell Metab.* **2018**, *28*, 737–749.e4. [CrossRef]
38. Abenavoli, L.; Di Renzo, L.; Boccuto, L.; Alwardat, N.; Gratteri, S.; De Lorenzo, A. Health benefits of Mediterranean diet in nonalcoholic fatty liver disease. *Expert Rev. Gastroenterol. Hepatol.* **2018**, *12*, 873–881. [CrossRef]
39. Golonka, R.M.; Yeoh, B.S.; Vijay-Kumar, M. Dietary Additives and Supplements Revisited: The Fewer, the Safer for Gut and Liver Health. *Curr. Pharmacol. Rep.* **2019**, *5*, 303–316. [CrossRef]
40. Rivero-Gutiérrez, B.; Gámez-Belmonte, R.; Suárez, M.D.; Lavín, J.L.; Aransay, A.M.; Olivares, M.; Martínez-Augustin, O.; de Medina, F.S.; Zarzuelo, A. A synbiotic composed of *Lactobacillus fermentum* CECT5716 and FOS prevents the development of fatty acid liver and glycemic alterations in rats fed a high fructose diet associated with changes in the microbiota. *Mol. Nutr. Food Res.* **2017**, *61*, 1600622. [CrossRef]

41. Park, S.; Kang, J.; Choi, S.; Park, H.; Hwang, E.; Kang, Y.; Kim, A.; Holzapfel, W.; Ji, Y. Cholesterol-lowering effect of Lactobacillus rhamnosus BFE5264 and its influence on the gut microbiome and propionate level in a murine model. *PLoS ONE* **2018**, *13*, e0203150. [CrossRef] [PubMed]
42. Natividad, J.M.; Lamas, B.; Pham, H.P.; Michel, M.-L.; Rainteau, D.; Bridonneau, C.; da Costa, G.; Van Hylckama Vlieg, J.; Sovran, B.; Chamignon, C.; et al. Bilophila wadsworthia aggravates high fat diet induced metabolic dysfunctions in mice. *Nat. Commun.* **2018**, *9*, 2802. [CrossRef] [PubMed]
43. Li, J.; Sung, C.Y.J.; Lee, N.; Ni, Y.; Pihlajamäki, J.; Panagiotou, G.; El-Nezami, H. Probiotics modulated gut microbiota suppresses hepatocellular carcinoma growth in mice. *Proc. Natl. Acad. Sci. USA* **2016**, *113*, E1306–E1315. [CrossRef] [PubMed]
44. Purohit, V.; Bode, J.C.; Bode, C.; Brenner, D.; Choudhry, M.A.; Hamilton, F.; Kang, Y.J.; Keshavarzian, A.; Rao, R.; Sartor, R.B.; et al. Alcohol, intestinal bacterial growth, intestinal permeability to endotoxin, and medical consequences: Summary of a symposium. *Alcohol* **2008**, *42*, 349–361. [CrossRef]
45. Khanna, S.; Walia, S.; Kondepudi, K.K.; Shukla, G. Administration of indigenous probiotics modulate high-fat diet-induced metabolic syndrome in Sprague Dawley rats. *Antonie van Leeuwenhoek* **2020**, *113*, 1345–1359. [CrossRef]
46. Hibberd, A.; Yde, C.; Ziegler, M.; Honoré, A.; Saarinen, M.; Lahtinen, S.; Stahl, B.; Jensen, H.; Stenman, L. Probiotic or synbiotic alters the gut microbiota and metabolism in a randomised controlled trial of weight management in overweight adults. *Benef. Microbes* **2019**, *10*, 121–135. [CrossRef]
47. Vallianou, N.; Christodoulatos, G.S.; Karampela, I.; Tsilingiris, D.; Magkos, F.; Stratigou, T.; Kounatidis, D.; Dalamaga, M. Understanding the Role of the Gut Microbiome and Microbial Metabolites in Non-Alcoholic Fatty Liver Disease: Current Evidence and Perspectives. *Biomolecules* **2021**, *12*, 56. [CrossRef]
48. Lu, X.; Jing, Y.; Zhou, X.; Zhang, N.; Tai, J.; Cao, Y. Bacillus licheniformis Zhengchangsheng®Inhibits Obesity by Regulating the AMP-Activated Protein Kinase Signaling Pathway. *Probiotics Antimicrob. Proteins* **2021**, *13*, 1658–1667. [CrossRef]
49. Kang, J.-H.; Yun, S.-I.; Park, M.-H.; Park, J.-H.; Jeong, S.-Y.; Park, H.-O. Anti-Obesity Effect of Lactobacillus gasseri BNR17 in High-Sucrose Diet-Induced Obese Mice. *PLoS ONE* **2013**, *8*, e54617. [CrossRef]
50. Fåk, F.; Bäckhed, F. Lactobacillus reuteri Prevents Diet-Induced Obesity, but not Atherosclerosis, in a Strain Dependent Fashion in Apoe−/− Mice. *PLoS ONE* **2012**, *7*, e46837. [CrossRef]
51. Mularczyk, M.; Bourebaba, Y.; Kowalczuk, A.; Marycz, K.; Bourebaba, L. Probiotics-rich emulsion improves insulin signalling in Palmitate/Oleate-challenged human hepatocarcinoma cells through the modulation of Fetuin-A/TLR4-JNK-NF-κB pathway. *Biomed. Pharmacother.* **2021**, *139*, 111560. [CrossRef] [PubMed]
52. Chen, J.; Wang, R.; Li, X.-F.; Wang, R.-L. *Bifidobacterium adolescentis* supplementation ameliorates visceral fat accumulation and insulin sensitivity in an experimental model of the metabolic syndrome. *Br. J. Nutr.* **2011**, *107*, 1429–1434. [CrossRef] [PubMed]
53. Álvarez-Mercado, A.I.; Navarro-Oliveros, M.; Robles-Sánchez, C.; Plaza-Díaz, J.; Sáez-Lara, M.J.; Muñoz-Quezada, S.; Fontana, L.; Abadía-Molina, F. Microbial Population Changes and Their Relationship with Human Health and Disease. *Microorganisms* **2019**, *7*, 68. [CrossRef] [PubMed]
54. Haro, C.; García-Carpintero, S.; Rangel-Zuñiga, O.A.; Alcala-Diaz, J.F.; Landa, B.B.; Clemente, J.C.; Pérez-Martínez, P.; Lopez-Miranda, J.; Pérez-Jiménez, F.; Camargo, A. Consumption of Two Healthy Dietary Patterns Restored Microbiota Dysbiosis in Obese Patients with Metabolic Dysfunction. *Mol. Nutr. Food Res.* **2017**, *61*, 1700440. [CrossRef] [PubMed]
55. Gøbel, R.J.; Larsen, N.; Jakobsen, M.; Mølgaard, C.; Michaelsen, K.F. Probiotics to Adolescents with Obesity. *J. Pediatr. Gastroenterol. Nutr.* **2012**, *55*, 673–678. [CrossRef]
56. Leber, B.; Tripolt, N.; Blattl, D.; Eder, M.; Wascher, T.C.; Pieber, T.R.; Stauber, R.; Sourij, H.; Oettl, K.; Stadlbauer, V. The influence of probiotic supplementation on gut permeability in patients with metabolic syndrome: An open label, randomized pilot study. *Eur. J. Clin. Nutr.* **2012**, *66*, 1110–1115. [CrossRef]
57. Tripolt, N.J.; Leber, B.; Triebl, A.; Köfeler, H.; Stadlbauer, V.; Sourij, H. Effect of Lactobacillus casei Shirota supplementation on trimethylamine-N-oxide levels in patients with metabolic syndrome: An open-label, randomized study. *Atherosclerosis* **2015**, *242*, 141–144. [CrossRef]
58. Tripolt, N.; Leber, B.; Blattl, D.; Eder, M.; Wonisch, W.; Scharnagl, H.; Stojakovic, T.; Obermayer-Pietsch, B.; Wascher, T.; Pieber, T.; et al. Short communication: Effect of supplementation with Lactobacillus casei Shirota on insulin sensitivity, β-cell function, and markers of endothelial function and inflammation in subjects with metabolic syndrome—A pilot study. *J. Dairy Sci.* **2013**, *96*, 89–95. [CrossRef]
59. Tenorio-Jiménez, C.; Martínez-Ramírez, M.J.; Del Castillo-Codes, I.; Arraiza-Irigoyen, C.; Tercero-Lozano, M.; Camacho, J.; Chueca, N.; García, F.; Olza, J.; Plaza-Díaz, J.; et al. Lactobacillus reuteri V3401 Reduces Inflammatory Biomarkers and Modifies the Gastrointestinal Microbiome in Adults with Metabolic Syndrome: The PROSIR Study. *Nutrients* **2019**, *11*, 1761. [CrossRef]
60. Tenorio-Jiménez, C.; Martínez-Ramírez, M.J.; Tercero-Lozano, M.; Arraiza-Irigoyen, C.; Del Castillo-Codes, I.; Olza, J.; Plaza-Díaz, J.; Fontana, L.; Migueles, J.H.; Olivares, M.; et al. Evaluation of the effect of Lactobacillus reuteri V3401 on biomarkers of inflammation, cardiovascular risk and liver steatosis in obese adults with metabolic syndrome: A randomized clinical trial (PROSIR). *BMC Complement. Altern. Med.* **2018**, *18*, 306. [CrossRef]
61. Depommier, C.; Everard, A.; Druart, C.; Maiter, D.; Thissen, J.-P.; Loumaye, A.; Hermans, M.P.; Delzenne, N.M.; de Vos, W.M.; Cani, P.D. Serum metabolite profiling yields insights into health promoting effect of A. muciniphila in human volunteers with a metabolic syndrome. *Gut Microbes* **2021**, *13*, 1994270. [CrossRef] [PubMed]

62. Kassaian, N.; Feizi, A.; Aminorroaya, A.; Amini, M. Probiotic and synbiotic supplementation could improve metabolic syndrome in prediabetic adults: A randomized controlled trial. *Diabetes Metab. Syndr.* **2018**, *13*, 2991–2996. [CrossRef]
63. Rahimi, F.; Pasdar, Y.; Kaviani, M.; Abbasi, S.; Fry, H.; Hekmatdoost, A.; Nikpayam, O.; Sohrab, G.; Rezaei, M.; Nachvak, S.M.; et al. Efficacy of the Synbiotic Supplementation on the Metabolic Factors in Patients with Metabolic Syndrome: A Randomized, Triple-Blind, Placebo-Controlled Trial. *Int. J. Clin. Pract.* **2022**, *2022*, 2967977. [CrossRef] [PubMed]
64. Eslamparast, T.; Zamani, F.; Hekmatdoost, A.; Sharafkhah, M.; Eghtesad, S.; Malekzadeh, R.; Poustchi, H. Effects of synbiotic supplementation on insulin resistance in subjects with the metabolic syndrome: A randomised, double-blind, placebo-controlled pilot study. *Br. J. Nutr.* **2014**, *112*, 438–445. [CrossRef] [PubMed]
65. Rabiei, S.; Hedayati, M.; Rashidkhani, B.; Saadat, N.; Shakerhossini, R. The Effects of Synbiotic Supplementation on Body Mass Index, Metabolic and Inflammatory Biomarkers, and Appetite in Patients with Metabolic Syndrome: A Triple-Blind Randomized Controlled Trial. *J. Diet. Suppl.* **2018**, *16*, 294–306. [CrossRef] [PubMed]
66. Cicero, A.F.G.; Fogacci, F.; Bove, M.; Giovannini, M.; Borghi, C. Impact of a short-term synbiotic supplementation on metabolic syndrome and systemic inflammation in elderly patients: A randomized placebo-controlled clinical trial. *Eur. J. Nutr.* **2021**, *60*, 655–663. [CrossRef]
67. Chang, B.J.; Park, S.U.; Jang, Y.S.; Ko, S.H.; Joo, N.M.; I Kim, S.; Kim, C.-H.; Chang, D.K. Effect of functional yogurt NY-YP901 in improving the trait of metabolic syndrome. *Eur. J. Clin. Nutr.* **2011**, *65*, 1250–1255. [CrossRef]
68. Mohammadi-Sartang, M.; Bellissimo, N.; de Zepetnek, J.O.T.; Brett, N.; Mazloomi, S.M.; Fararouie, M.; Bedeltavana, A.; Famouri, M.; Mazloom, Z. The effect of daily fortified yogurt consumption on weight loss in adults with metabolic syndrome: A 10-week randomized controlled trial. *Nutr. Metab. Cardiovasc. Dis.* **2018**, *28*, 565–574. [CrossRef]
69. Rezazadeh, L.; Gargari, B.P.; Jafarabadi, M.A.; Alipour, B. Effects of probiotic yogurt on glycemic indexes and endothelial dysfunction markers in patients with metabolic syndrome. *Nutrition* **2019**, *62*, 162–168. [CrossRef]
70. Barreto, F.M.; Simão, A.N.C.; Morimoto, H.K.; Lozovoy, M.A.B.; Dichi, I.; da Silva Miglioranza, L.H. Beneficial effects of Lactobacillus plantarum on glycemia and homocysteine levels in postmenopausal women with metabolic syndrome. *Nutrition* **2014**, *30*, 939–942. [CrossRef]
71. Bernini, L.J.; Simão, A.N.C.; Alfieri, D.F.; Lozovoy, M.A.B.; Mari, N.L.; de Souza, C.H.B.; Dichi, I.; Costa, G.N. Beneficial effects of Bifidobacterium lactis on lipid profile and cytokines in patients with metabolic syndrome: A randomized trial. Effects of probiotics on metabolic syndrome. *Nutrition* **2016**, *32*, 716–719. [CrossRef] [PubMed]
72. Bellikci-Koyu, E.; Sarer-Yurekli, B.P.; Karagozlu, C.; Aydin-Kose, F.; Ozgen, A.G.; Buyuktuncer, Z. Probiotic kefir consumption improves serum apolipoprotein A1 levels in metabolic syndrome patients: A randomized controlled clinical trial. *Nutr. Res.* **2022**, *102*, 59–70. [CrossRef] [PubMed]
73. da Silva Ghizi, A.; de Almeida Silva, M.; da Andrade Moraes, F.; da Silva, C.; Endringer, D.; Scherer, R.; Lenz, D.; de Lima, E.M.; Brasil, G.A.; Maia, J.F.; et al. Kefir improves blood parameters and reduces cardiovascular risks in patients with metabolic syndrome. *PharmaNutrition* **2021**, *16*, e100266. [CrossRef]
74. Shapiro, H.; Suez, J.; Elinav, E. Personalized microbiome-based approaches to metabolic syndrome management and prevention. *J. Diabetes* **2017**, *9*, 226–236. [CrossRef]
75. Duseja, A.; Acharya, S.K.; Mehta, M.; Chhabra, S.; Shalimar; Rana, S.; Das, A.; Dattagupta, S.; Dhiman, R.K.; Chawla, Y.K. High potency multistrain probiotic improves liver histology in non-alcoholic fatty liver disease (NAFLD): A randomised, double-blind, proof of concept study. *BMJ Open Gastroenterol.* **2019**, *6*, e000315. [CrossRef]
76. Reinehr, T.; Roth, C.L. Fetuin-A and Its Relation to Metabolic Syndrome and Fatty Liver Disease in Obese Children Before and After Weight Loss. *J. Clin. Endocrinol. Metab.* **2008**, *93*, 4479–4485. [CrossRef]
77. Gorenjak, M.; Gradišnik, L.; Trapečar, M.; Pistello, M.; Kozmus, C.; Škorjanc, D.; Skok, P.; Langerholc, T.; Cencič, A. Improvement of lipid profile by probiotic/protective cultures: Study in a non carcinogenic small intestinal cell model. *New Microbiol.* **2014**, *37*, 51–64.
78. Llévenes, P.; Rodrigues-Díez, R.; Cros-Brunsó, L.; Prieto, M.I.; Casaní, L.; Balfagón, G.; Blanco-Rivero, J. Beneficial Effect of a Multistrain Synbiotic Prodefen®Plus on the Systemic and Vascular Alterations Associated with Metabolic Syndrome in Rats: The Role of the Neuronal Nitric Oxide Synthase and Protein Kinase A. *Nutrients* **2020**, *12*, 117. [CrossRef]
79. Wang, L.; Liu, K.; Gao, D.; Hao, J. Protective effects of two *Lactobacillus plantarum* strains in hyperlipidemic mice. *World J. Gastroenterol.* **2013**, *19*, 3150–3156. [CrossRef]
80. Oh, J.-H.; Schueler, K.L.; Stapleton, D.S.; Alexander, L.M.; Yen, C.-L.E.; Keller, M.P.; Attie, A.D.; van Pijkeren, J.-P. Secretion of Recombinant Interleukin-22 by Engineered Lactobacillus reuteri Reduces Fatty Liver Disease in a Mouse Model of Diet-Induced Obesity. *mSphere* **2020**, *5*, e00183-20. [CrossRef]

MDPI
St. Alban-Anlage 66
4052 Basel
Switzerland
Tel. +41 61 683 77 34
Fax +41 61 302 89 18
www.mdpi.com

Nutrients Editorial Office
E-mail: nutrients@mdpi.com
www.mdpi.com/journal/nutrients

www.ingramcontent.com/pod-product-compliance
Lightning Source LLC
Chambersburg PA
CBHW061438210926
51850CB00008B/1545